1973

ADVERSE EFFECTS OF COMMON ENVIRONMENTAL POLLUTANTS

Papers by
Kingsley Kay, M. M. Hipskind, M. Schafer, et al

MSS Information Corporation
19 East 48th Street New York, N.Y. 10017

TABLE OF CONTENTS

Toxicity of Pesticides to Man

Effects of Noise on Man

CREDITS AND ACKNOWLEDGEMENTS

Beasley, Oscar C.; and Keith R. Long, "Toxic Effects From the Use of Agricultural Chemicals: A Review with Special Attention to Problems in Iowa Agriculture," *Journal of Iowa Medical Society*, January, 1966, 56:45-48.

Booker, Harold E.; Francis M. Forster; and Hallgrim Kløve, "Extinction Factors in Startle (Acousticomotor) Seizures," *Neurology*, December, 1965, 15:1095-1103.

Bunting, John J., "Space Medicine," *Medical Record and Annals*, June, 1964, 57:405.

Casarett, L. J.; G. C. Fryer; W. L. Yauger, Jr.; and H. W. Klemmer, "Organochlorine Pesticide Residues in Human Tissue–Hawaii," *Archives of Environmental Health*, September, 1968, 17:306-311.

Dickson, R. C., "Some General Impressions and Some Thoughts on the Problem of Industrial Noise," *Proceedings of the Mine Medical Officers Association*, January-April, 1964, 43:88-93.

Dille, J. Robert; and Stanley R. Mohler, "Drug and Toxic Hazards in General Aviation," *Aerospace Medicine*, February, 1969, 40:191-195.

Dille, J. Robert; and Paul W. Smith, "Central Nervous System Effects of Chronic Exposure to Organophosphate Insecticides," *Aerospace Medicine*, May, 1964, 35:475-478.

Edmundson, Walter F.; and John E. Davies, "Occupational Dermatitis From Naled: A Clinical Report," *Archives of Environmental Health*, July, 1967, 15:89-91.

Edmundson, Walter F.; Joseph J. Freal; and John E. Davies, "Identification and Measurement of Dichloran in the Blood and Urine of Man," *Environmental Research*, 1967, 1:240-246.

Fairbanks, Wendell L.; and Kenneth C. Hoffman, "Organic Phosphate Intoxication," *The Nebraska State Medical Journal*, August, 1967, 52:376-379.

Gallaher, George L., " 'Low Volume' Insect Control and Parathion Poisoning," *Texas Medicine*, October, 1967, 63:39-41.

Hipskind, M. M., "Hearing Tests in Industry," *Industrial Medicine and Surgery*, June, 1967, 393-402.

Jenkins, R. B.; and J. F. Toole, "Polyneuropathy Following Exposure to Insecticides," *Archives of Internal Medicine*, May, 1964, 113:691-695.

Kay, Kingsley, "Pesticides and Associated Health Factors in Agricultural Environments: Effect on Mixed-Function Oxidizing Enzyme Metabolism, Pulmonary Surfactant and Immunological Reactions," *Industrial Medicine*, January, 1969, 38:52-63.

Laws, Edward R.; August Curley; and Frank J. Biros, "Men With Intensive Occupational Exposure to DDT: A Clinical and Chemical Study," *Archives of Environmental Health*, December, 1967, 15:766-775.

Laws, Edward R.; Francisco Ramos Morales; Wayland J. Hayes, Jr.; and Charles Romney Joseph, "Toxicology of Abate in Volunteers," *Archives of Environmental Health*, February, 1967, 14:289-291.

Milby, Thomas H.; Fred Ottoboni; and Howard W. Mitchell, "Parathion Residue Poisoning Among Orchard Workers," *Journal of the American Medical Association*, August 3, 1964, 189:351-356.

Mohler, Stanley R.; and Charles R. Harper, "Protecting the Ag Pilot," *Federal Aviation Agency*, September, 1966, AM66-30.

Nalbandian, Robert M.; and James F. Pearce, "Allergic Purpura Induced by Exposure to *p*-Dichlorobenzene: Confirmation by Indirect Basophil Degranulation Test," *Journal of The American Medical Association*, November 15, 1965, 194:238-240.

Quinby, Griffith E., "Feasibility of Prophylaxis by Oral Pralidoxime: Cholinesterase Inactivation by Organophosphorus Pesticides," *Archives of Environmental Health*, June, 1968, 16:812-820.

Radomski, J. L.; W. B. Deichmann; and E. E. Clizer, "Pesticide Concentrations in the Liver, Brain and Adipose Tissue of Terminal Hospital Patients," *Food and Cosmetics Toxicology*, 1968, 6:209-220.

Sataloff, Joseph; Lawrence Vassallo; Joseph M. Valloti; and Hyman Menduke, "Long-Term Study Relating Temporary and Permanent Hearing Loss," *Archives of Environmental Health*, November, 1966, 13:637-640.

Sataloff, Joseph; Lawrence Vassallo; and Hyman Menduke, "Temporary and Permanent Hearing Loss: A Ten-Year Follow-Up," *Archives of Environmental Health*, January, 1965, 10:67-70.

Sataloff, Joseph; and John A. Zapp, Jr., "The Environment in Relation to Otologic Disease," *Archives of Environmental Health*, March, 1965, 10:403-415.

Schafer, M., "Pesticides in Blood," *Residue Reviews*, 1968, 24:19-39.

Selby, Lloyd A.; Kenneth W. Newell; Carmel Waggenspack; George A. Hauser; and Gladys Junker, "Estimating Pesticide Exposure in Man As Related to Measurable Intake; Environmental Versus Chemical Index," *American Journal of Epidemiology*, 1969, 89:241-253.

Silverio, John, "Organic Phosphorus Insecticide Poisoning in Children," *The Journal of School Health*, November, 1969, 39:607-611.

Teichner, Warren H.; and Ernest Sadler, "Loudness Adaptation as a Function of Frequency, Intensity, and Time," *The Journal of Psychology*, 1966, 62:267-278.

Weiner, Betsy P.; and Robert M. Worth, "Insecticides: Household Use and Respiratory Impairment," *Hawaii Medical Journal*, March-April, 1969, 28:283-285.

West, Irma, "Lindane and Hematologic Reactions," *Archives of Environmental Health*, July, 1967, 15:97-101.

West, Irma, "Pesticide-Induced Illness: Public Health Aspects of Diagnosis and Treatment," *California Medicine*, October, 1966, 105:257-261.

Wyckoff, David W.; John E. Davies; Ana Barquet; and Joseph H. Davis, "Diagnostic and Therapeutic Problems of Parathion Poisonings," *Annals of Internal Medicine*, April, 1968, 68:878-882.

Zavon, Mitchell R., "Treatment of Organophosphorus and Chlorinated Hydrocarbon Insecticide Intoxications," *Modern Treatment*, July, 1967, 4:625-632.

Toxicity of Pesticides to Man

Toxic Effects From the Use of Agricultural Chemicals

A Review With Special Attention to Problems in Iowa Agriculture

OSCAR C. BEASLEY, M.D., and
KEITH R. LONG, Ph.D.

A NEW FIELD IN MEDICINE is appearing because of the universal utilization of toxic chemicals in the growth, preservation, and preparation of human food. This field deserves and is receiving study by members of many disciplines, including the economist, the entomologist, the botanist, the agronomist, the geneticist, the industrial hygienist, the toxicologist, members of government, and the practicing physician. Laymen in all walks of life have become excited about the possible ultimate effects of these materials on "nature," in large part because of Rachel Carson's book entitled, SILENT SPRING.[1]

Belatedly, doctors of medicine as a group are becoming aware of the human illnesses that agricultural chemicals can produce. It seems that in many hazardous industries the interest and help of the medical profession have arrived late in the day, and only after extensive evidence of human disease has been compiled. Thus it was in mining industries, where workers were endangered by silicosis and tuberculosis in the 1920's; in industries that used radioactive materials before protective devices and methods came into general use; and in the beryllium industry before protection was defined and effected for the workers. In all these instances and in agricultural-chemical toxicity, a large body of basic knowledge existed many years before it was applied by industry and by practitioners of medicine.

From the standpoint of the practicing physician and the industrial hygienist, the psychologic ill-effects of agricultural chemicals on workers must be considered. The newspapers are replete with articles, often prominent editorials, on the ill-effects of chemicals on "nature." They are also replete with hints and discussions of frank human poisoning with agricultural chemicals. One editorial in the DES MOINES REGISTER advised the farmers in Iowa who must fend off a new variety of corn-root infestation with a recommended new chemical to notify their doctors in advance about their likelihood of exposure.

Additional examples of public concern are evident in government circles. For example, the President's Science Advisory Committee met in the spring of 1963 and compiled a report entitled "The Use of Pesticides,"[2] which was then published and released to the general public. In the winter of 1963-64, a Congressional bill to allocate funds for an investigation of the hazards of agricultural chemicals attracted wide attention. It is apparent that the public is being aroused over the problem. The medical profession must assert itself and decide the following points:

(1) Is there a significant health problem from the use of agricultural chemicals?

(2) If there is, how can it be prevented?

(3) If poisoning occurs, can victims be treated effectively?

Our knowledge of the toxic effects of agricultural chemicals in human beings is quite limited. It is a purpose of this paper to outline broadly and to comment upon the toxicity of those agricultural chemicals that are in popular use in Iowa.

Although the number of individual pesticidal agents which organic chemists may contrive is almost infinite, it is fortunate for the student of this subject that the many compounds fall into a few broad classes, each of which has more or less specific biologic effects. For a pesticide to work at practicable dosage levels, it must disrupt or alter some vital process of the plant or animal organism. Presently-useful poisons block the cell's transformation of energy by interfering with the enzymes of energy transfer, interrupting communications between cells (e.g., nervous system), blocking transport of vital material across cell membranes, preventing synthesis of cellular constituents for growth or reproduction, and impeding vital bio-physical phenomena involved in locomotion or in circulation of body fluids.[3]

Obviously, man has no immunity to these cellular poisons, and hence his safety must depend on the physical separation of these substances from his body's internal environment. Many of the newer pesticides are rapidly absorbed through human skin; others are not. An intelligent avoidance of exposure requires first a knowledge of the relative toxicity of the various compounds, and then the implementation of physical devices that will enable him to handle these chemicals without absorbing harmful quantities.

This year, farmers will be using not only the usual organophosphorus and chlorinated hydrocarbon pesticides but also some combinations of them.

Organic phosphorus compounds in use will be:

Product	*Manufacturer*
Diazinon	Geigy Chemical Corporation
Di-Syston	Chemagro Corporation
Niran-10G	Monsanto Company
Thimet	American Cyanamid
Sta-Thion	Woodbury Chemical Company
4072	Shell Chemical Company
P-Ten-G	Ortho Division, California Spray Company

Chlorinated hydrocarbon pesticides in use will be:

Aldrin	Shell Chemical Company
Heptachlor	Velsicol Chemical Corporation
DDT	Geigy Chemical Corporation

Combinations of organophosphorous and chlorinated hydrocarbon compounds will be:

Aldrex (Aldrin-Parathion)	Shell Chemical Company
Hepathion (Heptachlor-Parathion)	Woodbury Chemical Company
Parahep (Parathion-Heptachlor)	Velsicol Chemical Company
AP-10-10-G (Aldrin-Parathion)	Ortho Division, California Spray Company

I. THE ORGANIC PHOSPHORUS INSECTICIDES

All organic phosphorus compounds are derivatives of phosphoric acid and have an identical physiologic action, namely, the inhibition of the enzyme cholinesterase. They therefore permit the accumulation of excessive quantities of acetylcholine. The symptoms and signs of organic phosphorus poisoning are neurologic in origin, thus accounting for the term "nerve gas"[4] that was applied to the organic phosphorus war gas that was developed but never used in World War II. The whole gamut of cholinergic effects, from the slightest symptoms to coma, are dependent upon the dose. In man, the usual symptoms are headache, lightheadedness, anxiety, blurred vision, weakness, nausea, cramps, diarrhea and tightness in the chest. Signs include sweating, constriction of pupils, weeping, salivation. excessive bronchial secretion, vomiting, cyanosis. papilledema, muscle twitching, convulsions, coma. loss of reflexes, and loss of sphincter control.[5]

In animals experimentally poisoned with the organic phosphate insecticides, various degrees of heart block and cardiac arrest are seen. In suicide cases, death has occurred within five minutes of ingestion.[5]

These insecticides are absorbed by three routes: skin, respiratory tract, and gastrointestinal tract. Skin absorption is enhanced by warm temperature and by the presence of dermatitis. Therefore, a dose so small as to be inocuous for intact skin may produce frank poisoning when significant skin abrasions exist.[5]

The cholinesterase of various tissues and blood can be quantitatively measured.[6] The degree of illness produced by organic phosphorus poisons correlates only grossly with the degree of cholinesterase reduction. Although these poisons act more or less as irreversible inhibitors of acetylcholine. the organism is able to replenish the enzyme continuously. Therefore rapid absorption is usually much more detrimental than slower absorption of an identical quantity. Also, a given degree of depression of the blood cholinesterase level may be associated with mild to severe chemical poisoning. depending upon the rapidity of absorption. Thus. a subject may show marked depletion of blood cholinesterase in cases of repeated exposure to small doses, and yet not appear ill. Critical poisoning usually does not occur in the absence of a marked reduction of the enzyme level. Once poisoned, however, the patient may take as long as three months to regain his normal enzyme level. Therefore in an agricultural worker, re-exposure to the poison after clinical recovery but before laboratory recovery holds the danger that an additional dose that would be clinically harmless in an undepleted subject may reduce his enzymes to a critical level.[5]

The organic phosphorus insecticides vary in their rates of absorption, in their potency, in their physical properties, and in their biochemical action. Some are direct inhibitors of cholinesterase but others first undergo a chemical alteration in the body, and are therefore called indirect inhibitors.[5]

The poisoned subject may show leukocytosis. albuminuria, glycosuria, and hemoconcentration. By a special technic, the cholinesterase activity of the blood, plasma and solid tissue can be determined. A value of less than 0.5^{Δ} pH/hr. is believed to be significant, although in workers who have been chronically exposed, levels of 0.2^{Δ} pH/hr. or less may be noted without evidence of illness.[5]

The end products of organic phosphorus absorption can be estimated in the urine. Paranitrophenol has been used as an index of the amount of parathion absorbed. Other end products have also been measured.[5]

There are no gross or microscopic changes in the tissues of animals that have been poisoned with the organophosphates. Effective treatment of victims is available, but because there are no definitive pathognomonic findings, three considerations in differential diagnosis deserve special emphasis: (1) In the absence of an immediate laboratory determination of cholinesterase, various unrelated conditions including cerebral hemorrhage, heat stroke, hypoglycemia, gastroenteritis and pneumonia have been confused with poisoning. (2) In the presence of mild (and perhaps asymptomatic) poisoning, asthenia and various psychogenically induced reactions such as hysteria are likely, especially in the co-workers of poisoning victims. (3) Men exposed to some of the less toxic organic phosphorus insecticides may actually exhibit symptoms resulting from unrecognized exposure to more toxic insecticides such as mercurial or arsenical agents.[5]

In the great majority of cases treatment of organic phosphorus poisoning can be successful if certain principles are kept in mind. The illness is a physiologic disturbance resulting from a reversible enzyme depletion, and no structural damage occurs except by way of complication. The pharmacologic derangements are directly counteracted by atropine and other newer pharmacologic agents.[5, 7, 8] If the vegetative functions—specifically respiration and tracheobronchial toilet—can be maintained artificially until the critical enzymatic depletion has been sufficiently overcome, recovery is usually assured.

Artificial respiration should be the first step in severe cases. Then large doses of atropine (two to four milligrams) should be given intravenously, followed by 2-PAM (2-pyridine aldoxime methylchloride and 2-pyridine aldoxime methyliodide). Decontamination of the skin, stomach and eyes should follow.

Although this summary is not intended to be complete, several points in treatment merit additional emphasis: (1) Organophosphorous insecticides are usually much more soluble in alcohol than in water. Hence, cleaning contaminated skin with ethyl alcohol is worthwhile. (2) Sedatives, tranquilizers and narcotics are contraindicated. (3) Any victim who is sick enough to need one dose of atropine merits at least 24 hours of close medical observation because of the possibility of progression to a more serious state of toxicity.

II. THE CHLORINATED HYDROCARBON INSECTICIDES

DDT is the best known of the chlorinated hydrocarbon group of insecticides. They have two features in common, namely (1) the chemical composition implied by their name and (2) a principal effect on the human central nervous system. The chemical structures of these compounds vary widely, and there are definite quantitative and qualitative differences in their effects. At least to some extent, they are all absorbed via the respiratory and gastrointestinal tracts, and the skin. In man and higher animals, skin absorption is minimal for some of these compounds including Dilan and DDT in powder form. Others such as Dieldrin and Heptachlor are rapidly absorbed through the skin, even when the chemical is in the dry form.[3, 5]

Pathologic studies of the tissues of poisoned subjects usually reveal no structural abnormality. In the case of chronically poisoned animals, microscopic changes in the liver and kidney may be seen.[5] The intracellular biochemical reaction produced by the chlorinated hydrocarbon insecticides has not as yet been elucidated. However, accruing evidence suggests that these compounds interfere with the enzyme system which generates high-energy phosphate compounds.[9]

The symptoms and signs produced by these compounds are varied, but serious poisoning always culminates in generalized convulsions. In DDT poisoning the characteristic early symptoms are paresthesias around the mouth and in the extremities, malaise, headache and fatigue. Ataxia, confusion and tremor are noted. Except in the more severe cases, recovery is well underway within 24 hours. Therefore in patients who do not start recovering promptly, a misdiagnosis must be considered.

The chlorinated-hydrocarbon insecticides, being central-nervous-system affectants, produce specific electroencephalographic abnormalities, namely bilateral synchronous spikes, spike-and-wave complexes, and slow theta waves. These changes may be evident in subclinically-poisoned subjects. These compounds and some of their degradation products are stored in the fat, where they can be measured on biopsy or at autopsy. This storage apparently sequesters the material and renders it more or less harmless, since the total amount that is found stored in a chronically poisoned animal may be greater than the amount that would be lethal when administered in a single dose. These insecticides, with their degradation products, are excreted in the urine. It is practical to examine urine for these products in cases of suspected poisoning.[5] There are no practical laboratory tests for diagnosis.

The treatment of chlorinated-hydrocarbon insecticide poisoning is limited to the control of the convulsions. Barbiturates are preferred, but calcium gluconate has been effective in experimental animals. Barbiturates and calcium gluconate may be used concurrently because the mechanics of their action are different.

SUMMARY

The toxicology of two frequently used classes of agricultural poisons has been reviewed. Attention is called to the fact that regardless of long-term broad biologic considerations, the use of these substances is permanently entrenched in and economically essential to modern agriculture. The important place of agricultural poisons in the spectrum of human disease is a neglected area in the physician's training. Because of the psychologic manifestations which may occur in patients who think that they have been poisoned, it is just as important for physicians to know what effects these materials do *not* produce as to recognize those which they do cause.

There is much to be learned, especially in the area of accumulated effects and acute subclinical effects. At present, it appears that with proper precautions man is well able to utilize the new scientific advances in agricultural chemicals, and to reap the greater food abundance resulting therefrom without suffering significant detriment.

REFERENCES

1. Carson, Rachel L.: Silent Spring. Houghton Mifflin Co., Boston, 1962.
2. The President's Science Advisory Committee: Use of pesticides. Washington, D. C., U. S. Government Printing Office, (May 15) 1963.
3. Smith, P. D.: Toxic hazards in aerial application: 62-8 Federal Aviation Agency, Civil Aeronautical Research Institute, Oklahoma City, Oklahoma, (April) 1962.
4. Nerve Gases, (Editorial). J.A.M.A., **178**:755-756, (Nov. 18) 1961.
5. Hayes, W. J., Jr.: Clinical Handbook on Economic Poisons. Atlanta, Georgia, Communicable Disease Center, U. S. Public Health Service, 1963.
6. Michel, H. O.: Electrometric method for determination of red blood cell and plasma cholinesterase activity. J. Lab. & Clin. Med., **34**:1564-1568, (Nov.) 1949.
7. Quinby, G. E.: Further therapeutic experience with pralidoximes in organic phosphorus poisoning. J.A.M.A., **187**:202-206, (Jan. 18) 1964.
8. Verhulst, H. D.: Organic phosphate anticholinesterase poisoning; its occurrence and treatment. Presented at USPHS Clinical Society Meeting, April 7-8, 1962.
9. Daughterty, J. W., Lacey, D. E., and Korty, P.: Problems in aerial applications: effects of chlorinated hydrocarbons on substrate-linked phosphorylation. 63-4. Federal Aviation Agency, Civil Aeromedical Research Institute, Oklahoma City, (March) 1963.

Organochlorine Pesticide Residues in Human Tissue —Hawaii

L. J. Casarett, PhD; G. C. Fryer, MD; W. L. Yauger, Jr.; and H. W. Klemmer, PhD,

In RECENT years, considerable interest has been evoked in the levels of organochlorine pesticides in human body tissues. This interest arises from a continuing entry of these chemicals into the environment and the need for an assessment of potential effects of human health. Among the many evaluations of levels of pesticides in tissues are reports from several parts of the world including the United States,[1-7] and from a number of regional populations of the United States.[8-14]

The data presented in this paper have been gathered as part of a survey in Hawaii. Although this is a continuing program and the data presented have certain limitations, the report is presented because analytical techniques have been modified in current studies. None of the subjects of this report was known to have had an unduly high exposure to pesticides and, insofar as can be ascertained, this group appears to be reasonably representative of that segment of the population coming to autopsy with regard to ethnic and socioeconomic factors and, presumably, in potentiality for pesticide exposure.

From the Pacific Biomedical Research Center, University of Hawaii, Honolulu.

The views expressed herein are those of the investigators and do not necessarily reflect the official view of the Public Health Service.

Reprint requests to 2538 The Mall, Honolulu 96822 (Dr. Casarett).

Methods and Materials

Autopsy tissues were supplied by two major Honolulu hospitals from 44 subjects ranging in age from 28 weeks gestation to 88 years. Specimens of the following tissues were requested: perirenal, mesenteric and panniculus fat, liver, lung, tracheobronchial lymph nodes, brain, bone marrow, adrenal, kidney, gonad, and spleen. A majority of these tissues were supplied for most autopsy sets. Full pathology reports were examined carefully to segregate findings which might have had a relationship to residue levels.

Tissue samples were prepared for analysis essentially by methods described by Mills et al[15] and deFaubert Maunder et al.[16] Following homogenization in dry ice and extraction with hexane in the presence of anhydrous sodium sulfate the solutions were evaporated and the residue weighed to determine the lipid extracted. Redissolved in hexane, the sample was divided, one part for cleanup in a column (Florisil) for gas chromatographic determination of all organochlorines reported in this paper, except dieldrin. The second portion of the sample was further treated with 25% potassium hydroxide, extracted with hexane, placed on a column and eluted with a mixture of 25% diethyl ether in petroleum ether. This fraction was used for the determination of dieldrin.

Gas chromatography was carried out on an instrument (MT-220), equipped with four inlets with glass inserts, flow combiners and two electron-capture detectors. Columns were 6 × ¼ inches packed with either 5% trifluoropropylmethyl silicone (QF-1) or 10% silicone oil (Dow Corning), each on 60/80 mesh chromatographic supports (Chromosorb), treated with dimethyldichloro silane. Each sample was put through both columns. Column tempera-

Table 1.—Mean Organochlorine Values of Extracted Lipid

Tissue			Organochlorine Residue (ppm)						
Type	No. Analyzed	% Lipid Content	Heptachlor-epoxide	DDE	DDD	DDT	Dieldrin	Total	s†
Bone marrow	30	20.6	0.02	18.47	1.45	4.48	0.41	24.8	(24.9)
Liver	42	2.1	0.08	12.93	2.43	1.75	0.43	17.6	(26.1)
Perirenal fat	30	55.7	0.04	11.47	0.02	3.14	0.08	14.8	(21.8)
Mesenteric fat	29	54.2	0.05	10.26	0.08	2.85	0.17	13.4	(17.6)
Panniculus fat	30	60.6	0.05	9.86	0.03	2.81	0.06	12.8	(16.6)
TBLN‡	11	8.6	<0.01	8.31	0.03	3.73	0.11	12.2	(12.9)
Spleen	27	0.6	<0.01	6.05	0.49	2.62	1.80	11.0	(19.8)
Lung	25	0.7	0.02	8.78	0.05	1.82	0.16	10.8	(21.4)
Adrenal	18	10.5	0.01	8.57	0.94	0.98	0.10	10.6	(15.4)
Gonad	36	1.3	0.01	7.78	0.18	1.53	0.62	10.1	(16.7)
Kidney	38	3.2	0.01	6.32	0.12	1.66	0.66	8.9	(10.4)
Brain	32	7.9	<0.01	1.59	0.22	0.12	0.04	2.0	(4.5)

*Expressed as parts per million extracted lipid for residues of organochlorine pesticides detected in 12 different human tissues from 44 autopsies.
†s = standard deviations for total organochlorines per tissue.
‡Tracheobronchial lymph nodes.

tures were maintained at 180 C with nitrogen flow rate of 100 ml/min.

Confirmatory analyses were run periodically on the MT-220 equipped with a microcoulometer (Dohrmann C-200) and occasionally by thin-layer chromatography.

Results

Residue Analysis.—Mean values for five organochlorine residues detected for all of the tissues examined are presented in Tables 1 and 2. Table 1 presents the data expressed in parts per million of extractable lipid material while data in Table 2 are expressed in parts per million of whole tissue. In each Table, values are ranked in order of the total mean organochlorine residue.

There was a great deal of variation among the tissues which is clearly evident from the standard deviations. Analysis of variance of the total organochlorines in each tissue in each autopsy indicated significant differences among the autopsies and tissues as expressed in terms of lipid, and among tissues when data were expressed on a whole-tissue basis. Applying stringent criteria to the data suggested that the differences lay in the extremes of the distribution of values.

Table 1 shows that the highest values for all of the organochlorines except dieldrin were found in bone marrow or liver. The high value for dieldrin in the spleen was due to a single extremely high value which is in some doubt, so that the highest levels of dieldrin can be considered to be in the kidney and gonad. Of particular interest were the values for the liver and brain. The liver contained relatively high levels of organochlorines despite the fact that the lipid content was low and, conversely, the brain with an appreciable lipid concentration contained the smallest total amount of organochlorines. No attempt was made to identify the lipids extracted.

As indicated in Table 2, the fat samples contained the highest concentrations of organochlorines. The three samples of fat do not differ significantly from one another in their pesticide content and can be considered as adipose tissue without regard to the source. The total residue values corresponded generally to the relative amount of lipid in the tissue with only one major exception, the brain. The relative ranking of the tissues in Tables 1 and 2, disregarding the spleen, changed in only two major organs, viz, the liver and the lung.

A selection of data was made to compare autopsies from which comparable tissues had been analyzed. Table 3 presents data from five organs and three fat samples (combined as adipose) ranked in order of the total residue values for each autopsy. Analysis of variance indicates significant differences ($P < 0.01$) among tissues and among autopsies. Application of stringent criteria after Tukey[17] indicates the source of the variance. Liver and adipose differ from brain, but not from each other; autopsies 45 and 15 are different from one another and all other autopsies.

The last five autopsies in Table 3 (No. 21, 36, 41, 68, and 37) were from individuals of 36 weeks gestation to 2.5 months of age. Al-

Table 2.—*Mean Organochlorine Values of Whole Tissue**

Tissue			Organochlorine Residue (ppm)						
Type	No. Analyzed	% Lipid Content	Heptachlor epoxide	DDE	DDD	DDT	Dieldrin	Total	s†
Perirenal fat	30	55.7	0.0220	4.64	0.0110	1.33	0.0300	6.03	(5.30)
Mesenteric fat	29	54.2	0.0320	4.40	0.0470	1.35	0.0630	5.89	(4.98)
Panniculus fat	30	60.6	0.0270	4.48	0.0180	1.16	0.0270	5.71	(5.25)
Bone marrow	19	20.6	0.0040	2.08	0.0760	0.411	0.0620	2.63	(2.21)
TBLN‡	11	8.6	0.0001	1.38	0.0100	0.892	0.0190	2.30	(4.52)
Adrenal	18	10.5	0.0012	0.875	0.0570	0.125	0.0060	1.06	(1.31)
Kidney	38	3.2	0.0009	0.209	0.0022	0.0827	0.0056	0.300	(0.651)
Liver	42	2.1	0.0019	0.200	0.0326	0.0467	0.0037	0.285	(0.369)
Brain	32	7.9	0.0002	0.0831	0.0020	0.0105	0.0031	0.0989	(0.171)
Gonad	36	1.3	0.0001	0.0688	0.0015	0.0150	0.0021	0.0875	(0.103)
Lung	25	0.7	0.0003	0.0585	0.0009	0.0147	0.0022	0.0766	(0.125)
Spleen	27	0.6	Trace	0.0305	0.0031	0.0112	0.0021	0.0469	(0.074)

*Expressed as parts per million of whole tissue for residues of organochlorine pesticides detected in 12 different human tissues from 44 autopsies.
†s = standard deviations for total organochlorines per tissue.
‡Tracheobronchial lymph nodes.

Table 3.—*Total Organochlorine Residues as Parts per Million of Extracted Lipid in Comparable Autopsy Sets*

Autopsy No.	Age	Total Organochlorines (ppm)						
		Liver	Adipose	Spleen	Kidney	Gonad	Brain	Mean
45	36 yr	151.99	101.28	33.82	46.00	80.70	7.38	70.20
15	63 yr	56.20	19.85	97.10	20.16	14.66	0.97	34.82
35	80 yr	28.56	23.47	2.72	39.95	19.54	1.70	19.32
32	68 yr	23.61	18.28	16.60	8.06	6.71	0.68	12.32
44	67 yr	16.81	26.49	4.40	10.32	9.44	1.00	11.41
33	72 yr	16.11	16.65	8.69	6.93	3.44	1.58	8.90
64	50 yr	10.21	12.01	11.18	10.16	2.96	5.59	8.69
40	51 yr	24.56	12.12	2.21	4.67	4.28	0.48	8.05
22	50 yr	5.96	8.04	14.22	7.39	8.37	1.06	7.51
18	65 yr	5.02	12.77	9.14	9.31	5.40	1.16	7.13
63	53 yr	13.63	8.29	3.79	14.60	0.49	0.84	6.94
23	54 yr	16.59	9.34	<0.01	6.86	6.20	0.76	6.63
34	56 yr	19.61	5.96	3.24	5.48	5.00	0.02	6.55
20	71 yr	17.20	6.95	3.00	1.61	5.35	0.25	5.73
16	46 yr	4.19	4.65	0.65	5.31	11.43	0.54	4.46
21	3 days	6.29	1.71	3.36	2.46	2.02	1.01	2.81
36	36/40	6.95	6.79	0.03	<0.01	0.47	<0.01	2.38
41	40/40	1.75	4.04	2.90	0.52	3.20	0.14	2.09
68	38/40	3.84	4.18	<0.01	3.44	<0.01	0.55	2.01
37	2½ mo	1.51	3.83	<0.01	0.22	0.40	0.15	1.02

though the individual values were not consistently deviant from many of the adult tissues, they can be separated on the basis of age. The mean total residue level for these five autopsies was 12.4 ppm while the mean for the adult tissue was 87.5 ppm, significantly different from one another ($P < 0.01$).

Correlation With Pathology

A summary of autopsy findings is presented in Table 4, in which autopsies are listed in descending order of total organochlorines expressed on the basis of extractable lipid material. The summary lists major pathologic findings, the pathologic condition of the liver when present, and an indication of body state.

The five subjects with the highest total organochlorine levels for all tissues (Autopsy No. 45, 24, 66, 1, and 15) had three characteristics in common. They were emaciated,

Table 4.—*A Comparison of Summarized Pathologic Findings**

Autopsy No.	Age	Organochlorine Level in ppm	Pathologic Findings
45	36 yr	70.5	Nasopharyngeal carcinoma with liver metastases; cachexia
24	63 yr	53.0	Bronchogenic carcinoma; Laennec's cirrhosis; emaciated
66	57 yr	35.6	Adenocarcinoma of stomach; widespread metastases including liver; emaciation and fat atrophy
1	56 yr	34.1	Nasopharyngeal carcinoma; focal fatty change in the liver; emaciated
15	63 yr	30.6	Carcinoma of stomach; hepatoma in cirrhotic liver; emaciated
26	65 yr	24.0	*S albus* septicemia; acute and subacute bacterial endocarditis
35	80 yr	20.2	Bleeding diathesis, c. cerebral hemorrhage; emaciated
19	51 yr	17.8	Ruptured intracranial aneurysm; well nourished
44	67 yr	15.2	Adenocarcinoma of colon; liver failure; emaciation and edema
32	68 yr	14.1	Myocardial infarct—ventricular fibrillation; obese
9	60 yr	12.5	Epidermoid carcinoma of piriform fossa
22	50 yr	10.7	Septicemia and liver failure; Laennec's cirrhosis; well nourished
33	72 yr	10.3	Occlusion R. int. carotid (thrombus left atrial appendage)
64	50 yr	9.8	Carcinoma of cervic uteri; thrombosis inferior vena cava; not emaciated
18	65 yr	9.7	Ruptured liver—adenocarcinoma; not emaciated
40	51 yr	8.9	Ruptured abdominal aorta
65	62 yr	8.6	Myocardial infarct; centrilobular necrosis of liver
67	53 yr	8.5	Myocardial infarct
17	88 yr	7.7	Bronchopneumonia following trochanteric fracture (aged, chronic chest infection)
23	54 yr	7.4	Sudden death; suspected pheochromocytoma
63	53 yr	7.4	Carcinomatosis following adenocarcinoma of the kidney
14	57 yr	7.3	Ruptured intracranial aneurysm; fatty change in liver; moderate obesity
34	56 yr	7.0	Myocardial infarct
16	46 yr	6.9	Carcinoma of pancreas; pulmonary embolus; not emaciated
12	82 yr	6.8	Acute pneumonia on chronic emphysema
13	77 yr	6.5	Aspiration pneumonia; emaciated—unexplained anorexia
8	86 yr	5.9	Subdural hemorrhage
4	52 yr	5.3	Catastrophic hemorrhage from peptic ulcer
20	71 yr	5.3	Pontine infarct
5	57 yr	4.7	Disseminated carcinoma
2	52 yr	4.5	Myocardial infarct
6	56 yr	3.9	Myocardial infarct
7	53 yr	3.6	Myocardial infarct
10	57 yr	3.0	Chronic lymphocytic leukemia
21	3 days	2.8	Pneumonia; immaturity
3	64 yr	2.7	Myocardial infarct
11	68 yr	2.6	Myocardial infarct; obese
36	36/40	2.6	Stillbirth, erythroblastosis
68	38/40	2.5	Neonate; respiratory distress
25	2 yr	2.2	Head injury
41	40 yr	2.1	Fetal death in utero (infection)
38	66 yr	1.8	Pulmonary tuberculosis; emaciated
37	2½ mo	1.1	Suppurative encephalitis (meningomyelocele)
39	28/40	1.0	Fetal death in utero, unknown cause

*Total organochlorine levels expressed as parts per million of extracted lipid for 44 autopsies.

they had a variety of carcinoma, and they had widespread focal or generalized abnormalities of the liver. Other subjects with carcinoma did not show as high levels of total organochlorines in their tissues. Subjects 45 and 15 which were significantly different from other subjects in residue values differed also from the others with the triad of carcinoma, cachexia, and generalized liver abnormalities, in that they were the only subjects with metastatic carcinomata in the liver.

When a similar comparison was made with total organochlorine residues expressed on a whole-tissue basis, there was no apparent preponderance of any particular abnormality with organochlorine level. When autopsy cases were arranged in this way, the five showing highest organochlorine levels had, as major pathologic finding, *Staphylococcus albus* septicemia, carcinoma of the pyriform fossa, carcinoma of the stomach, ruptured intracranial aneurysm, and bleeding diathesis with cerebral hemorrhage. It was noted, irrespective of which method of expression was used, those dying a sudden death, such as by myocardial infarct, seemed to be clustered near the center of the distribution of organochlorine levels compared to others who had undergone prolonged periods of illness before death.

Comment

The range and magnitude of the organochlorine residues found in this study were not markedly dissimilar from those reported elsewhere in the United States or in other countries with few exceptions. The concentration in adipose tissue as with other parts of the United States was somewhat higher than those reported for Germany,[6] the United Kingdom,[18] and Australia,[1] but smaller than values reported for Hungary,[19] and India.[3]

Levels in other organs and tissues tended to be comparable to those found in other parts of the United States. For example, in Florida,[10] fat values expressed in parts per million of whole tissue were somewhat higher than those reported here, but concentrations found in the liver, kidney, gonad and brain were of the same order of magnitude in the Hawaii population as in the Florida study. Other tissues reported here were not reported in the Florida group. Although no strict parallel was found between pesticide level and age, infant and fetal tissue contained significantly lower levels than adult tissues in this limited study in contrast to that found in the Florida study.

Expressed as parts per million of whole tissue, organochlorine levels generally related to the amount of extractable lipid in the tissue, except for the brain. It should be noted, however, that this is not a strict proportionality. Compared to adipose pesticide concentration, the visceral organs contained inordinate concentrations of organochlorines. This is especially true for the liver and lung, and is illustrated by the fact that in changing modes of expression from whole tissue to lipid basis, the only marked changes in tissue ranking are accounted for by these organs.

It is also of interest to comment on the relative significance of the lipid levels. Material contained in adipose tissue can be considered to be "stored" pesticide, the concentration at the time of sampling being dependent on the steady state existing in the body. Pesticide levels in viscera, on the other hand, are likely to be a part of lipoid elements of parenchymal tissue. In this light, the concentrations of organochlorines in these tissues, even at low levels, should be considered as being of greater functional significance. This is particularly true for organs which are sites of metabolic activity or sites of known effects of the pesticides.

The discrepancy between storage and functionally significant levels is not always readily apparent, but can be quite marked. For example, in Table 2, if the concentration in adipose tissue is taken as a storage level, the ratio of what is actually present to what would be predicted on a simple storage basis is about 16 and 33 for liver DDT and dichlorodiphenyl dichloroethylene (DDE) respectively and 50 and 68 for lung on the same basis.

Such considerations become of some importance for at least two reasons. In assessing probable routes of entry and disposition of pesticides, the relative levels of individual organochlorines in the lung might well provide indications of recent exposure rather than steady-state levels. As occurred in this series, high levels in pulmonary lymph nodes associated with elevated lung values suggest that this is a possible pathway of clearance of material deposited in the lung by inhalation as described for many other materials.[20,21] This suggestive evidence is strengthened in some instances where, in individuals showing elevated lung or lymph node values, the ratio of DDE to DDT is markedly reduced or even reversed compared to the tissues from the same individual. A second obvious reason for viewing tissue levels in terms of "non-stored" pesticides is in assessing possible effects more realistically. This is of importance in a metabolically active site such as the liver, a site of entry such as the lung or gastrointestinal tract, or a possible site of effect with lasting consequences (eg, the possible production of chromosomal aberrations).

The fact that individuals with the highest total organochlorine levels (Table 4) had pathologic conditions which included the triad of cachexia, carcinoma, and liver abnormalities is of some import. One can speculate that weight-loss, particularly if sudden, might raise the pesticide levels in the tissues as storage depots are depleted. Although admittedly remote, if weight-loss occurred suddenly, one might expect that some evidence of pesticide poisoning could occur from tissue levels which had been below the range of effective levels before the loss of weight.

Carcinoma, if accompanied by wasting, might also be expected to raise tissue levels. Although it is impossible to support from present data, the possibility that carcinoma might enhance sequestration of pesticides in the body cannot be disregarded. Liver abnormalities on the other hand, can be expected to influence the distribution and metabolism of the organochlorines.

Two cases in which tracheobronchial lymph nodes were quite high (43 and 27 ppm) are worthy of comment. These are the only two cases out of 11 in which nodes were analyzed in which septicemia deaths occurred. Although this is suggestive at best in the context of this report, other unpublished cases support the possibility of a relation between lymph node levels and the occurrence of infection via the lung.

This study was supported by grant PH 86-65-79 from the Pesticides Program, National Communicable Disease Center, Bureau of Disease Prevention and Environmental Control, Public Health Service, Department of Health, Education, and Welfare.

Drs. James G. Bennett and Grant N. Stemmerman, provided the specimens and allowed access to their very detailed autopsy reports.

References

1. Bick, M.: Chlorinated Hydrocarbon Residues in Human Fat, *Med J Aust* **54**:1127-1130, 1967.

2. Dale, W.E., and Quinby, G.E.: Chlorinated Insecticides in Body Fat of People in the USA, *Science* **142**:593-595 (Nov 1) 1963.

3. Dale, W.E.; Copeland, F.M.; and Hayes, W.J., Jr.: Chlorinated Insecticides in the Body Fat of People in India, *Bull Wld Hlth Org* **33**: 471-474, 1965.

4. Hayes, W.J., Jr.; Dale, W.E.; and Le Breton, R.: Storage of Insecticides in French People, *Nature* **199**:1189-1191 (Sept 21) 1963.

5. Hunter, C.G.; Robinson, J.; and Richardson, A.: Chlorinated Insecticide Content of Human Body Fat in Southern England, *Brit Med J* **1**:221-224, 1963.

6. Maier-Bode, H.: DDT in Körperfett des Menschen, *Med Exp (Basel)* **1**:146-148, 1960.

7. Quinby, E., et al: DDT Storage in the US Population, *JAMA* **191**:175-179 (Jan 18) 1965.

8. Durham, W.F., et al: Insecticide Content of Diet and Body Fat of Alaskan Natives, *Science* **134**:1880-1881 (Dec 8) 1961.

9. Hayes, W.J., Jr.; Dale, E.W.; and Burse, V.R.: Chlorinated Hydrocarbon Pesticides in the Fat of People in New Orleans, *Life Sciences* **4**:1611-1615 (Aug) 1965.

10. Fiserova-Bergerova, F., et al: Levels of Chlorinated Hydrocarbon Pesticides in Human Tissues, *Indus Med Surg* **36**:65-70 (Jan) 1967.

11. Robinson, J., and Hunter, C.G.: Organochlorine Insecticides, *Arch Environ Health* **13**:558-563 (Nov) 1966.

12. Hoffman, W.S.; Fishbein, W.I.; and Andelman, M.B.: Pesticide Content of Human Fat Tissue, *Arch Environ Health* **9**:387-394 (Sept) 1964.

13. Hoffman, W.S.; Fishbein, W.I.; and Andelman, M.B.: Pesticide Storage in Human Fat Tissue, *JAMA* **188**:819 (June 1) 1964.

14. Zavon, M.R.; Hine, C.H.; and Parker, K.D.: Chlorinated Hydrocarbon Insecticides in Human Body Fat in the United States, *JAMA* **193**:837-839 (Sept 6) 1965.

15. Mills, P.A.; Olney, J.N.; and Gaither, R.A.: Rapid Method for Chlorinated Pesticide Residues in Non-fatty Foods, *J Assoc Offic Ag Chemists* **46**:186-191 (Feb) 1963.

16. deFaubert Maunder, N.J., et al: Clean-up of Animal Fats and Dairy Products for Analysis of Chlorinated Pesticide Residues, *Analyst* **89**:168-174 (March) 1964.

17. Tukey, J.W., as cited by R.G.D. Steel and J. Torrie, *Principles and Procedures of Statistics*, New York: McGraw-Hill Book Co., 1960.

18. Robinson, J., et al: Organochlorine Insecticide Content of Human Adipose Tissue in S.E. England, *Brit J Indust Med* **22**:220-229, 1965.

19. Denes, A.: Lebensmittel Chemische Probleme von Rückständen Chlorieter Kohlenwasserstoffe, *Nahrung* **6**:48-52, 1962.

20. Casarett, L.J.: Some Physical and Physiological Factors Controlling the Fate of Inhaled Substances, *Health Physics* **2**:379-386 (May) 1960.

21. Morrow, P.E., and Casarett, L.J.: *Inhaled Particles and Vapours*, C.N. Davis (ed.) Oxford: Pergamon Press, 1961, p 167.

Central Nervous System Effects of Chronic Exposure to Organophosphate Insecticides

J. Robert Dille, M.D., and Paul W. Smith, Ph.D.

THE ORGANOPHOSPHATE cholinesterase-inhibitors, which were first introduced during World War II as potential agents of chemical warfare, have yielded a variety of derivatives which are now among the most widely used of all pest-control agents.

As early as 1941 it had been established that the high toxicity of these compounds was due to the persistence of their anticholinesterase action, although the mechanism by which they inactivated the enzyme was not known until later. Certain of these compounds are still used in the treatment of such clinical states as glaucoma and myasthenia gravis, where persistence of action can be an advantageous feature.

While the symptoms of organophosphate poisoning vary in rapidity of onset, severity and duration depending on the specific compound involved and a variety of other factors, they are readily divided into two categories, peripheral and central, depending upon their point of origin.

From the Civil Aeromedical Research Institute, Oklahoma City, Oklahoma.

The peripheral symptoms are, in general, more dramatic and immediately incapacitating than are the central effects, and until about ten years ago the latter had received little attention. At about that time it began to be realized that cholinesterase inhibitors could produce sensorimotor, behavioral and personality changes, but there was no general agreement as to their potential severity and importance.[1]

Little has been written about this problem in the intervening years, but recent reports, particularly that of Gershon and Shaw,[3] have firmly established the psychiatric sequelae of organophosphate poisoning as a toxicologic entity. Moreover, although the patients in these studies were ultimately considered to have recovered completely, it is now apparent that this group of chemicals is capable of causing reversible symptoms of surprisingly long duration.

We have recently encountered two cases, both aerial-applicator pilots, who present the typical complex of such psychiatric symptoms. There seems to be little doubt of the causal relationship between exposure to organophosphate pesticides and their behavioral changes. Both cases are considered to be of special interest because of the long duration of their symptoms after their last contact with agricultural chemicals.

REPORT OF CASES

Case #1.

At the time of his physical examination for a Federal Aviation Agency second class medical certificate, a 35-year-old farmer-pilot with 4775 flying hours revealed a history of previous treatment for a psychiatric condition. Because a diagnosis of psychosis had at one time been made, he was denied certification under Section 29.3 (d) (i) (ii) of Part 29 of the Civil Air Regulations (now Section 67.15 (d) (i) (ii) of the Federal Aviation Regulations). A current psychiatric evaluation was obtained and accompanied his appeal to the Civil Air Surgeon for reconsideration.

Review of the record revealed that he had undergone eight months of psychotherapy for an anxiety reaction with tension headaches 3½ years prior to this examination. Just prior to this treatment, he had failed to accept realistically the death of his 93-year-old father and was greatly upset when one of his children ingested a foreign body one month later. He was actively engaged in the aerial application of organophosphate insecticides immediately preceeding and during this treatment. When little improvement occurred he was hospitalized with a diagnosis of neurotic reaction with obsessive-compulsive, phobic and depressive symptoms.

At the time of admission he complained of angina and was unable to work despite the assurance of physicians that no disease existed; he was fearful of being alone, depressed, and despondent with crying spells. He had no suicidal impulses, insomnia, ideas of guilt or punishment, or depersonalization. The past medical history, family history, and physical examination were not significant.

Laboratory studies were as follows: hemtocrit, 43 per cent; hemoglobin, 15.4 gm.; white blood count, 7,700, with normal differential; BUN, 10.3 mg. per cent; fasting glucose, 114 mg. per cent; total serum proteins, 7.60 gm. per cent; albumin, 4.27 gm. per cent; globulin, 3.33 gm. per cent; alkaline phosphatase, 3.3 units; bromsulfalein test, 12 per cent retention in 45 min.; cephalin flocculation, 24 hrs.: 2 +, 48 hrs.: 2 +; prothrombin time, 80 per cent activity; Kline, negative. No cholinesterase determination was performed.

Baseline and double Master's test electrocardiograms were normal. No electroencephalogram was performed.

A chest x-ray was normal as were upper and lower gastrointestinal series and oral cholecystography.

He received 19 electroshock treatments and was discharged after two months on iproniazid, 35 mgs. daily, and pyridoxine, 150 mgs. daily. The final diagnois was psychotic depressive reaction.

This hospitalization was not reported until the third annual physical examination for a second class medical certificate following discharge. At this time he denied any commercial flying during the two years immediately following his hospitalization. He reported 600 hours flying time in the preceding year and denied any symptoms referable to the nervous system.

The medical record including the current psychiatric evaluation was reviewed by the Civil Air Surgeon's Medical Review Board. Since there was no disqualifying physical defect present at this time and since the correct diagnosis may have been that of "acute brain syndrome probably resulting from exposure to toxic chemicals," a medical certificate was issued with the stipulation that a current psychiatric evaluation be obtained at the time of his examination for renewal of his medical certificate.

Case #2.

A 37-year-old Latin American agricultural pilot visited CARI in February 1963 seeking help for recurrent bouts of acute anxiety, which he believed to have resulted from long exposure to organophosphate insecticides.

He began aerial application in Texas ten years ago after serving ten years as a pilot in the Chilean Air Force. One year later he moved to Guatemala where he still resides. Review of his role as an owner-operator reveals the use of a wide variety of aircraft to spray mostly cotton with methylparathion, DDT, toxaphene, endrin, and dieldrin. The normal season is from July through December; the normal workday is from 6:00 a.m. to 2:00 p.m., five to seven days a week depending upon the weather. The farmer selects and purchases the insecticide and furnishes the landing strip and mixing facilities. The owner-operator employs a mixer and a flagman and, when business warrants, may employ a second pilot. The pilot usually helps with the mixing about every third day. The insecticide mixture for each of the 30 to 50 daily flights is loaded in 1½ minutes from an overhead tank. Spillage is common and results in discoloration of the left wing and the ground. Review of his personal habits revealed that he cleaned his plane, inside and out, at the end of each day and wore a respirator in which he changed the filters every day and the cartridge every two or three days.

In January 1955 he began applying a mixture of methylparathion and DDT on a plantation in Peru. The living quarters were quite near the mixing and loading area and the odor of insecticide was said to be constant. He reported that nausea and diarrhea were prevalent among the twenty pilots employed. In March he sought medical advice, was told that he had liver and kidney damage and was advised to stop spraying.

He resumed spraying in July 1955. During the next six seasons he experienced gastrointestinal complaints about every ten days. He reported that this is a common experience of aerial applicators in his country and that the use of a mixture of atropine, sugar and charcoal for relief of symptoms is widespread. During 1957 he sought medical advice for nausea, chills, and tightness in the chest. He was advised to rest a few days before he returned to flying.

On September 26, 1961, his clothing was saturated with a mixture containing methylparathion, toxaphene and Dipterex when the tank of his aircraft was accidentally overfilled. He washed his face, shoulders and hands and continued flying. Three trips he was "too sick to fly" with nausea, abdominal cramps, dizziness, chest tightness, extreme weakness of hands and legs, and a rash on his wrists. He was hospitalized for five days. He reports that his cholinesterase was 20 per cent, his liver was enlarged and liver tests were "abnormal." His ability to stand erect with his eyes closed was tested daily; he consistently fell to the left until the sixth day, the day he was discharged. He became nauseated and dizzy in a nearby restaurant. This led to another cholinesterase determination which he was told showed 12 per cent activity. The following day he returned to flying but experienced a recurrence of symptoms on the third day. He was advised against further aerial application and received social security for the next six months (through March 1962). An EEG obtained in January 1962 appeared to be normal although he experienced dizziness during photic stimulation.

In May 1962 he was re-evaluated. At this time his cephalin flocculation was + (48 hrs.), his thymol turbidity was 3 units, and his bromsulfalein was 5.5 per cent retention after 45 minutes.

He returned to aerial application in July intending to work only two days per week. He experienced nausea and weakness on the second day and went to the hospital outpatient department where he was given the mixture of atropine, sugar and charcoal. En route home he experienced dizziness. He returned to the hospital and obtained a blood count which he reports showed 11,500 WBCs per cubic millimeter. He was denied any medical support for further social security benefits. He was informed that a tonsillitis could be responsible for the elevated white count; he, therefore, underwent a tonsillectomy. After recovery he attempted a return to working one day per week but experienced the same symptoms. He stopped work in late August 1962.

At the time of his examination here his chief complaint was of recurring episodes of acute anxiety. His usual geniality has been marred by sudden outbursts of temper the past two years. He feels "closed in" and experiences sweating and difficulty breathing when alone in a car or airplane. On one trip he had to land because of his symptoms. While he once enjoyed flying at altitudes up to 12,000 feet, he now becomes symtomatic above 5,000 feet. He is sympton-free in the presence of another person, even a non-pilot. Ingestion of alcohol is also noted to relieve his symptoms. He admits fear of being alone in a closed space as a child but denies any recurrence of these symptoms until the past two years. He is now extremely sensitive to the odor of even low concentrations of chemicals and experiences nausea and chest tightness when he smells benzene, gasoline, ether, or similar compounds.

The past medical history is noncontributory except for an uncomplicated appendectomy at age 20. The family history and review of systems are similarly noncontributory. There were no abnormal findings on physical examination.

Laboratory studies were as follows: red blood count, 5,340,000 per cubic millimeter; hematocrit, 52 per cent; hemoglobin, 17.2 gm.; white blood count, 7,100 with normal differential; total cholesterol (non-fasting), 318 mg. per cent; thymol turbidity (non-fasting), 5.2 units; urinalysis, normal.

A chest x-ray and an electrocardiogram were normal.

An electroencephalogram showed paroxysmal 5-7 per second slow waves (theta waves) of approximately 50 microvolts present most prominently in leads F^{P1}, F^{P2}, C^3, and C^4; less prominently in leads O^1, O^2, T^3, and T^4; and absent in leads T^7 and T^8. Occasional high voltage bursts (above 100 $_{\mu}$V) appear in leads F^{P1}, F^{P2}, C^3, and C^4. These were no significant changes with hyperventilation and photostimulation.

DISCUSSION

It is commonly accepted that the toxic actions of the organophosphate pesticides result from their inhibition of the cholinesterase enzyme. Unlike the carbamates and a variety of other chemicals and drugs which exert a temporary or transient effect on the activity of this enzyme, the organophosphates exert a persistent action because of the formation of a chemical bond between the phosphorus-containing radical and a receptor site on the enzyme itself. The rate at which phosphorylation takes place following contact between organophosphate and enzyme varies from compound to compound, as does the firmness of the bond. The rate of this reaction, along with many other factors such as size of dose, route of exposure, enzymatic transformation of the toxic material in animal tissues, and the rate of excretion, contributes to the rate of depression of cholinesterase activity and the time of onset and severity of such symptoms as occur. The firmness of the phosphate bond is the controlling factor in determining whether some spontaneous reactivation of cholinesterase can occur or whether recovery of full activity depends entirely on *de novo* synthesis of a new complement of enzyme.

The firmness of the bond increases with time, and this factor greatly influences the success of efforts to reactivate the enzyme by means of 2-PAM or other, similar antidotes.

Following a single, massive dose of an organophosphate compound given orally or parenterally, the activity of plasma (pseudo) cholinesterase can fall to near zero in one hour, and that of red cell (true) cholinesterase will reach a similar level in two hours. The red cell activity is believed roughly to parallel that in nerve, muscle and gland. A rapid fall of enzyme activity usually coincides with a rapid onset of acute symptoms. Following a single exposure, recovery of from 13 per cent to 19 per cent of the pre-existing enzyme activity can occur within 24 hours. Plasma activity thereafter returns to normal in 30 to 40 days, and the erythrocyte level in about 90 days.

A considerable lack of precision is inherent in the methods most commonly used for routine measurement of cholinesterase activity. Therefore the values of 20 per cent and 12 per cent reported by the second pilot as having been obtained in that sequence, five or six days apart, do not necessarily indicate a real decrease in activity during this period, especially since the determinations seem to have been made in different laboratories.

The peripheral autonomic and somatic motor responses to acutely toxic doses of organophosphates are in actuality, therefore, the responses of these systems to accumulated acetylcholine. Generally, the responses correspond to the predictable effects of systemically administered acetylcholine, the differences being primarily quantitative, but there may be exceptions. While the heart is usually slowed by acetylcholine, organophosphates may cause acceleration with palpitation and discomfort in the precordial region instead. This may be due to nausea sufficient to overcome the expected bradycardia, or to stimulation of the adrenal glands resulting in release of catecholamines. Dilation of the pupil may occur instead of the predictable myosis.

A more detailed account of the symptomatology of organophosphate poisoning is to be found in the recent review by Durham and Hayes[2] and the article by Gershon and Shaw.

Each of the subjects represented in this report experienced many of these acute symptoms during their active periods as agricultural pilots. Case 2 in particular recounts many episodes of nausea, retching, vomiting and abdominal cramps, and the muscular weakness which indicates an advanced stage of acute toxicity. Case 1 was reported at one time to have experienced anginal pain, but this may well have been identical with the sensation of "tightness" in the chest of which the second subject complained. Atropine is an effective antidote against the majority of these acute effects, a major exception being the initial tremor and later weakness of skeletal muscle.

With or without treatment, the acute signs and symptoms usually subside within one to six days although the activity of both plasma and red blood cell cholinesterase may still be low. It is also true that repeated exposures to small doses of cholinesterase inhibitors may reduce enzyme activity to a low level without the appearance of acute symptoms. The mechanism of adaptation to low levels of enzyme activity is not known.

The nature of the central nervous system symptoms depends upon the manner and degree of exposure. Acute poisoning has been reported to produce ataxia with stumbling gait, headache, restlessness, irritability, impaired ability to concentrate, depersonalization, indifference to environment, wild dreams, slurred speech, tremor, convulsions, medullary-center depression with dyspnea, cyanosis, coma and high-voltage, low-frequency electroencephalographic waves which are more pronounced after hyperventilation.

Since both of our cases are typical of chronic exposure, it is not surprising that neither of them presented an appreciable number of these symptoms at any time. Only headache was prominent in Case 1, and Case 2 showed only irritability, ataxia and slow EEG waves which were of moderate amplitude and not appreciably altered by hyperventilation.

Chronic exposure is associated with anxiety, uneasiness, giddiness, insomnia, somnambulism, lassitude, owsiness, tinnitus, nystagmus, dizziness, pyrexia, pa-lysis, paresthesias, polyneuritis, mental confusion, notional lability, depression with weeping, schizo-renic reaction, dissociation, fugue, inability to get ong with family and friends, and poor work perform-ce.

Of these, Case 1 presented the symptoms of anxiety, easiness, and depression with weeping. Case 2 com-ained of dizziness, anxiety, emotional lability, fre-ent and severe disagreements with family and as-ciates and an inability to perform familiar tasks.

Since only close questioning of a non-medically-ained individual would reveal the occurrence of cer-in of these symptoms, particularly if they were tran-nt, it is possible that some may have been missed these histories.

Other reports have stated that recovery from schizo-renic and depressive symptoms requires from six to elve months following removal from further contact th the toxic agents. Sufficient recovery for discharge om medical observation occurred in ten months in our ase 1, but commercial flying was not attempted for an-her two years. Symptoms which prevent commercial ing still persist in the second pilot six months after his st exposure. Symptoms that persist after 90 to 120 days nnot be associated with low cholinesterase levels in e brain. There is some indication in these and other stories that recovery may be a relative term and a gree of recovery which would permit the resumption one occupation might not permit the resumption of other.

The EEG pattern found in the second pilot (Fig. 1)

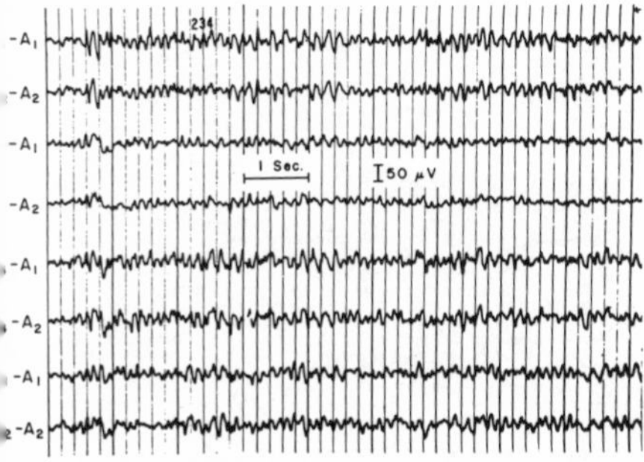

Fig. 1. Resting EEG showing paroxysmal 5-7 per second slow ves and occasional high voltage bursts (above 100 $_{\mu}$V) in ds Fp_1, Fp_2, C_3, and C_4.

not inconsistent with the findings of other au-ors [4, 6, 7] despite lower amplitude and the absence of anges with hyperventilation. The principal difference ncerns the duration of the effects, which are reported other studies as seven days,[4] eleven to eighteen days,[5] d four weeks.[3] If the EEG changes observed here are deed residual from organophosphate exposure, their rsistence exceeds six months. Since experts differ in eir interpretation of slow-wave activity, it is not cer-in whether the EEG obtained one year earlier and onounced normal was different from the current one. In the light of present knowledge, it must be con-sidered more likely that the mild liver damage found in both pilots was cause by the chlorinated hydrocarbons to which they were also chronically exposed.

It is interesting to speculate that our cases were the victims of a combination of circumstances almost perfectly designed to produce their central nervous system involvement. They operated in regions with long growing seasons. The nature of the crops and the insect pests demanded the use of organophosphate insecticides. Prolonged periods of good weather and a great demand for their services led to long, uninterrupted periods of flying. All of these factors combined to cause continuous contact with the toxic materials for protracted periods, and encouraged the practice, common among agricultural pilots, of suppressing acute symptoms with atropine and continuing their work. In the meantime, the more insidious central nervous effects were developing.

These circumstances are not unique, and it is almost certain that more such cases will appear. Physicians in Latin America and the agricultural regions of the United States are becoming alert to the acute peripheral signs and symptoms of organophosphate toxicity, and recognize them readily. It is now apparent that they must be watchful for central nervous system symptoms as well. They should be especially emphatic in warning pilots to avoid the prolonged, chronic exposure which may lead to psychiatric difficulties.

The case of the second pilot of this report indicates that it is scarcely possible to take excessive precautions against such exposure, since he customarily used a respirator which he kept in good condition, had flown a closed-cockpit aircraft since 1959 and was careful about the cleanliness of both his aircraft and his person.

Even this need not be surprising. Of 23 subjects in one study,[4] all of whom used carbon-filter respirators and wore rubber gauntlets and coveralls, 18 had mild to moderate symptoms, four had severe symptoms but recovered, and one died.

ACKNOWLEDGMENTS

The authors are deeply indebted to Harry L. Gibbons, M.D., for the first case and to P. C. Tang, Ph.D., who performed the electroencephalographic studies on the second case.

REFERENCES

1. Conference of Human Pharmacodynamics of CW Compounds, Chemical Corps Medical Laboratories Special Report No. 72, Sept. 1955.
2. Durham, W. F., and Hayes, W. J., Jr.: Organic phosphorus poisoning and its therapy. *Arch. Environ. Health,* 5:21, 1962.
3. Gershon, S., and Shaw, F. H.: Psychiatric sequelae of chronic exposure to organophosphorus insecticides. *Lancet,* 1:1371, 1961.
4. Grob, D., Garlich, W. L., and Harvey, A. M.: The toxic effects in man of the anticholinesterase insecticide parathion (p-nitrophenyl diethyl thionophosphate). *Bull. Johns Hopkins Hosp.,* 87:106, 1950.
5. Grob, D., and Harvey, A. M.: The effects and treatment of nerve gas poisoning. *Amer. J. Med.,* 14:52, 1953.
6. Grob, D., and Harvey, A. M.: Observations on the effects of tetraethyl pyrophosphate (TEPP) in man, and on its use in the treatment of myasthenia gravis. *Bull. Johns Hopkins Hosp.,* 84:532, 1949.
7. Rowntree, D. W., Nevin, S., and Wilson, A.: The effects of diisopropyl fluorophosphonate in schizophrenia and manic depressive psychosis. *J. Neurol. Neurosurg. Psychiat.,* 13:47, 1950.

Drug and Toxic Hazards in General Aviation

J. Robert Dille & Stanley R. Mohler

DRUGS AND ALCOHOL are frequently cited as major causes of automobile accidents. Drugs are rarely reported as a cause of aircraft accidents. This is due to the relatively infrequent search for their presence and to unprecise definition of levels at which most of them begin to impair pilot performance. Alcohol has been found present in up to 35.4 percent[4] of fatal general aviation accidents but a causal role has been ascribed in only about seven percent.[6] This discrepancy is due to the absence of experimentally determined and generally accepted blood alcohol levels where pilot performance is significantly impaired. In practice, the legal level of intoxication for driving an automobile is frequently used. Most authorities agree that the 150 mg % level, accepted in most states, is too high for safely operating even an automobile. Most aviation authorities agree that piloting an aircraft is more complex than driving an automobile. The effects of combinations of alcohol and drugs are of even greater concern.

While improved somewhat over previous years, aerial application still has the highest accident rates among commercial categories of aviation. Evidence of pesticide poisoning has been found in about one-third of fatal aerial applicator accidents, but on two small samples. Here, again, the level at which a causal role should be assigned is uncertain.

Carbon monoxide is the only other toxic material which is occasionally found, and frequently discussed, in general aviation. It is the determined cause of about three fatal accidents per year; most are due to faulty exhaust manifolds and heater assemblies. Carbon monoxide from smoking tobacco products can significantly affect hypoxia tolerance and night vision but this is rarely, if ever, found as an accident cause.

Toxic products of combustion of cabin interior materials and electrical insulation are primarily of concern in air carrier operation and will not be discussed here.

DRUGS

Drug usage in this country is almost universal and apparently increasing each year. Many potent drugs can be purchased over the counter, others are prescribed by physicians who do not inquire about or understand the demands of various occupations, and still others are passed around by well-meaning relatives, friends and neighbors.

It is assumed that drugs have frequently been taken by victims of general aviation accidents but their presence is rarely detected and, even if it is, rarely warrants assignment of a causal role based upon present knowledge. A previous report[2] cited two cases where ATR pilots with 3000 and 8000 hours had human factors accidents and liver barbiturate levels of 0.7 mg % and 0.4 mg %. One of the compounds taken also contained d-amphetamine. Despite known possible effects on

gment of both drugs, no causal role was felt as-
nable.

All known "side effects" are now routinely listed for ry drug. The problem lies in predicting the occur-ce of these undesirable effects. One approach, the itary one, is to ban the use of all drugs by aircrew mbers. This is neither practical nor enforceable in l aviation. Supervised test doses are useful to de-nine individual responses. The fact that the condi-ı for which the drug is taken may, in itself, contra-icate flying should be remembered. The fact that formance with drug-relieved symptoms may exceed formance with unrelieved symptoms (but not per-nance without symptoms or drugs) must also be con-ered.

Jntil more is known, a cautious approach to drug by airmen is urged. Studies of drug effects on the formance of aviation tasks and expanded investiga-ı of aircraft accidents are expected to provide use-information.

ome of the medications of greatest use and concern, some recent accidents involving drugs, are dis-sed here.

nalgesics—Probably no over-the-counter drug is d more often or more indiscriminately than acetyl-cylic acid. Toxic effects are relatively rare and are ost always associated with large doses. However, trointestinal hemorrhage, acute renal failure, blood crasias and idiosyncratic reactions (such as urticaria angioneurotic edema) are possible. Hemorrhage, en it occurs, is usually due to a competitive antago-n with vitamin K and a decreased circulating pro-ombin. A reduced tolerance to hypoxia has been nd with salicylates, mostly because of an increase he metabolic rate.

nalgesic compounds containing aniline derivatives y cause methemoglobinemia if used indiscriminately. essive use of bromide-containing compounds may se psychosis or dermatitis. Quinine-containing parations may cause vertigo, tinnitus, deafness or sea.

)f greatest concern is the frequent combination of lgesics, antihistamines and decongestants in com-nds which may be taken for analgesic purposes y. The roles of these added ingredients are discussed ow.

ntihistamines—Undesirable effects which are pos-e with the use of antihistamines are drowsiness, tention, confusion, mental depression, dizziness, de-ased vestibular function and impaired depth per-tion.

Because of the adverse effects of these symptoms on safety of flight, airmen generally should not take ort-acting" antihistamines during the 8 hours before ht or take the "long-acting" preparations within 16 rs of flying.

ndividual consideration should be given to allergic ents who have taken the same drug and dosage for g intervals with a good symptomatic response and noted side effects.

Nonsedating" antihistamines (for example, phenindamine tartrate) should be evaluated for their suitability for safe use.

A 38-year-old pilot was killed in a crash of his helicopter due to the improper operation of the powerplant and flight controls. He had two blood alcohol level determinations reported as 80 and 100 mg % and a "significant" level of an antihistamine believed to be diphenhydramine. The combined effects of these two agents are believed to have caused impaired efficiency and judgment and are given as the cause of the accident.

Antihistamines have also been implicated in a recently reported British accident.[1]

Nasal Decongestants—Since these compounds can occasionally be used to advantage topically during flight (usually for the relief of a blocked eustachian tube during descent), their proper use in flight is not contraindicated. Indiscriminate use of these compounds, particularly by the systemic route, can cause tachycardia, nervousness, tremor, incoordination and mydriasis with visual disturbances.

Motion Sickness Medications—Several types of drugs are used for the relief of motion sickness. Scopolamine, a parasympathetic depressant sometimes used for this purpose, is effective but has sufficient side effects to limit its use. Antihistamines are used widely but often cause drowsiness and dizziness. The sedative antihistamines, such as promethazine, are particularly likely to produce drowsiness. Cyclizine and meclizine also can produce drowsiness, and blurred vision may occur with cyclizine. However, side effects are less common with these two drugs. Barbiturates have been used but are seemingly of less value and are definitely contraindicated during flight.

Most pilot trainees who become airsick will have no difficulty by the tenth flight. Therefore, cyclizine or meclizine may be temporarily given for motion sickness before training is discontinued, but only under medical supervision, after a test dose, on dual flights and with the consent of the flight instructor. Otherwise, the use of these preparations during flight or within 8 hours to 24 hours (with meclizine) before flight is generally contraindicated.

Amphetamines—Amphetamines diminish a sense of fatigue, can delay its onset up to four hours, and tend to force the body beyond its natural capacities. Nervousness, impaired judgment and euphoria are sometimes reported, particularly with overdosage or unusual susceptibility.

Many patients find the stimulation produced by amphetamines pleasant and complain when the physician suggests discontinuance. They can be habit forming and excessive use is common.

When amphetamines are taken in conjunction with a weight reduction program hypoglycemia may be present. The effects of hypoglycemia are additive to those of hypoxia.

Amphetamines should not be used during flight except in unusual situations (mostly military) in which mission completion is paramount and fatigue represents a greater hazard than drug use during a critical, relatively brief phase of the flight.

A 35-year-old pilot with 63 hours flying time was killed when he crashed while buzzing a tavern at night. He had a blood alcohol level of 200 mg % and a blood amphetamine level of nine micrograms percent. A 27-year-old pilot with 19.8 hours flying time was killed in an unexplained accident. He had a kidney and liver amphetamine level of 0.1 mg % and tissue amobarbital levels of 0.20 to 1.00 mg %. Causal roles were not ascribed to amphetamine in either accident but the frequent presence of more than one drug is demonstrated.

The monoamine-oxidase inhibitors, which are also psychic stimulants, are rarely indicated for persons who are well enough to fly. Significant side effects, particularly hypotension, blurred vision and excitement, also make flying contraindicated. The possible altered response to other drugs and alcohol should be recognized for this group.

Tranquilizers and Sedatives—In most cases flying is contraindicated by a condition which requires tranquilizers. However, experience has shown that they are prescribed quite liberally and without due concern for the patient's occupation or the undesirable effects of these drugs. While it may not be readily apparent, even the nonsedating tranquilizers usually have some measurable effects on alertness, judgment, efficiency and over-all performance.

Sedatives have been used under controlled conditions by the military to guarantee adequate rest before flight and alertness during flight. Secobarbital sodium, often used for this purpose because it is considered a short-acting sedative, has been found to have effects for up to eight or nine hours when given in 200 mg doses under controlled experimental conditions.

Pilot duties are contraindicated for 12 to 24 hours after the use of sedative agents.

A 50-year-old pilot with 108 hours of flying time had a fatal accident when he departed with one passenger on a VFR flight under IFR conditions. He had blood levels of 0.7 mg % of long acting barbiturate and 40 mg % of alcohol.

A 51-year-old pilot, with 1136 hours, and four passengers were killed when their aircraft went into a spin at low altitude and crashed. A blood barbiturate level of "0.2 to 0.4 mg %", blood alcohol level of 8 mg %, COHb of 12% and fatigue after a four hour flight are felt possibly to have been additive and causal.

A 24-year-old pilot with 2203 hours was involved in a fatal helicopter crash of undetermined cause. He had a blood phenobarbital level of 0.2 mg % which was not felt to be clinically significant. The reason for drug use was not determined.

A 49-year-old physician was known to use narcotics, barbiturates, and tranquilizers. He had lost his medical license. Erratic behavior had been noted for over two years. He had passed out while driving and had had several automobile accidents. A relative, who stopped flying with him, had considered reporting him to the FAA! He had a fatal accident after a stall and spin. He had blood levels of 3.54 mg % of phenobarbital and 0.95 mg % of meprobamate, a combined effect near the stupor level.

A 49-year-old pilot with 100 hours of flying tim was killed when his plane dived into a reservoir. H had been hospitalized for depression due to alcoholis two years before. He had blood levels of 100 mg of alcohol and 11 mg % of bromide.

Attention is drawn here to the involvement of mu tiple drug, alcohol and physiologic factors, and th family (and sometimes local official) apathy that observed too frequently.

Cardiac Agents—Hypertension and most hypotensi agents are disqualifying for flying duties. While med cation is the simplest treatment for a physician render, most authorities agree that, where indicate weight reduction is the treatment of choice. A patie with benign essential hypertension without demon trable eye, kidney or electrocardiographic changes ma not require medication if salt intake is restricted an a controlled weight reduction program is institute preferably with exercise. Thereafter, the patient ca probably fly safely.

Intermittent drug therapy is not indicated becau of the long duration of effects, particularly with rese pine.

Because of the common knowledge that drug reli from disqualifying hypertension is available, a milita medical evaluation center analyzes the urine of referre airmen for these agents. Because of the likelihood a reduction in G tolerance with these drugs, partic larly with the ganglionic blocking agents, their use of greater concern in military airmen.

In civil aviation, medical certification may be grante for pilots who have taken a hypotensive agent (pa ticularly thiazide diuretics) for a prolonged peri and who have demonstrated good control of a relative stable condition with no significant side effects. Su actions must, of necessity, be on an individualiz basis.

A 45-year-old pilot with 91 hours entered an une plained bank and dive just after take-off and he ar his passenger were killed. A 0.05 mg % liver level "quinine or quinidine" raises the question of a possib medical condition under treatment and inflight i capacitation.

Pilots have had pargyline and other drugs prescrib simultaneously; one such pilot has had an accide due to alcohol.[7] Flying is contraindicated with parg line and most of the agents which augment its effec and pilots should be advised of this by their physician *all patients* should be warned of the possible combin effects.

Muscle Relaxants—These agents, with or witho analgesic and tranquilizer actions, cause sufficient wea ness, sedation and vertigo to contraindicate pilot duti within at least 12 hours after their use.

Steroids—These compounds are often used syster cally for the relief of arthritic, allergic, dermatolog and inflammatory conditions which may not, in ther selves, contraindicate the performance of pilot duti Flying is considered contraindicated for three da after the systemic use of steroids because of the pc sible mental changes and other undesirable effec

he topical use of these preparations is not expected
compromise flying safety.

Drugs for Hyperuricemia—These agents are fre-
uently discussed but there is no unanimity of opinion
garding their acceptability for use by flying person-
el. The authors are not opposed to the supervised use
uricosuric agents by asymptomatic pilots with hyper-
ricemia. Headache and gastrointestinal effects, while
ndesirable, are not usually incapacitating. An attempt
ould be made to determine the presence and severity
these or other symptoms when these patients are
amined for medical certification.

The xanthine oxidase inhibitors are also reported to
e effective and relatively free from side effects.

In metabolic diseases like hyperuricemia and diabetes
ellitus, where the treatment and not the disease may
e disqualifying for flying, pilots and cooperative
ysicians will often shun treatment. The relative haz-
ds, both immediate and long term, of the treatment
. the untreated condition must be considered. Long
scussions have resulted from such considerations but
ese are equally inconclusive.

Anticholinergics—Anticholinergic compounds are fre-
ently used and occasionally found associated with
iation accidents and incidents. Since they are fre-
ently combined with sedatives and tranquilizers,
eir side effects can include, not only blurred vision
d ataxia, but sedation, muscle weakness and altered
dgment. Their use is contraindicated with flying.

And Finally, "The Pill"—There are approximately
,000 active female pilots under the age of 45. We
not know, nor will we probably ever know, the
umber who take oral contraceptive preparations.
veral pilot-physicians who have several patients on
ese medications have expressed concern over the
mptoms of tension noted by "many" of their patients.
hile there is insufficient evidence of significant un-
sirable effects at this time to consider stricter con-
ol on their reporting or use by female pilots, we must
ep abreast of evolving knowledge on thrombo-
nbolic, or other, problems with these products and
nsider these data, age, past medical history, and
ysical findings in the counseling and certification of
ese pilots.

THYL ALCOHOL

Everyone interested in aviation safety agrees that
inking and flying do not mix and that alcohol should
tensify the effects of hypoxia. However, despite many
udies of the subject, there is no generally accepted
ood alcohol level, or time after recovery, for the safe
eration of an aircraft. The degree to which alcohol
creases the effects of hypoxia is similarly inconclusive.

Nystagmic and EEG changes can be noted after the
ood alcohol level returns to zero. Decrements in
sion and hearing have been reported at levels as low
10 mg % (though for abstainers and probably not
alistic). Judgment, comprehension and fine atten-
on are important in flying, and are reduced at low
ood alcohol levels, but are most difficult to measure.
A significant decrement in performance is often not
seen at levels below 50 mg %.

A complete review of the literature on this subject
is beyond the scope of this paper. The high incidence
of alcohol involvement in fatal accidents and the reporting discrepancies were mentioned earlier. The incidence is clearly too high. Further research is indicated to better identify the *legal* level of intoxication
with flying. From a practical point of view, considering the high percentage of human factors accidents,
individual variation of effects, the frequent presence
of other factors such as drugs and fatigue, and residual
effects, no alcohol level should be considered com-
patible with flying. Grounding for three to four hours
per drink is suggested as a "rule of thumb" for most
situations.

Education and, if unsuccessful, attempted enforce-
ment of existing regulations should be used to reduce
the incidence of flying while under the influence of
ethyl alcohol.

PESTICIDES

The high accident rate in aerial application and the
high incidence of poisoning in the few accidents investigated were mentioned earlier. We have previously reported several of these cases.[2,3,5]

We are disturbed that several cases of poisoning have
been admitted to hospitals, and even to university medical centers, and undiagnosed and/or not treated with
available effective drugs. In extensive educational activities with the "ag" operators and pilots, we discour-
age self-medication. While their use of drugs to reduce
symptoms and increase tolerance to the pesticides is
dangerous and therefore contraindicated, it must be
admitted that most safety-minded aerial applicator
pilots know more about the symptoms of poisoning
and the correct drugs and dosages than does the average physician.

For physicians who may be called upon to treat any
patient with exposure to insecticides, an annual review
of the signs, symptoms and treatment is recommended.
Purchase of the Clinical Handbook on Economic Poisons is also recommended (available from the Super-
intendent of Documents, U. S. Government Printing
Office, Washington, D. C. 20402, for 55 cents).

In cases of mild poisoning, miosis, rhinorrhea, and
salivation are seen in about 90 percent of the cases;
constriction of the chest is reported by about 80 percent; and dimness of vision and cough are noted by
60 percent. Some report headache, fatigue, anorexia,
nausea, increased perspiration, dizziness, twitching,
increased dreaming, irritability and mood changes. In
severe poisoning, all symptoms are increased with
vomiting and uncontrollable muscular twitching having
the largest proportional increase in incidence. The
most important signs are cyanosis, convulsions, coma,
loss of reflexes, and loss of sphincter control.

Diagnosis depends upon the signs and symptoms, the
label from the container, reliable information (not from
the patient), the blood cholinesterase level, and, in
poisoning by parathion and its congeners, the urine

paranitrophenol level.

Treatment, which is started before the laboratory values are received, consists initially of establishing an airway, artificial respiration, oxygen and suction when necessary.

Atropine, given intravenously, is the initial drug of choice. The success of treatment is directly proportional to the promptness of administration and adequacy of dosage of atropine.

Atropine is of little or no value in the treatment of poisoning by pesticides other than the organophosphate compounds. It is absolutely contraindicated for pentachlorophenol poisoning, another insecticide which produces symptoms similar to those of the organophosphates, may be contraindicated in chlorinated hydrocarbon poisoning. The importance of the label and reliable information in making the correct diagnosis is therefore emphasized.

Atropine should be administered intravenously in dosages of 1 to 2 mg (for mild cases) to 2 to 4 mg (for severe cases) every 5 to 10 minutes until atropinization occurs (dry flushed skin, pulse rate of 140 per minute, and pupillary dilatation).

One gram of 2-PAM chloride should be slowly administered intravenously. Another 500 mg should be given in one-half hour if muscular weakness persists. This is a specific chemical antidote which releases cholinesterase inactivated by the phosphate ester but its effectiveness decreases after 24 hours. Best results are obtained when 2-PAM and atropine are both given.

The clothing should be removed and the skin, hair, nails and eyes should be decontaminated.

Symptomatic treatment should be continued for 24 to 48 hours.

A blood sample for cholinesterase determinations should be drawn before 2-PAM is administered.

No morphine, tranquilizers, or barbiturates should be given.

Since there are over 55,000 trade name pesticide products in the United States representing formulations of over 300 different chemicals, problem cases may be encountered. A telephone call to the nearest Poison Control Center is advised for perplexing problems.

The chlorinated hydrocarbon insecticides vary widely in chemical structure and activity. They act primarily on the central nervous system but the exact mechanism of action is not known. They also produce some gastrointestinal effects and, at least in animals, liver and kidney damage. They are stored in fat and released very slowly. The fat levels are felt to be mostly inactive materials but convulsions many months after exposure have been noted in animals.

Signs and symptoms usually consist of headache, nausea, dizziness, hyperexcitability, tremor, and, classically with severe poisoning, convulsions which may be followed by coma.

Laboratory diagnosis is difficult. The insecticide or its derivatives can be identified in body fat, urine and stomach contents by properly equipped laboratories. The electroencephalogram may show bilateral synchronous spikes, spike and wave complexes and slow theta waves. These changes may persist for a f
months after the last exposure. A good history is
portant.

Treatment, too, is less specific than for organoph
phate poisoning. Removal of the poison from the s
or gastrointestinal tract, high dosages of sedatives a
sometimes, calcium gluconate are the principal effect
measures.

CARBON MONOXIDE

Carbon monoxide constitutes up to 2.5 percent
the volume of cigarette smoke and more in ci
smoke. If the smoke of three cigarettes is inhaled
rapid succession at sea level, a carboxyhemoglo
saturation of 4 percent may result. This can red
visual acuity and dark adaptation to the extent of
mild hypoxia of 8000 feet. Smoking at 10,000 feet
produce effects equivalent to those seen at 14,000 f
without smoking. The effects of smoking at 20,
feet are equivalent to the effects otherwise expected
22,000 feet. With heavy smoking, carboxyhemoglo
concentrations as high as 8 percent are possible.

While the adverse effects of smoking have been, a
should continue to be, included in airman educat
material, this is not likely to be a determined cause
an aircraft accident. This is because of the necess
combined effects and the usual lack of objective e
dence.

Carbon monoxide has been found as the cause
death in both general aviation and air carrier accide
with fire.

The greatest threat is from carbon monoxide le
into the cabin from worn or defective exhaust st
slip joints, exhaust system cracks or holes, openings
the engine firewall, defective gaskets in the exha
manifold and defective mufflers.

In one aerial application fatal accident, a 40 perc
carboxyhemoglobin level was found. In modifying
light aircraft for crop dusting, the cabin heater w
removed and the exhaust aperture ineffectively capp

While this is a relatively rare cause of accidents,
can be prevented by periodic inspection of the exha
manifold and heater assembly, awareness of the ea
symptoms of poisoning and use of a carbon monox
detector.

REFERENCES

1. British Board of Trade: Medical Factor Studied in Bella Crash. *Aviation Week and Space Technology* 87: (December 11) 1967.
2. DILLE, J. R., and MORRIS, E. W.: Human Factors in G eral Aviation Accidents. *Aerospace Med.* 38:1063, 19
3. DILLE, J. R., and SMITH, P. W.: Medical Advances Aerial Application Problems. *Flying Physician* 9:6, (tober) 1965.
4. HARPER, C. R., and ALBERS, W. R.: Alcohol and Gene Aviation Accidents. *Aerospace Med.* 35:462, 1964.
5. MOHLER, S. R., and HARPER, C. R.: Protecting the Pilot. Federal Aviation Agency Office of Aviation Me cine Report 66-30, September 1966.
6. National Transportation Safety Board: Annual Review U.S. General Aviation Accidents Occurring in Calen Year 1966, November, 1967.
7. SMITH, P. W.: Unpublished data.

Occupational Dermatitis From Naled

A Clinical Report

Walter F. Edmundson, MD, and John E. Davies, MD, MPH,

NALED is a contact and stomach insecticide and acaricide which has the chemical name of 1,2-dibromo-2,2-dichloroethyl dimethyl phosphate and the following structural formula:

```
CH3O     O
    \   /
      P          Br   Br
    /   \        |    |
CH3O      O ---- C -- C -- Cl
                 |    |
                 H    Cl
```

This is rather unusual organophosphate pesticide in that it is brominated. The acute oral 50% lethal dose (LD_{50}) to rats is about 430 mg/kg and the acute dermal LD_{50} to rats is about 1,100 mg/kg, relatively low values of toxicity compared to highly toxic organophosphates, such as parathion, but higher than others, such as malathion. Naled is insoluble in water in which it rapidly hydrolyzes, and it is highly soluble in aromatic solvents (eg, xylene).[1]

A report was received by this office that nine of twelve female workers who had been making cuttings of tops of unflowered chrysanthemum plants for replanting had dermatitis shortly after performing this piece of work at a chrysanthemum grower's farm. This was the first instance known in the area of such an occurrence.

Four of these workers were examined four days after the occurrence; the following history of exposure was obtained and dermatologic findings observed.

On the day of the occurrence, several of the women complained of burning and itching of the arms, face, neck, and abdomen within a few hours of the cutting operation. It was immediately suspected that one of the many chemicals that are used on the plants was the cause of the trouble. It was determined that within two hours of the time that the women began the cutting that the field had been sprayed with a mixture of naled, an insecticide; captan, a fungicide (N-trichloromethylthio-4-cyclohexene-1,2-dicarboximide); and dicofol, an acaricide (4,4′-dichloro-α-trichloromethylbenzhydrol). The spray mixtures consisted of 0.9 lb naled, 0.5 lb captan, and 0.2 lb dicofol, calculated as active material per 100 gallons of water. It is conjectured that some of the mixture was still wet on the leaves when the women were in the fields. None of these chemicals were known as trouble makers to the flower growers. Chrysanthemum flowers were not present on the plants and the women had been working regularly (before and after the episode) in the flower fields on which these same chemicals had been used many times. The women dressed similarly in working clothes consisting of a short sleeved, open-necked, loose cotton shirt, slacks (with one exception); all wore glasses; and all were exposed identically. Gloves were not habitually worn. All were white.

Report of Cases

CASE 1.—This 44-year-old woman stated that within one hour after starting to make cuttings

From the Community Studies on Pesticides—Dade County, Florida State Board of Health, Miami.

Reprint requests to 1390 NW 14th Ave, Miami, Fla 33125 (Dr. Edmundson).

Table 1.—*Patch Test Results (24 Hours) on Reported Cases*

Case No.	1	2	3	4
Naled (60% in xylene)*	1+	3+	...	2+
Xylene	...	...	...	...
KB†	...	...	...	...
Dicofol (18.5% in xylene)	...	...	...	...
Dicofol (99% crystals)	...	...	...	...
Captan (50% powder)	...	...	...	...
p, p'-DDT (99% crystals)	...	...	...	...

* 1+ indicates erythema; 2+, erythema and edema; 3+, vesicles, erythema, and edema.
† Saturated solution.

she noticed that her face was burning and "welts" were appearing on her face and neck. The next morning she had redness and swelling of her face, her right eyelid was swollen, and her neck was reddened. On the following day an eruption appeared on her abdomen above the belt line and an eruption appeared on the antecubital areas of her arms. All of the involved areas itched and burned. She has continued to experience the itching and burning sensations in the involved areas.

The worker denied previous attacks of skin trouble, but she does have symptoms of hay fever. She has been working on the farm for the last five years.

Upon examination, erythema of popliteal areas and the face was noted, there was mild swelling of the face. A maculopapular eruption was present on the abdomen above the belt line. The eruption appeared to be improving; she had been applying triamcinolone acetonide (Kenalog) cream and diphenhydramine hydrochloride (Benadryl) in a calamine lotion base, and taking tripelennamine hydrochloride (Pyribenzamine) tablets, 50 mg four times a day.

CASE 2.—This worker, aged 19, stated that within an hour after starting work her arms itched, and that in the evening of that day "welts" appeared on her abdomen. Two days later a mild eruption appeared on the inner sides of the knees.

This girl denied previous skin trouble or allergic disorders. She had been working in this occupation only one month. She wore a skirt rather than slacks, which probably accounts for the eruption about the knees.

An improving mild eruption was present in antecubital areas of the arms, on the abdomen, and inner sides of the knees. She had applied triamcinolone acetonide cream to the involved areas one or more times daily.

CASE 3.—This 25-year-old woman stated that on the night of the episode she "itched all over" and noted "welts" on her arms, face, and neck. The following morning she experienced burning and itching of the face (not eyelids), sides of the neck, and antecubital areas of the arms.

She denied previous skin or allergic disorders. She had been working at this occupation for two years.

An examination revealed a residual dermatitis consisting of a mild patchy erythema was present on the neck and on the right antecubital area.

CASE 4.—This woman, aged 41, stated that she noticed itching and burning of the external and internal surfaces of the upper right arm, the right side of her neck, and right ear about three hours after working in the cutting operation. About three days later, she noted a rash above the belt line on the abdomen and an irritated "spot" in the right groin.

This worker gave a history of having hay fever and chronic sinusitis. She has a brother who also has hay fever. She stated that she had eczema of the hand at about 9 years of age, and a mild rash on her neck and inner aspects of the arms about three weeks earlier, cause undetermined. She has worked at this occupation about nine years.

On examination there was a residual papular dermatitis of the right upper arm, and glazing of the skin on the lower right cheek, mild irritation of both sides of the neck, and a maculopapular patchy eruption of the abdomen above the belt line. She had been treating the areas with triamcinolone acetonide cream and diphenhydramine hydrochloride-calamine lotion. She complained that the areas affected itched more while she was picking flowers on the day of examination.

The clinical diagnosis of the above eruptions was contact sensitization type dermatitis.

Patch testing was carried out two weeks later on the inner surface of the forearms of the above patients after preliminary patch testing of the chemicals for primary irritation. At that time these women presented no evidence of dermatitis in spite of the fact they had contin-

Table 2.—*Patch Tests Results (24 hours) on Controls*

Control No.	Naled (60% in xylene)	Xylene	KBr
1	...	...	...
2	...	...	...
3	...	...	...
4	...	...	...
5	...	...	...
6	...	...	...
7	...	...	...
8	1+	1+	...

ued working in the same fields, which are known to have been treated with naled on several occasions.

In Table 1 it can be seen that three of four of these workers had reactive patch tests to the designated materials in 24 hours.

In Table 2 the results of patch testing controls, consisting of persons who were not engaged in occupations related to pesticide exposure, are recorded. These controls were nonreactive to naled (60% in xylene) in each of eight cases except one. This one is considered to be sensitive to xylene since he was nonreactive to a saturated solution of potassium bromide and reactive to pure xylene. Pure naled was not available so naled alone could not be used as patch test material in these cases, and it is questionable whether pure naled should be used in patch testing.

Comment

The above workers are considered to be sensitive to naled while it is in the nonhydrolyzed form. They were not reactive to the solvent xylene or to a saturated solution of pure potassium bromide by 24-hour patch testing. Realizing that negative patch test results are nonconclusive, however, the reactive results of three of four workers to patch testing provides strong evidence for the sensitizing property and dermatitis-producing potential of naled when workers are exposed to it before hydrolyzation of the compound has had time to occur after its use. The exposure experience of these workers since the occurrence testifies to its safety when adequate time for its breakdown has been allowed to pass. It is, therefore, recommended that after naled has been used on vegetation that workers not be exposed to the material for several hours after the application of the chemical.

This study was supported by the Pesticides Program, National Communicable Disease Center, US Department of Health Education, and Welfare, Atlanta, through the Florida State Board of Health, contract PH-86-65-26.

Reference

1. Burchfield, H.P.; Johnson, D.E.; and Storrs, E.E.: *Guide to the Analysis of Pesticides Residues,* Public Health Service, Office of Pesticides (BSS-EH), 1965, Vol 2, VIII. B. (nal).

Identification and Measurement of Dichloran in the Blood and Urine of Man[1]

WALTER F. EDMUNDSON, JOSEPH J. FREAL, AND JOHN E. DAVIES

With the expanding use of pesticides and the concomitant exposure of man to them, it has become increasingly important to identify pesticides in human tissues as a measure of absorption of these chemicals. This paper describes a method for the identification and measurement of a fungicide, dichloran, in the blood and urine of men industrially exposed to this pesticide.

Dichloran, 2,6-dichloro-4-nitroaniline (Fig. 1) is an agricultural fungicide

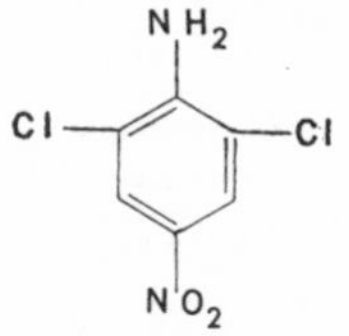

FIG. 1. Dichloran.

used either in soil or as a foliar dust or spray. It is practically insoluble in water and is less than 4% soluble in most organic solvents; it is stable to hydrolysis and relatively stable to oxidation. This chemical, Botran,[2] is said to have an oral LD_{50} in rats of more than 10000 mg and intraperitoneal LD_{50} in mice of about 5470 mg/kg (Tuco Products Co., 1965). No cases of human toxicity have been reported. This fungicide was encountered in the form of a 75% dichloran wettable powder and a 6% dichloran dust in our continuing field studies on the effects of pesticides on human beings.

An individual engaged in aerial dusting had been using the 6% dichloran dust on several occasions during the 3-week period before blood and urine samples were obtained. The level of dichloran in his urine was 41 ppb, and in his blood 6.6 ppb, which led us to the more intensive study reported here.

[1] This research was supported by the Pesticides Program, National Communicable Disease Center, Bureau of Disease Prevention and Environmental Control, Public Health Service, U.S. Department of Health, Education and Welfare, Atlanta, Georgia, through the Florida State Board of Health, under contract number PH-86-65-26.

[2] Trade name is provided for identification only. Its mention does not constitute endorsement by the Public Health Service or by the U.S. Department of Health, Education, and Welfare.

METHODS AND MATERIALS

At the formulating plant, from which the aerial duster received his supply, four operators were studied intensively over a period of 8 days. During this period of time these formulators furnished us with a urine sample at the beginning and end of 5 working days and a blood sample at the end of the same working days. The number of hours each man was occupied with formulating 6% dichloran dust from 75% dichloran wettable powder was recorded. No precautions were taken to avoid exposure to the material, and no respirators were used; the operators were themselves grossly dusted with the formulation, as was the working area. A comparison of the exposures of each formulator and his whole blood and urine values are presented (Table I).

The exposure data are presented in terms of "$[K]$ t", the time the worker was exposed to the material in an average concentration, between the highest concentration of the chemical used and the concentration of the chemical in the finished formulation. The exact amounts which the men handled on all occasions were not known. This expression of hours exposed (t) times the average percentage concentration $[K]$ is abbreviated as $[K]$ t and represents a rough, but practical index of his exposure to the chemical during his work day. (As examples: the time required to make a 75% dichloran mixture into a 4% dichloran formulation may be 1 hour, then $[K]$ t, $= (75 + 4)/2 \times 1 = 39.5$ $[K]$ t, dichloran; a 50% chemical formulated to 10% mixture in 4 hours would be expressed as $(50 + 10)/2 \times 4 = 120$ $[K]$ t, chemical.)

The Pesticides Manual (Food and Drug Administration, 1964) of the Food and Drug Administration contains two methods for the analysis for Botran; one is colorimetric, and the other is based on microcoulometric gas chromatography. Dichloran has also been determined using electron-capture gas chromatography on fruits (Kilgore *et al.*, 1962). Since neither the colorometric nor microcoulometric method is adequately sensitive to the parts to the billion level in small samples, a gas-chromatographic method was used for blood and urine, and, in a large volume of urine, confirmed by microcoulometry, partitioning, and infrared spectrophotometry.

In the analysis of human blood this laboratory uses a method of direct extraction with hexane (Dale *et al.*, 1966), and this extraction by hexane of blood and urine was carried out prior to these analyses for dichloran by the electron capture gas chromatograph. Results of these analyses, at the parts per billion level (ppb), are reported (Table I).

The method used was similar to that of Kilgore *et al.* (1962) with the exception of the substitution of hexane for benzene. A 1-ml sample of whole blood or urine was placed in a 5-ml glass-stoppered graduated centrifuge tube. One milliliter of Mallinckrodt Nanograde hexane was added. The mixture was shaken vigorously for 10 minutes on a mechanical shaker and then centrifuged for 5 minutes at 2000–3000 rpm or sufficiently to break any emulsion which may form. A 2–5 μl aliquot was injected into the gas chromatograph. The conditions on the gas chromatograph used were as follows: Microtek 220 with a tritium foil electron-capture detector at a temperature of 195°C. This laboratory used the following columns: 5% QF-1 on Chromport XXX at a temperature of 181°C and a nitrogen flow of 85 cc/minute. Both of these materials were in a 4-ft long stainless steel column with a ¼ inch ID. In the 10% Dow 200 on Chromport-XXX column, the column temperature was 184°C and nitrogen flow of 120 cc/minute. The 10% Dow-200 column was 4-ft long and ¼ inch ID glass column. At a sensitivity of 1.6×10^{-10} A a ½ FSD was obtained with 200 pg of dichloran.

A urine sample was run after a partition cleanup, as suggested in the FDA Pesticides Manual (Food and Drug Administration, 1964), on a Dohrman microcoulometer under the following conditions: 4% SE-39 plus 6% QF-1 on Chromport XXX at a temperature of 184°C and a nitrogen flow of 85 cc/minute, amplifier gain 200, sweep gas 10 cc/minute, oxygen flow 25 cc/minute, range 100 ohms. With these settings, 60 ng was required for a ½ FSD. The retention times (RRT) relative to aldrin were as follows:

Column	RRT
5% QF-1	1.39
4% SE-30 plus 6% QF-1	0.70
10% Dow 200	0.40

As a further check, a partition value of 0.059 was found between hexane and acetonitrile for Botran; and infrared spectra of the Botran standard and dichloran from urine were identical.

RESULTS

Figure 2 shows the occurrence of dichloran under conditions of Botran exposure. This is a tracing of a gas chromatograph obtained on the urine of an aerial duster and is presented together with a tracing obtained from a 99% Botran pure standard. The blood from this same individual was taken simultaneously with the urine sample and was also analyzed by GLC chromatography using a QF-1 column and electron-capture detector.

To further substantiate the identity of the peak obtained the urine was extracted with hexane and the hexane concentrate was cleaned up by partitioning with acetonitrile; a portion was then run on a 5% DEGS column using a Coulson detection system. The retention time of the standard Botran and the only peak from the urine extract matched exactly. Further, infrared spectrophotometric identification of dichloran in urine was carried out.

As in all analyses using electron-capture gas chromatography, care should be taken to avoid misinterpretation of data. In many of the urine samples in this study, a group of unidentified peaks occurred, including one at the position where aldrin elutes. A check with the records of exposure indicated that the formulators had been exposed to mixtures containing elemental sulfur. When elemental sulfur was dissolved in hexane and injected into the electron-capture gas chromatograph the peaks corresponded to the previously unidentified peaks. It should be noted that none of these peaks interfered with dichloran.

In Table I the results obtained in the blood and urine of the four formulators by serial sampling and analysis are presented; the relation of the values of

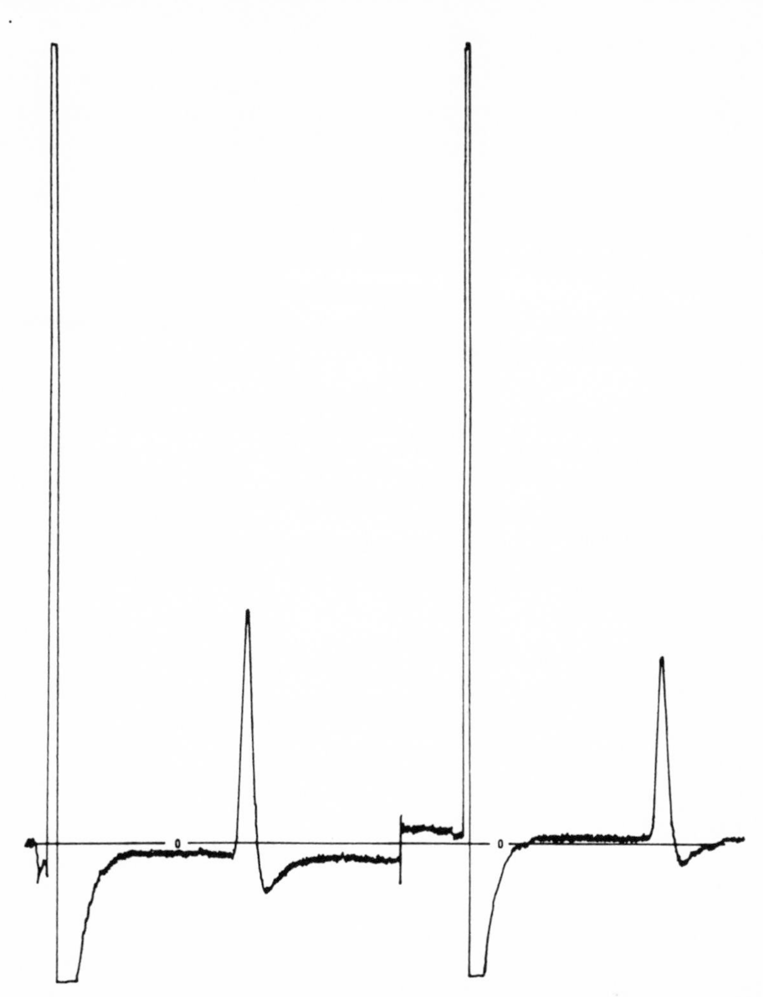

FIG. 2. The occurrence of dichloran in urine, under conditions of Botran exposure in Dade County, Florida, December 1966. Left-hand side of figure: (1) Sample, B. Urine; (2) Date, January 5, 1967; (3) Instrument, Microtek 220; Detector, E.C. type; (4) Temperature, 195°C; Column, 4%, SE 30 + 6% QF-1, temperature: 181°C; Carrier, N_2, Flow, 99cc/minute; Pressure, 39 psig; Sample, 3.7 μl; Attenuation, 10 × 16; Speed, ½ in./minute; and Recorder, Honeywell. One $\mu l \equiv 1$ mg, 41 ppb. Right-hand side of figure: Sample, Botran standard; Same conditions except as follows: Sample: 0.9 μl, 1 $\mu l \equiv 120$ pg, relative retention time, 0.71.

dichloran to their estimated exposures, $[K]$ t, Botran, on each of the days is also recorded.

It should be noted that two of the operators (Nos. 1 and 2) did the actual formulating of Botran dust, while the other two (Nos. 3 and 4) were exposed only to the considerable amounts of dust and spillage incident to the formulating operations involving Botran. These operators, (Nos. 3 and 4), were occupied in formulations with other chemicals in an adjacent area in the plant.

In Table I there can be seen an increase in the urine concentrations of operator No. 1 from a baseline of zero (0) on December 12 to a peak value within 24 hours after exposure to Botran dust, from which the peak concentration began to decline, and then rose again after another Botran exposure. In operator No. 2 the urine concentrations continued to rise in response to repeated exposures for 3 days and then declined 24 hours after the last Botran exposure. In operators (Nos. 3 and 4) blood and urine dichloran concentrations show rises in response to their exposure to dust in the plant, incident to the formulations, though they did not formulate Botran. The time required for blood concentrations to decline after reaching their peaks was not determinable since further exposures to Botran occurred before the declines became evident.

TABLE I
DICHLORAN IN PPB[a] IN BLOOD AND URINE OF FOUR FORMULATORS IN RELATION TO EXPOSURE, EXPRESSED IN $[K]t$[b]

Date	Operator no. 1			Operator no. 2			Operator no. 3			Operator no. 4		
	Blood	Urine	$[K]t$	Blood	Urine	$[K]t$	Blood	Urine	$[K]t$	Blood	Urine	$[K]t$
12/12 AM	NS[c]	NS	—	NS	0	—	NS	NS	—	NS	0	—
PM	3.0	0	0	3.4	10.2	20.0	9.4	5.6	0	5.4	0	0
12/13 AM	NS	0	—	NS	10.0	—	NS	18.7	—	NS	0	—
PM	2.3	8.6	39.5	6.1	16.5	39.5	5.3	6.6	0	3.3	±	0
12/14 AM	NS	20.0	—	NS	47.3	—	NS	5.9	—	NS	±	—
PM	8.9	24.2	0	15.4	34.3	0	6.6	±[d]	0	8.5	±	0
12/15 AM	NS	NS	—	NS	NS	—	NS	NS	—	NS	NS	—
PM	NS	NS	10.0	NS	NS	10.0	NS	NS	0	NS	NS	0
12/16 AM	NS	35.7	—	NS	71.6	—	1.9	0	—	NS	±	—
PM	13.3	19.0	0	17.0	40.4	10.0	NS	0	0	5.9	0	0
12/19 AM	NS	8.4	—	NS	24.8	—	NS	0	—	NS	NS	—
PM	10.9	15.6	30.0	18.4	11.8	30.0	NS	NS	0	5.0	±	20.0

[a] The ppb in urine corrected to osmolality 1000 milliosmols/liter.
[b] $[K]t$ = percentage pure chemical × hours formulated (per day).
[c] No specimen collected.
[d] Trace.

Dichloran at parts per billion levels appeared in the blood and urine within 6 hours of formulating Botran dust. The levels of the chemical in the grab samples of urine rose and fell more rapidly than did the blood levels, and seemed to rise precipitously after the blood concentration reached 9 ppb; below this blood level, after initial exposure, there was no measurable dichloran in the urine. All urinary levels of dichloran were corrected to an osmolality of 1000 milliosmols/liter.

Figures 3 and 4 are graphic representations of the results listed in Table I of the two operators (Nos. 1 and 2) who actually formulated Botran.

During this same period of time all of these formulators also worked with many other chemicals which included DDT, lindane, chlordane, toxaphene, carbaryl, phosdrin, parathion, zinc phosphide, and zineb.

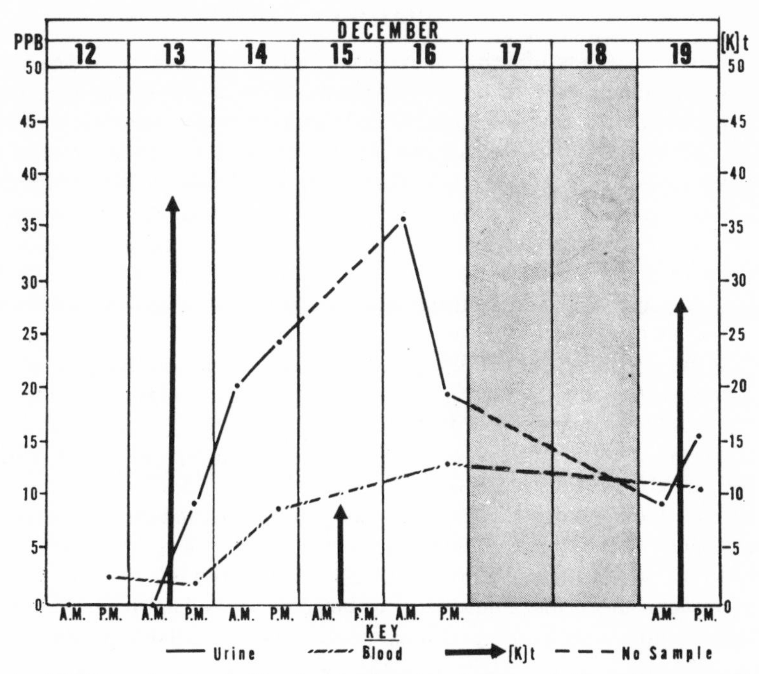

FIG. 3. Dichloran in ppb, in blood and urine, in relation to $[K]\ t$, of operator No. 1.

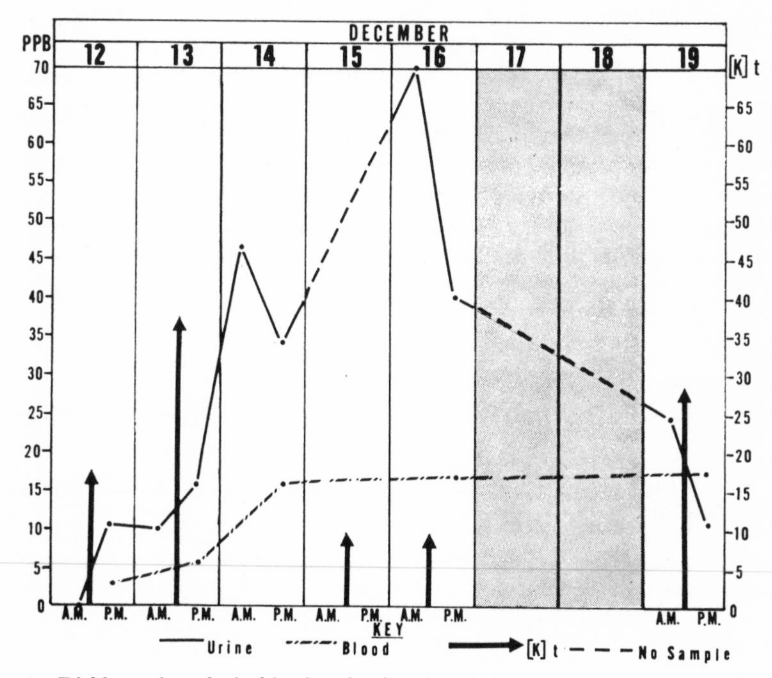

FIG. 4. Dichloran in ppb, in blood and urine, in relation to $[K]\ t$, of operator No. 2.

COMMENT

The maximum levels of urinary excretion, after an exposure of the kind and intensity described, occurred within 24 hours. Most of the material is probably excreted in about 72 hours if further absorption does not take place. However, further exposures, especially within 48 hours, are additive in their effect of raising blood and urine levels. The highest urinary level recorded in this study was 71.6 ppb, and the highest blood level was 18.4 ppb. Although no special effort was made to assess the toxicity of the chemical, the investigators encountered no evidence of any overt toxicity in their twice a day dealings with these volunteers to this or any other chemical during this study.

The finding of rising blood levels of dichloran in operators (Nos. 3 and 4), who were not involved in the actual formulation of the pesticide, but working in the plant at the time, points up the probable routes of entry as being by way of the respiratory tract and by dust swallowing.

This study, involving gross contamination and absorption of an exogenous chemical, illustrates the complete lack of concern that occurs daily toward the handling of materials which, although not known to produce acute or chronic toxicity under occupational conditions, should be and can be environmentally controlled. Efforts to avoid absorption of Botran and myriads of other potentially toxic exogenous chemicals, would seem to be good public health practice, whether toxicological information on specific contaminants is available or not.

SUMMARY

After occupational exposure to Botran, dichloran was found in the whole blood and urine of an individual engaged in aerial dusting, and in four pesticide formulators. The chemical procedure used for its identification and measurement is described, using electron capture gas chromatography. The identity of dichloran was confirmed by infrared spectrophotometry.

A study of the formulators was carried out over an 8-day period, comparing the intensity of exposure, $[K]\ t$, of the workers, with blood and urine dichloran values (in ppb). No evidence of overt toxicity was observed to this or any other chemical to which these formulators were exposed during the study period.

ACKNOWLEDGMENTS

We wish to thank Mr. James Carroll of the staff of the Pesticides Research Laboratory, PHS-NCDC, Perrine, Fla. for his assistance in running the extracted urine samples on the equipment in his laboratory i.e. Coulson detector, and Dr. Morris Cranmer for the infrared spectrophotometric identification of dichloran in urine.

REFERENCES

TUCO PRODUCTS Co., Division of Upjohn Co., Kalamazoo, Michigan (1965). Botran, Technical data brochure.

DALE, W. E., CURLEY, A., AND CUETO, C., JR. (1966). *Life Sci.* **5**, 47.

FOOD AND DRUG ADMINISTRATION, U.S. Department of Health, Education, and Welfare (1964). Pesticide Analytical Manual, vol. II.

KILGORE, W. W., CHENG, I. W., AND OGAWA, J. M. (1962). *Agri. Food Chem.* **10**, 399.

Organic Phosphate Intoxication

WENDELL L. FAIRBANKS, M.D.

KENNETH C. HOFFMAN, M.D.

THE purpose of this paper is to present two cases of significant intoxication with organic phosphorous insecticides and a discussion from the clinical as well as the chemical standpoint. The manner of diagnosis and some comments regarding treatment, as well as contamination by agricultural workers is considered.

Introduction

The organic phosphorous insecticides have been replacing the halogenated hydrocarbons in agricultural use in recent years because of resistance developed to these latter compounds by crop pests, and by the growing concern of residuals of the halogenated hydrocarbons on domestic and wild life. These chemical compounds are used for root worm and corn borer, but are also available to the general public in milder or more diluted concentrations for garden and household insects. As the use of these compounds grows, the possibility of seeing intoxications from them in general practice is increasing because of the carelessness in handling these potentially dangerous products. In the past two seasons, one case of intoxication (Case No. 1) with a probable hematologic complication, and several milder cases have been seen in general practice in a small farming community in Nebraska. One severe case (Case No. 2) and several mild cases have been seen in a pathology practice where the determinations of plasma and red blood cell cholinesterases are done for a large chemical plant producing these insecticides. In the production of these compounds, protective clothing, shoe covers, and in some areas of work, even separate air supplies are used by personnel. It has been estimated that 31% to 58% of the people working with organic phosphate pesticides show absorption and clinical symptoms. Absorption may be through inhalation of the dust or spray, through the gastrointestinal tract, or through any mucous membrane or intact skin. Skin absorption may be increased at higher temperatures or enhanced by an underlying dermatitis. The dose needed to produce symptomatology varies with the concentration and potency of the agent but may be quite small. When the material is absorbed through intact skin there is usually little or no irritation present. In spite of strict warnings concerning the potential danger in using these compounds, symptomatic agricultural workers we have seen have taken no special precautions other than occasional washing of their hands before eating. No thought had been given to decontamination of themselves or changes of clothing after accidental contamination, and all had smoked while applying the product. One of the patients presented here had been out of the field one week but was cleaning and repairing the spraying equipment when he became symptomatic.

Case No. 1 — A 35-year-old white male farm worker complained of abdominal pain, chills, fever, myalgia of the back and extremities, and headache for three days prior to admission. The epigastric abdominal pain radiated to the right upper quadrant, was continuous but mild, except for an intermittent increase in pain which was "something like a cramp." He was anoretic without nausea or vomiting, constipation, diarrhea, hematochezia, or melena. There was some frequency the past several days without dysuria. Several months before admission, the patient had a "rundown condition" but had felt well after undergoing a full mouth extraction. He stated that during a two-week period three days prior to the onset of pain, he had applied an organic phosphorous insecticide and a herbicide to a field. The patient had repaired the applicator during the three days before admission and admitted to taking no unusual medication.

Physical examination revealed a temperature of 102° with the patient appearing lethargic and ill. The pupils were constricted and reacted only slightly to light. The abdomen was flat and with mild epigastric tenderness, but less than would be expected from the history of abdominal pain. There was no rebound tenderness and there were no masses or organomegaly. Bowel sounds were active. The remainder of the examination was normal. The patient continued to spike temperatures to 103°, and over a 24-hour period, the WBC fell from 7,000 to 3,500. There was a trace of albumin in the urine. On the next day, the white blood count fell to 2,500, with 41% band forms and 38% segmented neutrophils. The remainder of the differential was normal. The patient was referred for bone marrow examination, which showed a maturation arrest of the neutrophilic series at the metamyelocyte level with no abnormalities of the erythrocytic or megakaryocytic series. Cholinesterase levels, done in a laboratory other than that of the author, revealed a below-normal plasma cholinesterase, but a normal red blood cell cholinesterase. The BUN and fasting blood sugar were both normal. During the hospitalization, the pupils became less constricted and the abdominal cramping which varied from the epigastrium to the lower quadrants became less severe. No therapy was instituted because the clinical picture was improving. The marrow recovered completely, and the patient progressively improved.

Case No. 2 — A 22-year-old, white, male, chemical industry worker developed abdominal cramping and severe vomiting with weakness a few hours after cleaning an area contaminated with organic phosphorous insecticide. The individual had blood drawn for a cholinesterase determination which was 0.330 Michel units/ml for the red blood cells and 0.135 Michel units/ml for the plasma. (Normal range varies with individuals, but the overall range is 0.390 to 1.020 Michel units/ml for red blood cells and 0.440 to 1.630 Michel units/ml for plasma). The patient continued to have severe vomiting with retching, abdominal colic, profound muscular weakness, and lethargy. Blood pressure on admission to the hospital was 128/80, and there was severe bilateral pupillary constriction with the pupils responding very poorly to light. Mild tremor and moderate muscular weakness as well as moderate dehydration was to be noted. The remainder of the physical examination was within normal limits. A complete blood count on admission showed HGB 13.6/100 ml, PCV 40 vol %, WBC 7,700 with a normal differential. The admitting urinalysis was unremarkable with a specific gravity of 1.037 and a 2+ acetonuria. The patient was given atropine gr. 1/150 IM with 1 gm of pralidoxime chloride IV in 250 ml of isotonic saline. Shortly after this, he felt improved and stated that his eyes were not blurring as they had previously. It was noted that his face was flushed, but the skin was warm and moist with the pupils showing some degree of dilatation. Cholinesterase levels the following morning revealed 0.786 Michel units/ml in the plasma and 1.100 Michel units/ml in the red blood cells.

During the second hospital day, he was given IM atropine and oral tablets of pralidoxime chloride. Late in the afternoon he became somewhat confused and had a semiliquid stool. A repeat cholinesterase study showed the plasma to contain 0.140 Michel units/ml and the red blood cells 0.680 Michel units/ml. The amount of pralidoxime chloride was increased to 2 gm and the following day the patient stated he felt well and had no complaints. His cholinesterase at this time, was 0.830 Michel units/ml for the red blood cells, and 0.270 Michel units/ml for the plasma.

Discussion

Organic phosphorous insecticides act by phosphorylation of the cholinesterases, and thus inactivate these enzymes. This allows an accumulation of acetylcholine at somatic and parasympathetic nerve endings. The parathion pesticide will inhibit cholinesterase only after being metabolized mostly by the liver. Many of the other compounds such as tetra-ethyl pyrophosphate (TEPP) and the 2-carbomethoxy-1-methyl vinyl dimethyl phosphate (Phosdrin) are direct inhibitors of the enzymes. Accordingly, they may also block sympathetic transmission at the ganglia and interrupt synaptic transmission in the central nervous system. The resulting inhibition is more or less permanent and the phosphorylated enzyme returns only very slowly to its active form without specific treatment. Some of these compounds are immediately active and accordingly are potent cholinesterase inhibitors which may cause death quickly, such as with the nerve gases and TEPP. Most common agricultural chemicals are sulfonated and metabolized to a more active cholinesterase inhibitor (oxygenated) after entering the body (Parathrone, Malathion). Other organic phosphorous insecticides are as follows: Carbophenothion, Chlorthion, Co-Ral, DDVP, Delnav, Demeton, Diasinon, Dicapthon, Dimethoate, Di-Syston, EPN, Ethion, Fanthion, Guthion, Malathion, Methylparathion, Methyl Trithion, NPD, Parathion, Phorate, Phosdrin, Phosphamidon, Ronnel, Schraden, TEPP, and Trichlorofon.

True acetylcholinesterase is found in its highest concentration both in nerve tissue and red blood cells. The cholinesterase found in plasma and some of that found in liver is considered nonspecific and probably is more properly called a pseudo-cholinesterase. In most individuals, symptoms are not common until the cholinesterase level is depressed to about 40% of normal. In any patient with intoxication, the plasma levels will be the first to fall, followed by the red cell cholinesterase levels. The latter will also be the last to recover, especially in cases without specific treatment.

It is to be realized that some individuals, due to liver disease or the taking of such drugs as antihistaminics, may have low plasma cholinesterases. Accordingly, any individual who is working with organic phosphorous insecticides should have two or three cholinesterase determinations prior to any contact with these compounds. The clinical symptoms may be predicted from the mechanism of action of the insecticide. In our mild cases, the patients complained of vague abdominal discomfort or actual cramping which was migratory. Giddiness, with occasional headache and some lethargy, was also noted frequently. In severe intoxication, there is blurring of vision, pupillary constriction, gastrointestinal pain and cramping which, with increased severity, leads to vomiting and retching. Weakness and lethargy, as well as discomfort in the chest and rhinorrhea may be seen. Physical signs will include sweating, tearing, excessive salivation, and excessive respiratory tract secretion with possible cyanosis and papilledema, as well as uncontrollable muscle twitches in addition to the miosis. If there is no treatment to disrupt the pattern, there can be further central nervous system depression with confusion or stupor, convulsions, coma, loss of reflexes and loss of sphincter control. Death is usually respiratory in nature when it occurs.

In the production and handling of these toxic pesticides, a rigid safety control program is constantly followed. All employees have two to three cholinesterase determinations made prior to their contact with the phosphate compounds and are checked at either weekly or bi-weekly intervals throughout the production time. Any employee who complains of even mild symptoms which may be due to intoxication is immediately examined by a physician, and has a cholinesterase determination. Also, if there is any decrease in the cholinesterase levels without symptomatology, the individual is removed from his duties and contact for a few days and warned of possible symptoms that may occur requiring specific treatment.

In severe cases, emergency treatment is necessary and is lifesaving. When the patient is in respiratory distress from bronchorrhea, skeletal muscle paralysis, or central nervous system depression, an airway must be maintained and any obstructing secretions removed by suction with artificial respiration being instituted. Oxygen may be used, especially to relieve cyanosis. After the cyanosis has passed, IV atropine in doses from 2-5 mg is begun (atropine in large IV doses may initiate cardiac fibrillation in a hypoxic patient). Atropine blocks the action of the accumulated acetylcholine at receptor sites, but has no effect on the nicotinic action of the accumulating acetylocholine. Accordingly, the muscular fasciculations as well as the respiratory distress from skeletal muscle paralysis may continue. Atropine in adequate doses is lifesaving in severe cases and can be given in large doses, if necessary. Doses of 2-4 mg of this drug every ten minutes is continued until the patient responds or signs of atropinization develop, such as dilated pupils, dry skin, or tachycardia. The use of metaraminal (Aramine) in 10 mg doses with the atropine has been proposed to decrease some of the latter drug's side effects. Any patient who is given atropine must be observed for a minimum of 24 hours for recurrence of possible clinical signs and symptoms.

The oxime drugs, of which pralidoxime chloride is available, are specific antidotes for anticholinesterase poisoning that is due to the organic phosphorous insecticides. It is not to be used where intoxication is due to carbamate insecticides which also are blood cholinesterase inhibitors. Pralidoxime chloride reactivates the inhibited cholinesterase, acts chemically with anticholinesterase, and depolarizes the myoneural junction by its own pharmacologic action. In severe cases, adults are given 1 gm IV at a rate not over 500 mg per minute or 1 gm in 250 ml of isotonic saline IV over a 30 minute period. The dose may be repeated in 30 minutes if muscular weakness persists. The pediatric dose is 25-50 mg per kilo. The amount of the drug is reduced by increasing time from exposure to treatment.

After the airway and artificial respiration have been established in severe cases, and the initial drugs given, decontamination must be promptly accomplished. The stomach is emptied, if the pesticide has been ingested, or the skin is washed with saline solution or alcohol if it is the site of contamination. The majority of the pesticides are more soluble in alcohol than in water. Several relatively common drugs are contraindicated because of their own anticholinesterase activity or because they lead to some degree of central nervous system depression. Among these are the narcotics, aminophyllin, theophyllin, phenothiazine derivatives and barbiturates. If a barbiturate must be used to control convulsions, it should be used very sparingly. In any severe case, it is recommended that in-hospital care be carried out for at least 48 hours.

In Case No. 1, the patient had a proven temporary arrest of the maturation in the myelogenous series at the metamyelocyte level. The etiology of this temporary alteration would appear to be most likely due to the organic phosphorous insecticide intoxication, although the herbicide or one of the mechanical lubricants used by the patient can not be absolutely ruled out. Mostromatte reviewed many cases of hemologic disorders associated with the use of insecticides and concluded that the development of blood dyscrasias to these agents was most unlikely. As with this case, most of the cases were exposed to many other agents that could also be responsible.

Case No. 2, showing severe intoxication, demonstrates the prompt recovery of the patient when aggressively treated as outlined. It also indicates the need for continuous observation because of possible recurrence of the symptoms and signs from agents still circulating and capable of depressing the cholinesterase even after the initial treatment.

Summary

The organic phosphorous pesticides are becoming progressively more important causes of accidental poisoning. Carelessness in handling these compounds is chiefly responsible for the intoxications. These cases will be seen more frequently as the use of these insecticides increases and it is necessary for all physicians to recognize them, obtain specimens for cholinesterase determinations and to treat accordingly. The organic phosphorous insecticides inhibit cholinesterase allowing accumulation of acetylcholine which is responsible for the symptoms and signs. Therapy includes recognition, respiratory resuscitation, and atropine in large doses.

'Low Volume' Insect Control And Parathion Poisoning

Until 1964, many pesticides used in agriculture, including the highly toxic, cholinesterase-inhibiting organo-phosphorus compounds, ethyl and methyl parathion, were applied in a diluted form; for example, one pint of the toxicant in three gallons of water.

In 1965 malathion, an organo-phosphorus insecticide with a low mammalian toxicity, was approved for use by the so-called low volume method. This new technique for insect control in fields was particularly welcomed by farmers because, in some instances, application rates were slashed by as much as 94 percent, and some costs were reduced by more than 50 percent. It was demonstrated that, in some instances, the cotton boll weevil could be controlled at rates as low as 5 oz. per acre.

During the 1965 cotton season, the low volume method was used extensively for aerial application of parathion in the lower Rio Grande Valley of Texas; and in 1966, most of the application of parathion was done in this manner. The degree of insect control achieved by the low volume method of applying this dangerous toxicant by air has been appreciably good. Many, perhaps most, Valley farmers have abandoned ground application methods because they can contract for aerial application more cheaply. However, low volume application of parathion has never been officially approved by the manufacturers of the material; and this procedure is carried out without sanction by federal or state departments of agriculture.

The Federal Aviation Agency is vitally interested in the aerial application of these and other agricultural poisons because of the possibility that airplane crashes due to pilot error may be the result of exposures to chemical substances which impair vision and other body functions. Also, the US Department of Agriculture, the Department of the Interior, farmers, chambers of commerce, manufacturers and formulators, state and federal health officials, physicians who treat cases of acute poisoning symptoms, and the aerial applicators themselves, are very much interested in the use and effects of these highly toxic materials in concentrated form on those who are exposed to or who handle them.

At the Poison Control Center at Valley Baptist Hospital in Harlingen, we have good records on cases of pesticide poisoning beginning with the year 1960.

During the four years 1960 through 1963, we reported a total of 70 human cases of poisoning due to pesticides. In 1964 we had a 400 percent increase in the amount of pesticides used in agriculture in this area and a corresponding, significant increase in the number of poisoning cases (68 cases). In other words, in 1964 the number of poisoning cases totaled 68 or approximately the same number of cases as in the combined previous four years. In 1964 all application was in the diluted form by air or by ground equipment.

During 1965, when approximately one-half of the total amount of pesticides used were applied by low volume aerial spraying, the number of acute poisoning cases was about equal to the previous year. During 1966, a year when most pesticide applications were done by low volume aerial spraying, there were 70 cases of poisoning. Therefore, according to our records, the use of low volume aerial spraying has not affected the seasonal number of cases of acute poisoning. However, there has been a change in the kinds of individuals poisoned. Formerly farm laborers who mixed and applied the pesticides were in the majority; whereas now, most cases occur in swampers, loaders, and flagmen who work for the aerial applicators.

This season, I predict that we will see a further decrease in the number of cases of poisoning. This prediction is based primarily on the fact that fewer people will be required for handling the same volume of pesticides. We should be able to indoctrinate them in safe handling procedures since, generally speaking, they are more intelligent and more cognizant of the very real need to take precautions against exposure to these highly toxic materials.

Another factor involved in fewer acute poisoning cases is that no mixing is done by the applicator, the material being transferred directly from the container in which it is purchased, to the spray plane. Mixing vats and defective hoses used in the dilution process have been the ultimate source of a large percentage of poisoning episodes.

There has been an apparent reduction in

the degree of severity observed in most cases of poisoning seen in recent years. I believe this is directly due to the education of employers and employees alike regarding the early symptoms of poisoning so that medical attention is more promptly sought.

In the area in which I practice, the Lower Rio Grande Valley of Texas, the bulk of pesticides are applied during the summer and the loading and flagging is done by teenagers. These young people are most difficult to train in the proper handling of toxic materials and, as a result, they are often acutely poisoned twice before they will follow instructions. Usually, by this time, their alarmed parents have made them quit and find other employment. Some of the aerial applicator companies are derelict in not providing adequate decontamination facilities, safe handling instructions, and protective clothing and equipment. One small company with only two aircraft had 12 cases of poisoning in 1965 and 9 cases in 1966. They did not have a shower and the hoses were saturated externally with parathion.

I wish to stress the fact that very few employees of the local formulating and manufacturing companies are ever poisoned. This is due to close supervision, adequate protective clothing, face masks, and equipment, and weekly or bi-weekly cholinesterase determinations. Routinely, if the cholinesterase level is depleted below 50 percent using the Caraway method, the employee is transferred to kinds of plant work where there is no exposure until his cholinesterase level has recovered to about three-fourths of its normal value.

Some observations that I have made in my practice are valuable and interesting. I treated two cases in which a connecting hose ruptured, squirting a concentrated organo-phosphorus directly into an eye. In both cases, adequate decontamination was accomplished by immediate showering, the shower being only 20 feet from the area where the accidents occurred. The second of these two poison victims had a plasma cholinesterase level of 0.48 delta pH one week before the accident; within 24 hours after the mishap it was reduced to 0.24 delta pH. The red blood cell cholinesterase was only slightly decreased. The pupil of the right eye, which had received the Bidrin, was exceedingly small and did not react; the left pupil was normal and light-reactive. The following day both pupils were constricted as a result of systemic response. In all likelihood this individual would have died had he not had rapid decontamination at the scene of the accident and prompt medical treatment.

We have found that, in treating poisoning cases, atropine in extraordinarily large doses is required. As much as 1/30th of a grain of atropine intravenously will cause no serious reaction, even in a person who is not poisoned but has only gastroenteritis with vomiting and abdominal cramps. I mention this to give courage to those who may be confronted with an emergency in which the diagnosis is uncertain. An intravenous injection of 1/30th grain of atropine administered to a person who is poisoned will usually result in prompt improvement, and this will suffice to confirm the diagnosis before the cholinesterase determinations can be made.

Last season, an aerial applicator crashed and was severely, but not fatally, burned. His cholinesterase was depleted but poisoning symptoms were not serious enough to cause him to seek medical attention. Later, it was learned from two pilots working with him that for about a week prior to the crash his handling of the aircraft was below par, that he had been making sloppy landings, and that on one occasion he had returned from a spraying job with the branch of a tree stuck in the landing gear. Undoubtedly, this pilot had an insidious decrease in cholinesterase due to frequent exposures over a considerable period of time which had affected his accommodation, probably the immediate cause of the crash.

The most common cause of chronic poisoning in these pilots for at least the past two years has been the practice of adjusting the heavily contaminated sprayer nozzles with their bare hands. Chronic poisoning in pilots is extremely hazardous because impairment of accommodation may be an early, perhaps unnoticed, symptom. Accommodation impairment precedes nausea which is one of the early recognizable symptoms and usually the one that causes the pilot to seek medical aid or to start self-medication with atropine tablets which may compound the risks involved in flying.

—George L. Gallaher, M.D.

Polyneuropathy Following Exposure To Insecticides

Two Cases of Polyneuropathy With Albuminocytologic Dissociation in the Spinal Fluid Following Exposure to DDD and Aldrin and DDT and Endrin

R. B. JENKINS, MB B Chir

AND

J. F. TOOLE, MD

A cause-and-effect relation between exposure to insecticides and subsequent development of polyneuropathy is very difficult to prove even when strongly suspected. However, in the two cases described below such a relation is so probable that we believe their full report is warranted. Our two patients had albuminocytologic dissociation in the spinal fluid and might have been considered to be suffering from a sporadic form of the Guillain-Barré syndrome (acute idiopathic polyneuritis) of cryptogenic origin had not the close association between exposure to insecticide and neurologic manifestations made an etiologic relationship probable. Despite the enormous quantities of insecticides used annually, case reports of polyneuropathy caused by them are few, perhaps because of their rarity, but perhaps also because of failure of physicians to consider insecticides as a possible causative agent.

CASE 1.—A 13-year-old boy was admitted to the North Carolina Baptist Hospital for evaluation of weakness and paresthesias. He was quite well until the first week of August, 1962, when he began to be awakened from his sleep by uncontrollable jerkings of his body of such severity that his father had to hold him in bed to keep him from injuring himself. At no time did he lose consciousness, so his local physician attributed the symptoms to nervousness and prescribed sedatives; however, they were probably myoclonic jerks. They continued almost nightly until Aug 28, when he first became aware of paresthesias in his feet. His father observed about this time that his son was unable to keep up with his work in the fields because of rapid fatigability. Within three days the boy had difficulty phonating and swallowing, and because of this he was admitted to the hospital for evaluation. No other member of his family or co-worker in the fields was similarly affected.

Inquiry showed that on May 10, 1962, hexachloro-*endo,exo*-hexahydrodimethanonaphthalene (aldrin) had been spread on the fields in which the boy worked. In June he began to work in the fields where, because of frequent rain, the pesticide-containing mud clung to his person for hours at a time.

it was his job to dust tobacco plants by hand with dichlorodiphenyldichloroethane (DDD), and, as a prank, he would frequently dust himself and his friends with 10% DDD mixture, so that he was exposed to this substance both by inhalation and through his skin.

Examination showed that he had a respiratory rate of 24, pulse 130, and rectal temperature of 99.4 F (37.4 C). He had bilateral facial paresis, but ocular movements and pupillary reactions were normal and his palate elevated normally on phonation. Weakness was more evident distally than proximally and greater in the lower than the upper extremities. Deep tendon reflexes were absent in the lower limbs and barely perceptible in his arms. The superficial reflexes were normal. Appreciation of pinprick, light touch, and position was normal although vibratory sensation was somewhat diminished in his toes. Despite the lack of objective sensory findings, he said that the tips of his fingers and toes did not feel right. The ulnar and peroneal nerves and the cords of his brachial plexus were tender to palpation, and severe pain was produced by flexion of his neck.

His muscle weakness progressed during the next three days to complete facial diplegia and loss of movement in the lower extremities. His respirations became shallower, and difficulty with secretions necessitated a tracheostomy on Sept 10. Beginning Sept 11, intravenous corticotropin 30μ in one liter of 5% dextrose in water was given over an eight-hour period daily for ten days. By Sept 15 his facial diplegia and muscle weakness showed signs of improvement. He complained bitterly of burning pain in his lower extremities throughout the last week of September. The tracheostomy tube was removed on Sept 29, and by Oct 5 subjective impairment of sensation had disappeared and he was able to move the toes of his left foot slightly.

Pertinent laboratory studies were: hemoglobin 13.7 mg%, WBC 6,000 with normal differential count, sodium 140 mEq/liter, potassium 4.4 mEq/liter, calcium 10.1 mg%, phosphorus 3.3 mg%, negative heterophil agglutination test. Three lumbar punctures showed clear colorless fluid with normal pressure, no cells, and proteins of 97 mg% on Sept 2, 240 mg% on Sept 15, and 302 mg% on Oct 1. No lead lines were seen on his gums or in x-rays of his wrists, and no punctate stippling was seen in the blood smears. Chest x-ray was normal. Several electrocardiograms showed nonspecific T wave changes.

Diagnostic tests with 10 mg edrophonium HCl (Tensilon) intravenously, atropine 1 mg intravenously, and pralidoxime (Protopam) chloride 1 gm were all without effect on his signs or symptoms.

Clinical Course.—Except for occasional slight elevations of temperature in the evenings, he remained afebrile with a persistent tachycardia throughout his hospital stay. No transverse striate leukonychia appeared. He was discharged Oct 10; by November he was able to stand without assistance. The peripheral nerves became less tender, and by December, 1962, the deep tendon reflexes had returned at his knees and ankles. His paresthesias had subsided, and the sensory examination was normal. By December, he was able to walk well enough to attend school, and when we examined him in May, 1963, his deep tendon reflexes were all brisk and symmetrical in both the upper and lower limbs, motor power was intact throughout, there were no sensory complaints, and it was difficult to determine that he had ever been ill.

In May, 1963, he was again put into the fields a week after his father had spread aldrin compound, and after several days he once more developed weakness, myoclonic jerks, and sensory disturbances similar to those he had had before. On this occasion there had not been any use of DDD. The boy was re-examined three days after the onset of his symptoms, and no neurologic deficit could be found. The family was warned again not to expose the boy to these materials in the future.

Case 2.—A 31-year-old male farmer was in good health until June 4, 1962, when he became contaminated with a mixture of hexachlorooctahydro-*endo,endo*-dimethanonaphthalene (endrin) and dichlorodiphenyltrichloroethane (DDT) made by tipping a one-gallon can of endrin concentrate and a one-gallon can of DDT concentrate into the tank of a power sprayer and adding water to make the volume up to 50 gallons. A blocked and defective spray nozzle came off in his hand and caused enough of the insecticide to blow onto him to wet his face and soak his upper clothing. Some of it entered his mouth, but he immediately spat this out without swallowing any. Because the concentrates were poured into the tank first and no mixing had yet been done, it is probable that he was contaminated with almost undiluted chemicals. Ignoring his wet shirt, he continued to work for the rest of the day and first cleaned himself off about eight hours later. The same night, he experienced malaise and slight aching in his limbs. Aching increased during the next few days and was more severe distally than proximally and in the upper more than in the lower extremities. On the sixth day, he saw his doctor because of weakness of his grip. He was re-examined at intervals and was hospitalized approximately 2½ weeks following exposure because of weakness in all four extremities. Lumbar puncture revealed clear acellular fluid with a protein of 136 mg%. Twenty-four-hour urine for lead was reported as containing 0.015 mg. One urine was reported positive and another as negative for coproporphyrin III. The patient was transferred to the North Carolina Memorial Hospital about four weeks after exposure.

Physical examination showed him to be of muscular build and bronzed by the sun. There was no lead line on the gums, and no transverse striate leukonychia. The distal portions of all four limbs were hypotonic and the forearm muscles were markedly flabby and weak, but the only wasting to be seen was in the left thenar eminence. The weakness was asymmetric, being greater in the left upper limb, where the extensors of the wrist could be overcome with one finger. Power was normal in the proximal muscles. His gait was slapping bilaterally but was not ataxic. The deep tendon reflexes were normal with the exception of the ankle jerks which were diminished and asymmetric, the right one being elicitable only with reinforcement. Superficial reflexes were normal. Sensation was normal with the exception of minimal impairment of appreciation of light touch distally. Co-ordination was normal. Romberg's test was negative. Examination of other systems revealed no abnormality.

Laboratory studies included a negative blood serology, hematocrit 45%, sedimentation rate 10 mm/hour, WBC 4,400 with normal differential count; no stippled red cells were seen, and three lupus erythematosus (LE) preparations were negative. Urine was negative for sugar and albumin, and three examinations for coproporphyrin III and three for porphobilinogen were negative. Blood urea was 15 mg%, fasting blood sugar 94 mg%, chest x-ray and electrocardiogram normal, and lumbar puncture (four weeks after exposure) showed no cells and a protein of 146 mg%. Sensory and motor nerve conduction studies showed a normal rate and amplitude of evoked nerve potentials in the left median and ulnar nerves.

Clinical Course.—By the time of the patient's discharge on July 2, 1962, the strength in his lower limbs was beginning to improve slightly. This improvement continued so that by the end of four weeks, he had started work again.

Comment

DDT (and also DDD, which closely resembles it) is a halogenated hydrocarbon. It is one of the least toxic of the insecticides, having an average median lethal dose (LD_{50}) for several animal species of 250 mg/kg. Despite the fact that the annual production of DDT is in excess of 100,000,000 lb in the United States alone, reports of poisoning in man are rare. The estimated lethal dose for man is 500 mg/kg, although this is probably lower when it is ingested in an oily vehicle.[1-4] Numerous instances are on record of large doses having been ingested by human subjects with slight or no harmful effects,[5-8] but distal numbness and weakness with wrist drop and loss of power in the hands, inability to stand, and an excited mental state have been described.[9]

Despite some negative experiments,[3,10] DDT is thought to be absorbed to some extent through the skin when the insecticide has been dissolved in oil or another organic solvent,[3,11] although this hazard is evidently not great, because innumerable members of spraying crews have suffered heavy contamination of the skin by oily solutions of DDT for long periods of time without harmful effect. Aching of the limbs with weak legs and generalized muscular tremors,[12] predominantly sensory neuropathy,[13] polyneuropathy and purpura,[13] polyneuropathy with paresthesias, fatigue, sensory loss, tremors, and ataxia of the legs,[14] fatigue and cramps,[15] aching and weakness in the limbs with diminution of deep tendon reflexes, patchy peripheral anesthesia, and apprehensive mental state,[16] monoparesis with swelling of the affected arm, numbness, and paresthesias,[8] and cerebellar signs [17] have been reported.

Reports of study of the central nervous system for possible pathology are scanty. Globus [18] administered large doses of DDT to cats, dogs, rats, and monkeys, some of which developed signs suggestive of a polyneuropathy; but no studies of peripheral nerve were performed, and no spinal fluid examinations were made. Histopathologic studies of the central nervous system, including the anterior horn cells, showed no significant abnormality.

Endrin and aldrin are members of the cyclodiene group of insecticides, which are highly chlorinated dimethanonaphalenes. Other members are dieldrin, isodrin, chlordane, heptachlor, and toxaphene.

Higher animals may absorb these insecticides by ingestion, by inhalation, or through the skin. The dermal route offers the most serious occupational hazard, for both toxic and lethal amounts have shown to be absorbed through the unbroken integument of experimental animals.[19,20] The cutaneous minimum lethal dose (LD_{00}) for endrin is between 60 and 94 mg/kg as compared with 2,500 mg/kg for DDT, but various solvents may increase the toxicity many times.[3,21] Species susceptibility varies greatly, monkeys being among the most susceptible animals, with a LD_{50} for endrin of the order of 3 mg/kg and a LD_{00} of 1 to 3 mg/kg.[1]

The mode of action of the cyclodiene compounds has not yet been elucidated, but they are thought not to act as cholinesterase inhibitors. Their fate in the animal body is not certain, but some are known to be very stable, being stored in the body fat for long periods, and are found in such body products as milk or eggs.[1,22,23] Similar acute toxic effects are produced by all of the various members of the group when they are administered experimentally to animals. They involve the central nervous system primarily, being manifested as hyperirritability, tremors, myoclonic jerks, epileptiform convulsions, and disturbances of respiration. Parasympathomimetic effects on the heart and intestine, probably the result of central stimulation of the vagus, have been described.[1,23] The effects in man are similar, with the addition of headache, vertigo, nausea and vomiting, generalized malaise, and disorders of consciousness.[24,25] Transient deafness has also been described.[26] No case of polyneuropathy secondary to endrin or aldrin intoxication has been found in the literature.

Of interest in our two cases is the similarity of the combination of insecticides to which exposure occurred, and the predominantly dermal route of absorption. It has been shown that aldrin, and to a lesser extent DDT, can exert effects on soil fauna for prolonged periods.[27] The occurrence of a relapse in case 1 followed re-exposure to soil treated that year by aldrin alone not only suggests the play of a personal idiosyncrasy or sensitivity effect, but tends to exonerate the noncyclodiene DDD from being a responsible agent. Although it is possible that residues of the latter may have played a part in pathogenesis, this seems unlikely in view of its low toxicity and low soil concentration. Convulsive phenomena may occur after exposure to any member of the cyclodiene group, but it is of interest that not only have grand mal seizures followed cutaneous exposure to aldrin,[25] but myoclonic jerks similar to those seen in our first case have also been reported after cutaneous exposure to dieldrin.[28]

In case 2 one cannot be certain whether DDT or endrin acted alone as the cause of the polyneuropathy, or whether a degree of synergy of their toxic effects occurred. Although cases of polyneuropathy following exposure to DDT have been described, they are very rare compared to the vast amounts of the chemical used: Exposure considerably greater than the present one had occurred, and the polyneuropathy was usually of a mixed or predominantly sensory variety. Endrin, on the other hand, is of more recent introduction. It is a much more highly toxic substance, and the danger of its absorption through the skin is far greater than that of DDT. It is possible that when the DDT is dissolved in aromatic naphtha emulsifier it may be absorbed through the skin as readily as is endrin,[29] but on balancing the probabilities it appears more likely that endrin was the responsible agent in this case.

Summary

Two cases of motor polyneuropathy resembling the Guillain-Barré syndrome following exposure to a mixture of a cyclodiene insecticide (aldrin or endrin) and DDD or DDT are described.

Polyneuropathy following exposure to DDT is known to occur but is excessively rare and is usually of the mixed sensory and motor variety. No previously reported case of polyneuropathy following exposure to aldrin or endrin has been found.

It appears that aldrin was the responsible agent in the first case; and it is probable that endrin was the responsible agent in the second case, although DDT may possibly have played a role either of synergy or as the primary agent if the patient possessed an idiosyncrasy to it.

In view of the enormous quantities of insecticides used, it is probable that other cases of polyneuropathy due to them are unrecognized. It is suggested that inquiries regarding exposure to them be made in every case of polyneuropathy of obscure etiology.

Dr. Weston M. Kelsey, Chairman of the Department of Pediatrics, Bowman Gray School of Medicine, gave permission to publish case 1.

R. B. Jenkins, MD, Department of Neurology, University of North Carolina School of Medicine, Chapel Hill, NC 27515.

REFERENCES

1. Negherbon, W. O.: Handbook of Toxicology, Philadelphia: W. B. Saunders Company, 1959, vol III.

2. Conley, B. E.: Pharmacologic and Toxicologic Aspects of DDT (Chlorophenothane), JAMA 145: 728, 1951.

3. Cameron, G. R., and Burgess, E.: The Toxicity of DDT in Man, Brit Med J 1:865, 1945.

4. Hill, K. R., and Robinson, G.: Oral Toxicity of DDT in Kerosene, Brit Med J 2:845, 1945.

5. Neal, P. A., et al: The Excretion of DDT in Man Together With Clinical Observations, Public Health Rep 61:403, 1946.

6. Stammers, F. M. G., and Whitfield, F. G. S.: The Toxicity of DDT to Man and Animals, Bull Ent Res 38:1, 1947; quoted by Brown.[26]

7. Lazar, T.: DDT Pancakes, Brit Med J 1:932, 1946.

8. Mackerras, I. M., and West, R. E. S.: Effect of DDT on Humans, Med J Aust 1:400, 1946.

9. Garrett, R. M.: Toxicity of DDT for Man, J Med Ass Alabama 17:74, 1947.

10. Dangerfield, W. B.: Toxicity of DDT to Man, Brit Med J 1:27, 1946.

11. Draize, J. H.; Nelson, A. A.; and Calvery, H. O.: The Percutaneous Absorption of DDT in Laboratory Animals, J. Pharmacol Exp Ther 82: 159, 1944.

12. Wigglesworth, V. B.: A Case of DDT Poisoning in Man, Brit Med J 1:517, 1945.

13. Campbell, A. M. G.: Neurological Complications Associated With Insecticides and Fungicides, Brit Med J 2:415, 1952.

14. Klingemann, H.: Die DDT Vergiftung, Aerztl Wschr 4:465, 1949.

15. Campbell, A. M. G.: DDT Poisoning in Man, Lancet 2:1178, 1949.

16. Case, R. A. M.: Toxic Effects of DDT in Man, Brit Med J 2:842, 1945.

17. Onifer, T. M., and Whisnant, J. P.: Cerebellar Ataxia and Neuronitis After Exposure to DDT and Lindane, Proc Mayo Clin 32:62, 1957.

18. Globus, J. H.: DDT Poisoning, J Neuropath Exp Neurol 7:418, 1948.

19. Treon, J. F., and Cleveland, F. P.: Toxicity of Certain Chlorinated Hydrocarbon Insecticides for Laboratory Animals, J Agric Food Chem 3:402, 1955.

20. Johnston, B. L., and Eden, W. G.: The Toxicity of Aldrin, Dieldrin and Toxaphene to Rabbits by Skin Absorption, J Econ Ent 46:702, 1953.

21. Treon, J. F.; Cleveland, F. P.; and Cappel, J.: Toxicity of Endrin for Laboratory Animals, J Agric Food Chem 3:842, 1955.

22. Sun, J. Y. T., and Sun, Y. P.: Microbioassay of Insecticides in Milk, J Econ Ent 46:927, 1953.

23. Metcalf, R. L.: Organic Insecticides, New York: Interscience Publishers, Inc., 1955.

24. Spiotta, E. J.: Aldrin Poisoning in Man, AMA Arch Industr Hyg 4:560, 1951.

25. Bell, A.: Aldrin Poisoning: A Case Report, Med J Aust 2:698, 1960.

26. Brown, A. W. A.: Insect Control by Chemicals, New York: John Wiley & Sons, Inc., 1951.

27. Edwards, C. A., and Dennis, E. B.: Some Effects of Aldrin and DDT on the Soil Fauna of Arable Land, Nature (London) 188:767, 1960.

28. Paul, A. H.: Dieldrin Poisoning: Report of a Case, New Zeal Med J 58:393, 1959.

Pesticides and Associated Health Factors in Agricultural Environments

Effect on Mixed-Function Oxidizing Enzyme Metabolism, Pulmonary Surfactant and Immunological Reactions

Kingsley Kay, M.A., Ph.D.

There have been rapid developments in knowledge of the effects of pesticides and associated health factors in agricultural environments since previous reviews by the author in 1965[1] and 1966.[2]

Early work in this laboratory[3] established that chlorinated hydrocarbon pesticides stimulated the activity of rat aliesterase (later included as microsomal drug-metabolizing) and reduced the toxicity of parathion sevenfold.[4] Carbon tetrachloride was also found to produce enhanced aliesterase activity but unaccountably failed to provide protection against parathion.[5] Frawley[6] observed potentiation between some organic phosphates. Later it was discovered that phenobarbital induced hepatic microsomal drug metabolism.[7] Induction and inhibition of the mixed-function oxidizing enzyme systems of the liver and other organs has been demonstrated to play an important role in pesticide toxicity although there is yet limited evidence of how these factors bear on human health.

Immunological study of pulmonary disease among agricultural workers has now led to the identification of the role of immune processes and the sensitizing exogenous agents in the agricultural environment. The capability of metabolites of DDT and malathion to form antigens with tissue proteins and induce an immune response has been demonstrated, raising the possibility of allergic tissue damage from these chemicals in certain persons predisposed, particularly by genetic factors such as atypical forms of pseudocholinesterase which were found by Kalow[8] in 1957. A full understanding of the haptens in pulmonary-renal disease has yet to be elucidated but there is current interest in an action status for some of the exogenous factors of the environment.

These considerations establish the outlines of a structure of associated factors affecting adjustment of the body to current agricultural stresses. It is the purpose of this paper to explore new boundary problems coming to light in this occupational milieu from the interactions within the structure outlined.

Enzymes In Metabolism Of Pesticides

The induction of hepatic aliesterases by chlorinated hydrocarbon insecticides in rats,[3] their role in increasing the metabolism of parathion[4] and the later discovery that phenobarbital effectuated similar induction[7] were the bases for a large number of (following) investigations. From these it has become clear that endogenous factors such as species, sex, age, hormonal status and exogenous factors such as nutrition, drugs, other foreign chemicals and ionizing radiation play a role as inducers or inhibitors in hepatic enzymology. These investigations were reviewed by the author in 1966[2] and in 1967 by Conney.[9] It is now understood that the induced hepatic enzyme systems activate the biotransformation of many foreign chemicals by processes such as oxidation, reduction, hydrolysis and conjugation and constitute a main aspect of defense against the toxic effects of pesticides to mammals as demonstrated to date on mice, rabbits, rats and the phylogenetically higher species, the squirrel monkey.[10] Of course, some hepatic enzymatic biotransformations increase toxicity at least through certain stages, for instance the epoxidation of aldrin to dieldrin. It is also important to stress that in our early work as shown in Fig. 1, aldrin was first an inhibitor and did not become an inducer for several hours. This pattern of activity was confirmed in 1962 by Kato et al.[11], [12] for SKF525A ordinarily an inhibitor but also a weak inducer. Phenobarbital, ordinarily an inducer, was also shown to be a weak inhibitor.

Another feature of the induction of the mixed function systems is that apparently this process occurs in extra-hepatic locations, at least in animals, as when 3,4-benzpyrene hydroxylase has been induced at the portals of entry, the lung and gastrointestinal tract.[13-15]

Evidence of Induction in Humans

Confirmation of the role of the exogenous inducers in humans has so far been limited. Kuntzman and coworkers[16] demonstrated *in vitro* that human liver contains enzyme systems catalyzing the ring hydroxylation of 3,4-benzpyrene, side-chain oxidation of pentobarbital, the b-dealkylation of acetophenetidin and the N-dealkylation of 3-methyl-4-monomethylaminoazobenzene. Creaven and Williams,[17] using autopsy human liver samples, demonstrated hydroxylation of biphenyl to 4-hydroxybiphenyl and coumarin to 7-hydroxycoumarin. A review of microsomal dealkylation of drugs by McMahon[18] lists a number of entities such as N-methyl phenobarbital, imipramine, chlorpromazine and mescaline which have been demonstrated in man. Other hepatic enzyme inductions in man have been reported by Remmer[19] for barbital, glutethimide and by Chen[20] for phenylbutazone which accelerated aminopyrine metabolism in human subjects.

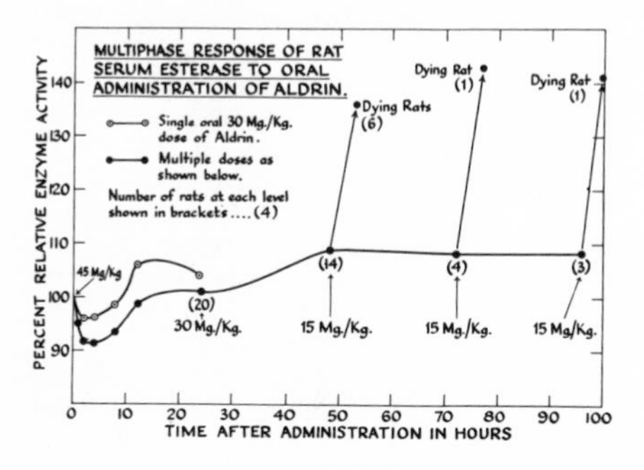

FIGURE 1. From Ref. 3.

Enhanced metabolism of the anticoagulant coumarin was found to be induced by barbiturates in 1961.[21] A recent report on inhibition of bishydroxycoumarin metabolism with resultant hypoprothrombrinemic effect in a patient exposed to CCl_4 concurrently with anticoagulant therapy[22] can retrospectively be attributed to the inhibitory action of CCl_4 on the metabolizing enzymes in light of early animal experiments from our laboratory (described later in this paper). More recently, the pharmacological effectiveness of bishydroxycoumarin has been shown to be reduced by phenobarbital,[23] chloral hydrate,[24] griseofulvin[25] and to be stimulated by phenyramidol hydrochloride.[26, 27] Warfarin metabolism in dogs has been shown to be stimulated by glutethimide, amobarbital, secobarbital, and meprobamate.[28]

Recently, Beckett and Triggs[29] have demonstrated increased metabolism of nicotine in man induced by tobacco smoking, a finding that confirmed previous work on dogs, rabbits and rats. Cigarette smoking has been found to stimulate metabolism of 3,4-benzpyrene by enzymes in human placenta.[30]

The clinical implications of these interactions undoubtedly apply to the potentialities of organochlorine pesticide exposure of humans under similar circumstances although there has not been confirmation of the role of the exogenous inducers, with the exception of a report this year from Poland where red cell acetylcholinesterase of a group of 51 workers exposed to chlorinated hydrocarbon insecticides was found to be increased 7% over controls, with no change in serum pseudocholinesterase.[31]

Although evidence in humans is slow to be recognized, it seems certain that the many hepatic enzyme-inducing and inhibiting drugs and chemicals must play an interacting role in homeostasis under appropriate combinations of agricultural environmental conditions. It is therefore important that laboratory experiments with other mammalian species be closely followed.

Recent Laboratory Findings on Exogenous Enzyme Inducers

Recent laboratory work on exogenous enzyme inducers has much potential significance in a consideration of health-related exogenous factors in the agricultural environments, as more and more chemicals prove to be inducers of mixed-function oxidizing enzymes, and some with considerable specificity.

Classification of Inducers

Attempts continue to classify inducers. In 1964 Fouts and Rogers[32] described differences in the stimulatory effects of phenobarbital, chlordane, 3,4-benzpyrene (BP) and 3-methylcholanthrene (3-MC) on hepatic microsomol enzyme activity and in the effects of these chemicals on the structure of the smooth endoplasmic reticulum (SER). Chlordane and phenobarbital stimulated metabolism of hexobarbital, aminopyrine, zoxazolamine and p-nitrobenzoic acid and brought about marked proliferation of SER, whereas BP and 3-MC stimulated only the metabolism of zoxazolamine, without pronounced increase in hepatic SER. It has just been reported by this group[33] from experiments with eight substrates that phenobarbital pretreatment had only insignificant effects on the metabolism of codeine, benzpyrene and amphetamine by adult male rabbits, but increased P-450 activity on the metabolism of zoxazolamine, hexobarbital, and aminopyrine. 3-MC, on the other hand, either increased or decreased the enzyme activity responsible for metabolism of the eight substrates.

It has been proposed by Kuntzman et al.[34] and Alveres et al.[35] that inducers can be classified into two groups: (a) compounds such as phenobarbital which induce the metabolism of a large number of exogenous substances, and (b) those exerting specificity as enzyme inducers which do not stimulate many reactions that are stimulated by phenobarbital. 3-MC and BP are given as examples of class b. These authors did obtain evidence that 3-MC causes an increase in a microsomal CO-binding pigment with a spectrum different from that of the pigment induced by phenobarbital. However the two-class proposition does not appear confirmed at the present stage of knowledge in light of the findings of Gram, Fouts and Rogers[33] described above. Uehleke[36] has also contributed to the problem, having found in rats multiform changes in specific activities after stimulation with phenobarbital, 3-MC and DDT singly and in combination. Creaven and Parke[37] and Davies and Creaven[38] have shown that three groups of compounds—carcinogenic polycyclic hydrocarbons, noncarcinogenic polycyclic hydrocarbons and a group including chlorinated hydrocarbon insecticides—can be differentiated on the basis of the induced metabolism of a series of alkoxybiphenyls.

It is evident from the foregoing review that systematic classification of exogenous chemicals of the agricultural environment, in terms of their metabolic position in hepatic processes, could be of great health-related importance but that a rational solution will have to await much further research.

It may also be noted that species-specific and strain-specific differences have yet to be accounted for; for example, phenobarbital will stimulate hepatic drug metabolism to a different extent in different strains of rabbits and within the same rabbit strain different enzymes can be stimulated to different degrees with some pathways unresponsive.[39] These workers also demonstrated[40] that DDT and γ-chlordane induced the metabolism of zoxazoalamine in two mouse strains, whereas the metabolism of hexobarbital was unaffected in either strain by DDT. In other studies by Lage et al., the metabolism of H^3-digoxin in rats[41] and squirrel monkey[42] was found to be unaffected by γ-chlordane or DDT pretreatment. Uehleke and Greim[43] found that phenobarbital stimulated the activity of various metabolic processes in both liver and kidney of rabbits, but in the case of rats cytochrome or oxidative drug reactions of kidney microsomes were not stimulated.

Environmental Temperature

An effect of cold on metabolism of certain drugs in normal rats had been demonstrated by Inscoe and Axelrod.[44] The mode of action of cold in microsomal drug metabolism has been further delineated by Furner and Stitzel[45,46] who have studied the effects of this stressor in normal and adrenalectomized animals. Cold and phenobarbital both increased the rate of p-hydroxylation of aniline and N-dealkylation of ethylmorphine. Adrenalectomy reduced the rate of metabolism of ethylmorphine aniline and hexobarbital. Cold—or the administration of hydrocortisone—increased the ethylmorphine and aniline metabolism of adrenalectomized rats but hexobarbital metabolism was further decreased. Phenobarbital pretreatment stimulated the metabolism of all three compounds in the adrenalectomized animals. Thus, both cold stress and phenobarbital affected drug metabolism independently of an intact adrenal, but not similarly. The incidental possibility is therefore raised that these two activators produced their changes in enzyme activity through different mechanisms. A role for extra-adrenal steroids, such as the sex hormones, in mediation of the enzyme activity induced by cold remains to be investigated.

An accelerating effect of cold environments on metabolism of pentobarbital in guinea pigs has lately been described by Sotaniemi,[47] who has also shown that this effect can be intensified by phenobarbital pretreatment and reduced by pretreatment with the mixed-function oxidizing enzyme inhibitor, SKF-525A. In hot environments, phenobarbital pretreatment failed to influence the rate of disappearance of pentobarbital from liver or brain, whereas SKF-525A retarded metabolism, but less effectively than at cold temperatures.

Nutrition

There have been a few recent contributions to the role of diet in hepatic enzyme induction, particularly with respect to sex differences. Kato[48] has found that sex differences in various drug-metabolizing processes were significant in rats fed on a high protein diet, but were minimal with a low protein diet. It was subsequently shown by Schenkman[49] et al. that sex differences in drug oxidase activity of liver microsomes from rats were related to difference in substrate affinity for the mixed function oxidase reaction. McLean et al.[50,51] have shown that a low protein diet reduced hepatic hydroxylating enzymes and P-450.

Wood-related Inducers

Vesell[52] has discovered, from bedding mice and rats on various types of wood shavings, that red cedar, white pine and ponderosa pine induced metabolizing enzymes for hexobarbital, ethylmorphine and aniline. These alterations were reversed by placing the animals on hardwood bedding such as beech, birch and maple.

Ethchlorvynol

A tentative report[25] that the sedative ethchlorvynol is a hepatic enzyme inducer in man was tested in terms of in vitro bishydroxycoumarin metabolism with livers from rats pretreated with ethchlorvynol and in vivo in dogs.[53] This drug was not found to be active as an inducer or inhibitor.

Caffeine, Coffee and Tea

Induction of microsomal mixed function oxidizing enzymes was affected by caffeine, coffee and tea as evidenced by metabolism of acetanilide, O-nitroanisole and aminopyrine by liver tissue from pretreated rats.[54]

Nicotine

Yamamoto et al.[55] have established in rats that pretreatment with 2-acetylaminofluorene (AAF) and 3,4-benzpyrene (BP), enhances the activity of nicotine-metabolizing enzymes and that nicotine pretreatment increases the activity of hepatic enzymes metabolizing AAF and BP.

3,4-benzpyrene Hydroxylase in Lung

Induction of increased 3,4-benzpyrene hydroxylase activity by phenothiazines, and other inducers of hepatic microsomal mixed function oxidase enzymes, has now been produced in a rat lung culture system.[56] The magnitude of the culture induction was substantially less than in the lung of the intact animal.

Late Work on Hepatic Enzyme Inhibitors

Last year Conney[9] reviewed hepatic drug-metabolizing enzyme inhibition by inhibitors of protein synthesis. In 1966 the author reviewed exogenous inhibitors of hepatic and serum enzyme activity.[2] Evidence of hepatic and serum enzyme inhibition from exogenous factors continues to accumulate, although the significance in biotransformations is not always clear. It must be reemphasized that some exogenous factors can affect both induction and inhibition[3] as demonstrated for aldrin in Fig. 1.

Aliphatic Chlorinated Hydrocarbons

Fig. 2, from studies on inhaled carbon tetrachloride in our laboratory,[5] demonstrates the reduction in serum B-aliesterase which occurs within a few hours of the introduction of carbon tetrachloride into the body. It shows also that as exposure continues, and particularly upon withdrawal of this chemical, either hepatic induction of the enzyme occurs, or the release of lysosomal esterases in necrotization; but whatever the source, the enhanced esterase activity did not reduce the toxicity of organic phosphates. Fig. 3 shows the reduction in B-aliesterase following a single dose of another aliphatic chlorinated hydrocarbon, ethylene dichloride, as well as by the chemically-unrelated compounds benzene and methanol. The significance of these findings in terms of damaged endothelial reticulum and later necrotization was discussed in an earlier publication[2] by the author.

Dingall and Heimberg[57] have just published results of studies on inhibition of hepatic drug metabolism following gastric intubation of carbon tetrachloride, chloroform and methylene chloride.

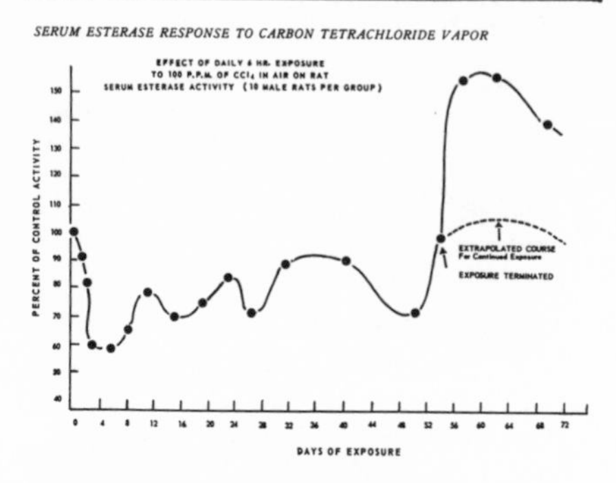

FIGURE 2. From Ref. 5.

It was demonstrated that CCl_4 impaired the metabolism of hexobarbital, evidenced by prolonged sleeping time. It also impaired the in vitro microsomal oxidation of this drug and aminopyrine, and reduced p-nitrobenzoic acid content. The observation of impaired aminopyrine metabolism confirmed work by Neubert and Maibauer[58] in 1959. Negative results were found by Dingell and Heimberg[57] with methylene chloride. Chloroform evoked only a moderate impairment of aminopyrine demethylation at the dosages employed. These authors support the view that the damage to the endoplasmic reticulum previously observed by electron microscopy, within a few hours of administration of CCl_4, is the explanation for the lack of microsomal drug-metabolizing enzyme activity.[59-61]

Another carbon tetrachloride study last year by Seawright and McLean[62] showed that metabolism of CCl_4 to CO_2 was reduced for 24 hours after intragastric administration to rats on a protein-free diet. This was also true when liver microsomal preparations were used in vitro. This group had earlier shown, by determining the LD_{50} of carbon tetrachloride, that the protein-free diets reduced the toxicity by more than half.[63] They subsequently proved that there was a loss of hydroxylating enzyme activity from the liver microsomes when diets were protein-free,[50] supporting the position that the toxic action of carbon tetrachloride is due to a metabolite formed in the microsomes.[50,63-65] In another test on starved sheep, toxicity of CCl_4 was greatly increased when phenobarbitone was preadministered, confirming the enhancing effect of induced hepatic microsomal enzymes on CCl_4 toxicity.[66] It now seems evident that CCl_4 was indeed a factor in the hypoprothrombinemia experienced by the patient on bishydroxycoumarin previously mentioned.[22] These findings could also bear on the ob-

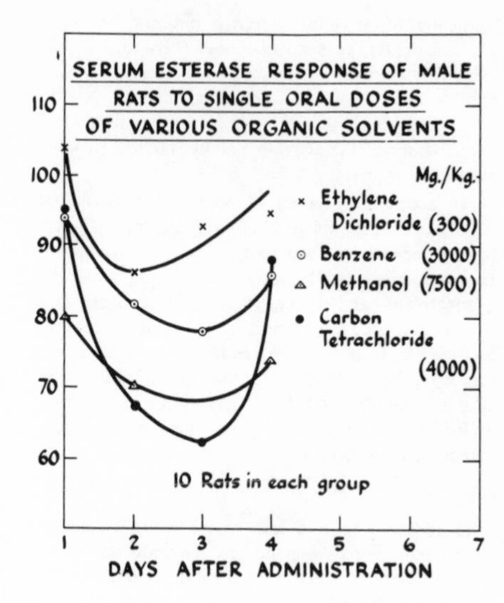

FIGURE 3. No other details necessary.

servations of Kondos and McClymont[67] who found that cold stress increased the toxicity of CCl_4 in newly-shorn sheep, if cold stress stimulated the induction of hepatic hydroxylating enzymes in the sheep as previously cited for some hydroxylating enzymes in rats.[45,46] On the other hand, Larson and Plaa,[68] and lately Gibb and Brody[69] have reported that liver damage after CCl_4 may be prevented, at least in part, by reducing body temperature. Because temperature stress is a feature of agricultural environments, it would appear a matter of some importance to have this aspect clarified by further research.

Heinrich and Klinger[70] have now shown a reduction in procaine-metabolising esterase in livers of rats pretreated with carbon tetrachloride.

Cytochrome P-450 and Carbon Monoxide

Studies on the mechanism of enzymic hydroxylations associated with TPNH oxidation established the role of Cytochrome P-450 as the oxygen-activating enzyme for many mixed function microsomal oxidations[71-73] which can be inhibited in vitro by carbon monoxide; for example, the O-demethylation of codeine,[72] the hydroxylation of aniline and 3,4 benzpyrene.[74] Inhibition of microsomal cytochrome P-450 by carbon monoxide in vivo appears to have been effected by Lewis[75] using the housefly Musca domestica (vicina strain) in carbon monoxide atmospheres where the independence of insects from haemoglobin for oxygen transport made survival possible. The epoxidation of aldrin to dieldrin, a hepatic microsomal reaction intensifying toxicity, was substantially reduced. A second experiment was made with carbaryl (1-naphthyl N-methylcarbamate) and a cyclodiene epoxide (1, 2, 3, 4, 9, 9-hexachloro-1, 4, 4a, 5, 6, 7, 8, 8a-octahydro-5, 6-epoxy-1, 4-methanonaphthalene). These compounds are detoxified in the liver and it was possible to demonstrate a striking increase in LD_{50} for both compounds when flies are maintained in a carbon monoxide atmosphere.

It has subsequently been established[76] that there exist two forms of the microsomal mixed function oxidase, called P-450 and P-446, which appear to be synthesized preferentially upon treatment of rabbits with the inducing agents phenobarbital or 3-MC. This finding may open the way to an understanding of the variability of enzymatic activities observed with different substrates of the mixed function oxidase reaction using microsomes from animals treated with various inducing agents.

Ionizing Radiation

The inhibitory effect of x-irradiation on the development of hepatic microsomal enzymes that metabolize organic phosphates was demonstrated by head-alone irradiation in 1964.[77] This proved to be largely an abscopal effect seeming to involve endocrine factors. Shielding liver or testes did not prevent the inhibitory effect on the development of phosphorothioate oxidase in the liver.[78] In utero exposure of rats to low doses of x-radiation (25 or 50R) proved inhibitory in the male offspring to the microsomal system that metabolizes hexobarbital.[79] Irradiation, total body or head alone, at 21 days and in adult rats, suppressed the hexobarbital-metabolizing system in comparison with normals but to a lesser extent than on the prenatal series.[80]

Studies on the effect of x-irradiation on resynthesis of the hexobarbital-oxidizing enzyme system after hepatectomy have shown that the radiation is more inhibitory if administered directly after the operation.[81]

Other investigators[82] exposed two-day-old male and female rats to 464R x-radiation and estimated acetyl and butyrylcholinesterase activity of whole blood and selected brain areas. Total esterase activity of whole blood was decreased at 10 days and declined as low as 60% below controls — without gross behavioral alterations but concurrent with disturbances in functional brain maturation as shown in earlier experiments.[83] Decreases in acetylcholinesterase and butyrylcholinesterase were observed in various parts of the brain.

It has been reported from the Karolinska Institute[84] that mice injected with 0.5 mg CCl_4 per gram of bodyweight, 72 hours before receiving 950 R from a cobalt source, had a substantially higher resistance to the radiation than controls. Carbon tetrachloride, as noted before, is primarily an inhibitor of mixed function oxidase reactions but as may be seen in Fig. 2, a rise in serum esterase activity to levels far above normal occurred around 72 hours after exposure of rats to CCl_4 vapor was terminated. Ionizing radiation, at the present stage of knowledge, is an inhibitor of esterases. One possible explanation for the antagonism between CCl_4 and ionizing radiation may therefore lie in the time relationship of the application of the other stresses; CCl_4 having perhaps provided excess esterase and other mixed-function oxidase throughout the body sufficient to replace that subsequently inactivated by the radiation from the cobalt source, so that detoxification and other metabolic processes related to the mixed function oxidases were able to continue.

Other New Inhibitions

The pentobarbital sleeping time in male and female Fischer strain rats was lengthened by pretreatment with the carcinogen N-hydroxy-N-2-fluorenyl-lacetamide.[85]

Netter et al.[86] have demonstrated that metyrapone inhibits a number of hepatic drug hydroxylations including O-demethylation of p-nitroanisole, N-demethylation of amidopyrine and ring hydroxylation of acetanilide. In vivo, metyrapone reduced urinary excretion of 4-amino-antipyrine in mice and rats dosed with amidopyrine.

Robertson[87] has estimated pseudocholinesterase changes in 33 women—before, and one month and three months after—taking anovulators, Ovulen (ethyndiol 1 mg., mestranol 0.1 mg.), Ortho-Novin (norethisterone 2 mg., mestranol 0.1 mg.), Norlestrin (norethisterone, 2.5 mg., ethinylestradiol 0.05 mg.), Gynovlar 21 (norethisterone 3 mg, ethinylestradiol 0.05 mg). There was a highly significant fall in serum pseudocholinesterase (20%) and serum albumin (16%). This investigator concluded that there is a steroid-induced depression of the liver's ability to synthesize protein or to secrete it into the circulation, which may represent a stage before hepatocellular damage or frank cholestasis.

Hickie and Kalant[88] have found that subcutaneously implanted Morris Hepatoma 5123tc will not metabolize hexobarbital in vitro; slices of non-tumorous liver from host rats have a reduced metabolism rate compared with liver slices from normal animals and a similar in vivo difference as measured by prolonged hexobarbital sleeping time when the hepatoma showed necrosis. Surgical removal of the tumor restored sleeping time to normal from which it was concluded that a diffusible product of tumor necrosis was active. Rogers et al.[89] also found that the presence of non-necrotic tumor tissue in rats did not affect drug metabolism in the liver.

Recently-Reported Interactions of Exogenous Substances

A number of new examples of toxicity modification, antagonism or potentiation, have lately been reported.

CCl_4 Organic Phosphates and Carbamates.

Massari and Querci[90] treated rats with carbon tetrachloride for three days prior to administration of parathion and showed that pseudocholinesterase levels in serum were reduced to half the level of controls treated only with parathion. This experiment constitutes the reverse of our experiment[4] in which aliesterase inducers in the form of chlorinated hydrocarbon insecticides maintained serum esterase levels of treated rats well above the parathion-treated controls.

Sakaguchi et al.[91] have delineated the relationship between chemical structure and protective effect of dithiocarbamate derivatives against hepatic injury induced by CCl_4 administration in rats. The mechanism of this protection was not explained in terms of hepatic metabolizing enzymes but it might be postulated that the esterase-inhibitory action of dithiocarbamates reduced the conversion of CCl_4 to its toxic metabolites.

CCl_4 and Phenothiazines

Phenothiazine derivatives have been shown to be in vivo inhibitors of a number of enzymes[92-97] as well as inducers of the mixed function oxidizing enzyme system. A Russian study[98] has claimed that promazine prevented the cellular escape of several enzymes in the liver maintaining normal limits of function when toxic doses of CCl_4 were administered to male rats. Guth et al.[99] showed that phenothiazine derivatives stabilized rat liver lysosomal membranes; thus it may be conjectured that the release of hydrolyzing enzymes which could convert CCl_4 to toxic metabolites was arrested. Experiments performed by Zimmerman et al.[100] and by Boyer et al.[101] confirmed these findings on the effect of phenothiazine derivatives as inhibitors of CCl_4 cytotoxicity. Slater[102] has just reported the results of investigations on the mechanism underlying the protective action of phenothiazines against CCl_4. This investigator has proposed that promethiazine inhibited the stimulation of lipid peroxidation produced in rat liver microsomes by low concentrations of CCl_4, through an inhibitory effect on the $NADPH_2$-ADP-Fe^{+2} system, although this system was admitted to share some components with the microsomal detoxication system.[103, 104]

CCl_4, SKF525-A and Antioxidants.

A recent report[105] has claimed that antioxidants d,1-tocopherol acetate and DPPD (N,N-diphenyl-p-phenylenediamine) do not prevent the inhibition, by necrotizing doses of CCl_4, of microsomal N-demethylation of ethylmorphine or P-450 activity. In contrast, SKF525-A completely prevented the loss of N-demethylating and P-450 microsomal enzyme activity in CCl_4-pretreated male rats, but within 24 hours its limited prevention of P-450 inhibition was dissipated. Histologically, both DPPD and SKF525-A mitigated changes caused by CCl_4, so it is concluded that these substances exert their protective effects differently. Presumably, pretreatment with SKF525-A as a hepatic microsomal enzyme inhibitor prevents the metabolism of CCl_4 to toxic products at least within 24 hours of administration, whereas DPPD acts through other metabolic pathways.

CCl_4 and Aliphatic Alcohols

Intubated methanol, ethanol and a number of other aliphatic alcohols administered 16 hours prior to exposure of rats to 1000 ppm CCl_4 for two hours, yielded elevated serum glutamic oxaloacetic transaminase (SGOT) levels assumed in this study to be evidence of hepatic damage. From this the experimenters concluded that potentiation of the effect of CCl_4 had occurred.[106] If the stimulation of ethanol metabolism by pretreatment with ethanol as reported by Ryan and Cornish[107] could be accepted as induction of hepatic mixed-function oxidases (possibly occurring along with alcohol dehydrogenase induction), then potentiation of CCl_4 might follow through its enhanced conversion to toxic metabolites. Recent experiments by Lind and Parker[108] demonstrated that chlorpromazine, pentobarbitone and some other drugs shortened ethanol sleeping time in mice whereas SKF525-A prolonged this phenomenon, which suggests that enhanced ethanol metabolism can be effected by the microsomol mixed-function oxidizing enzymes as well as by alcohol dehydrogenase. However Meksongee et al.[109] observed that ethanol pretreatment of mice increased the toxicity of carbaryl, and as shown in Fig. 3, it was found in our laboratory that methanol was an inhibitor of serum mixed-function esterase, at least following large single doses. It seems likely, therefore, that ethanol will be found to be primarily an inhibitor of microsomal mixed-function oxidases although a stimulator of alcohol dehydrogenase, in which case other metabolic pathways must exist to account for intensification of carbon tetrachloride toxicity if such, indeed, was the phenomenon behind the rise in SGOT observed.[106]

Further Observations on Ethanol and Drugs

In possible disagreement with the findings of Lind and Parker[108] who estimated metabolism of ethanol in terms of sleeping time, concurrent experiments by Vincenzi et al.[110] have revealed that liver triglyceride levels in ethanol-poisoned rats were lower than controls when pretreated with phenobarbital or SKF525-A. These results were interpreted as partial inhibition of the pro-oxidative effect of ethanol on the hepatic microsomes by phenobarbital and SKF525-A, both of which are known to bind to liver microsomes.[111, 112]

SKF525-A and Carbaryl

Pretreatment with SKF525-A caused a significant increase in the toxicity of carbaryl for mice[109] indicating that this chemical is metabolized by the microsomal mixed-function oxidases.

Drugs and Organic Phosphates.

Vardanis[113] has demonstrated that the microsomal drug-metabolizing enzyme stimulator phenobarbital increases the activation of OMPA (octamethylpyrophosphoramide) to its toxic oxidation product as previously shown by Kato[114] and McPhillips.[115] Vardanis also claims that the degradation of the toxic metabolite is stimulated.

Phenobarbital has recently been shown to enhance the protection afforded by atropine and 2-PAM (pyridine-2-aldoxime methiodide) against parathion poisoning in rats.[116]

Tri-O-cresyl phosphate, Phenobarbital and EPN

Tri-O-cresyl phosphate, an aliesterase inhibitor, abolished the protective effect of phenobarbital against malathion in rats, but did not affect the protection against EPN, from which the investigators conclude that aliesterase is not involved in EPN antagonism.[117]

Coumarin and 4-methylcoumarin

Liver enzyme studies with rats, by Feuer and Shilling,[118] indicated that 4-methylcoumarin induces hydroxylases which increase the rate of coumarin metabolism.

Aldrin and Anticholinesterases

Triolo and Coon[119] found that pretreatment with aldrin failed to protect against neostigmine and octamethylpyrophosphoramide (OMPA), poor brain penetrators, [120, 121], as opposed to other anti- cholinesterase agents which are effective inhibitors of brain cholinesterase. It is proposed that the protective mechanism of aldrin in pretreatment is to prevent excessive lowering of brain cholinesterase to which neostigmine and OMPA presumably failed to penetrate significantly.

Fat Storage and Chlorinated Hydrocarbon Insecticide Interactions

Street et al.[122] found that heptabarbital, aminopyrine, tolbutamide and phenylbutazone depressed dieldrin storage in rat adipose tissue. Subsequently,[123] DDT was found to exert a similar effect from which it was concluded that the phenomenon was due to induction of microsomal enzymes. Koransky[124] had earlier found phenobarbital an accelerator of α or γ BHC elimination, and lately others[125] found that it depressed dieldrin storage, as might have been anticipated.

Pulmonary Surfactant and Metabolism of Alveolar Cells

The significance of exogenous factors in agricultural environments on the integrity of the pulmonary surfaces is now in the initial stages of investigation and already there is evidence that this will be a fruitful area of research for establishing the etiology of some types of pulmonary disease occurring among agricultural workers.

The weed killer paraquat (1, 1'-dimethyl-4, 4'-dipyridilium dichloride) has been found to damage the lung both in experimental animals and man.[126] There is close resemblance between the lung pathology in paraquat poisoning and the respiratory distress syndrome (RDS) in infants. It had been shown by Avery and Mead[127] that pulmonary surfactant lipoprotein was deficient in cases of RDS. Manktelow[128] has examined pulmonary surfactant in relation to paraquat poisoning using specific pathogen-free mice from a Caesarian-derived colony in which naturally occurring pneumonia had never been detected. All mice that died had congested, firm, unexpanded lungs, which on histological examination showed intense congestion, collapse of alveoli with dilatation of alveolar ducts, and terminal bronchioles. Three out of 12 mice fatally poisoned showed the formation of hyaline membranes identical to those sometimes seen in infant respiratory distress syndrome. Pulmonary surfactant in all toxified mice was severely reduced or absent and appeared to be cleared from alveoli toward the major airways. No inflammatory lesions or fibrosis were found. However, Clark et al.[126] did note such changes. Delayed ocular burns from paraquat and diquat have also been reported.[129]

The effect of paraquat on the pulmonary surfactant and the integrity of the lung draw attention to recent studies on the metabolism of the alveolar cells. For instance, Tyler and Pearse[130] found evidence of a variety of oxidative enzymes in alveolar cells of rat lung. This was confirmed by Said et al.[131] in alveolar cells of dog lung. They found that when there was a deficiency of surfactant produced by pulmonary artery ligation, the enzyme-rich alveolar cells were absent. It was concluded that these cells are the site of elaboration of alveolar surfactant such as fatty acids and phospholipids long known to be important surface-active lining constituents. Goldfischer et al.[132] have found in Type II alveolar epithelial cells of rabbits, high levels of acid phosphatase, aryl sulfatase B, and B-glucuronidase-typical lysosomal enzymes capable of cytological digestion when released. Type II cells differ from alveolar macrophages in their levels of hydrolase activity and the fine structure of their active sites.

It is evident that there may be a heretofore unrealized role for exogenous factors possessing the power to alter lysosomal membrane stability and to inhibit oxidative enzymes of the alveolar cells such as NADP diaphorases, G6PD, cytochrome oxidase and the inducible mixed-function metabolizing enzymes such as benzpyrene hydroxylase.[13-15] As a subject for speculation in the pulmonary context, attention might be directed to a report just published[133] in which it has been found that after intraperitoneal injection of malathion and two pesticidal carbamates, rat liver lysosomes released arylsulfatase at a significantly higher rate than controls. The investigators propose that anticholinesterases are able to interact with non-specific esterases or pseudocholinesterase in structural components of the lysosomal membrane to alter permeability. Another report,[134] concerning aspergillosis, claims that cortisone, by stabilizing pulmonary lysosomes, prevented normal release of enzymatic and non-enzymatic protein in mice which resulted in a lesser tendency for insufflated aspergillosis spores to be killed outside and inside the pulmonary alveolar macrophages. It remains to be demonstrated whether further examples of such membrane interactions might apply to already documented pulmonary effects of pesticides and other exogenous factors in agricultural environments.

Immunological Reactions to Exogenous Factors in the Agricultural Environments

Investigation of immunological reactions of pesticides and other agricultural stressors is now developing into a major area of investigation, but it is already evident that this approach will be at least as productive as comparable investigations of drug idiosyncracies which were the subject of an informative review in 1966.[135] Immunopathology[136] has proved particularly useful in the agricultural health field as will be seen from the review to follow.

Reagin-dependent reactions constitute one phase of the subject, generally producing immediate effects such as oedema and irritation of the mucous membranes, urticaria, and bronchial constriction in asthma. There are also late allergic reactions of this type.

Then there are cytotoxic effects, generally delayed and involving reactions between antibodies and an antigenic component of membranes or with an antigen or hapten associated with the membranes. These are exemplified in allergic thrombocytopenic purpura and hemolytic anemias induced by certain drugs.

Another delayed category is the Arthus type reaction in which soluble antigen-antibody complexes precipitate in tissue spaces at site of formation or circulate and deposit elsewhere causing tissue damage, as in the agricultural pneumopathies and certain forms of glomerulonephritis.

The basis of immunological studies in the agricultural field is to identify the low molecular weight hapten that combines irreversibly with the tissue protein to form the antibody-inducing antigen. This is the key to immunization or other measures for protecting the affected individual. The haptens may be the exogenous chemical or a metabolic product of low enough molecular weight and bonding power to form antigens. Of course, haptens may also be endogenous as now implied in the concept of autoimmunity. The globulin antibodies in man are currently identifiable in four major classes and further systemization may be expected as classes and subgroups become linked with mediation of various allergic reactions. The human complement system at present is classified into nine plasma protein components, C'1 to C'9, and at least two inhibitors, those of C'1 esterase and C'3. In relation to exogenous factors such as the anticholinesterase insecticides which inhibit broad-band esterases, it is noteworthy that C'lp is a proesterase converted to an activated esterase, C'la, by the antigen-antibody complex.[137] Thus exogenous agricultural factors potentially capable of interfering with the complement reaction exist.

Among effective endogenous agents, the genetic have been prominent determinants of individual differences in immune response as well as in metabolism among drug idiosyncrasy cases, and it is already clear that, likewise, pesticides and other exogenous agricultural stressors will in the main produce significant immune response in only a limited proportion of exposed persons, those idiosyncratic to the extent that genetic factors dictate — unless esterase inhibitors are influencing complement fixation to an extent not fully comprehended at the current stage of knowledge.

Delayed Hypersensitivity Pneumopathies in the Agricultural Environments

A brief review of occupational pneumopathies related to agricultural environments was included in a publication by the author[1] in 1965. Recently there have been a number of significant developments in this field which, taken together, suggest that extrinsic allergic alveolitis or the delayed hypersensitivity pneumopathies are undoubtedly more incident in the agricultural environments than heretofore supposed. The findings related to the biochemistry of the lung lining already described indicate that certain agricultural chemicals challenge the integrity of the active surface of the lung. Furthermore, chemical pneumonitis has long been known to be a frequent accompaniment of overexposure to chlorinated hydrocarbon and organic phosphate insecticides to which agricultural workers in general are becoming increasingly exposed. Taking shape is a structure of exogenous lung irritants in the agricultural environments, arousing interest in the possibilities of uncovering heretofore obscured interactions—inductions, inhibitions, antagonisms and potentiations within the respiratory tract.

There have been many recent reports on hypersensitivity pneumopathies characterized by acute granulomatous interstitial pneumonitis. These cases have been occurring in the agricultural environments or among persons exposed to agricultural materials. The clinical similarities to farmer's lung are striking and exemplify antigen-antibody reactions to form precipitated complexes which cause tissue damage. Immunological mechanisms appear different from the atopic asthmatic where the hypersensitivity reaction, for instance, to vegetable dusts is immediate, takes place in the bronchi, and is mediated by non-precipitating reaginic antibody.[138] It may be noted that a brief review of occupational asthma was published in France in 1967.[139]

A clinical summary on farmer's lung with typical roentgenograms has recently been published by Katz.[140] This condition is still identified with sensitization to inhaled thermophilic actinomycetes, Thermopolyspora and Micromonospora vulgaris. IgE (IgND) specific antibody to castor-bean allergen has lately been detected by the red-cell-linked antigens-antiglobulin reaction.[141]

Bagassosis

Bagasse disease of the lungs with its clinical similarities to farmer's lung has now been established as a delayed hypersensitivity pneumopathy by Salvaggio et al. who have found a precipitating antibody against bagasse extracts in the sera of their patients with bagassosis.[142] The subject has recently been reviewed by Sodeman.[143] It appears that bagassosis is more common from bagasse stored outside for long periods with exposure to rain and hot weather. The presence of sensitizing fungi is suspected but not yet demonstrated.

Sequoiosis

Last year Cohen et al[144] described a case of granulomatous pneumonitis associated with redwood sawdust inhalation in a man who had worked 17 years in a redwood sawmill. The fungus provoking the antibody response with tissue damage was isolated.

Mushroom Workers

In 1959 Bringhurst[145] described a lung condition in mushroom workers strikingly similar to farmer's lung. Further cases with positive serum precipitin tests have now been reported from England.[146] These cases were found related to thermophilic actinomyces in compost. It was pointed out in a later communication that in some farmer's lung cases, serum precipitin tests are negative.[147] The desirability of differentiating between sarcoidosis and mushroom worker's lung by histological examination of scalene glands when serum precipitin tests are negative has been stressed, for Adams[148,149] has pointed out that granulomatous lesions in farmer's lung have not been found to extend beyond the lung.

Other Plant-associated Lung Irritants

Granulomatous pneumonitis has been found in sensitive individuals employed in removing the bark from maple logs.[150, 151] New Guinea natives exposed to organic dust from thatched roofs have been similarly affected.[152] Serum precipitins against extracts of thatch were demonstrated.

Avian-associated lung disease

There are now a sizeable number of cases of hypersensitivity pneumonitis reported among persons in contact with pigeons, budgerigars, and hens[153-163] since the first case was described by Reed et al.[164] in 1965. As with the plant-associated antigens, the main features of the disease are chills, fever with malaise, cough, and dyspnea five to eight hours after contact with the birds. These subside within 24 hours. There are high titers of precipitating antibodies and an Arthus type of skin sensitivity. Interstitial pneumonitis is present with impairment of pulmonary diffusion capacity but no bronchoconstriction. The disease may be coexistent with bronchial asthma and rhinitis, also avian-associated, and in these circumstances bronchoconstriction is present. Time lapse between contact and disease may range over years. There has been a recent review of the roentgenographic pulmonary findings in pigeon breeder's disease.[165]

Lung Reactive Antibodies in Pneumoconiosis

Burrell et al.[166] demonstrated in 1966 that serum of persons with chronic respiratory disease contained antibodies that were reactive with normal human lung tissue components among which collagen was one antigen. Further characterization of these antibodies[167] has disclosed that in patients with silicosis complicated by progressive massive fibrosis sera contained elevated IgA concentrations. Anti-lung reactivity was associated primarily with the purified IgA fractions and to a lesser extent with IgG and IgM serum immunoglobulin concentrations.

Studies on Antigenic Properties of Pesticides

To plant and avian-associated haptens, or hapten precursors in the agricultural environments, have now been added metabolities of DDT and malathion.[168] Investigators at McGill University have coupled the anhydride form of DDA (di-(p-chlorophenyl) acetic acid) to bovine serum albumin. DDT cannot be coupled. The conjugate was injected intradermally into rabbits on four occasions after which the antisera were titrated by hemagglutination (BDB technique). Titers up to 32,000 were observed. Specificity was tested with the benzylamine salt of N-(DDA)-aminocaproic acid which acting as an inhibitor, reduced the titer to 16. The technique was employed to detect and estimate antibodies to DDT in 7 human cases which presented allergic-type reactions when in contact with DDT. The maximum titer was 128 versus 16 in supposedly normal controls.

In the case of malathion, the metabolite used was the 0,0-dimethyl S-(1,2-biscarboxy) ethyl phosphorodithoate. Rabbit antisera with titers up to 64,000 were produced but no human studies have yet been possible.

Speculations on the Etiology of Pulmonary-Renal Immune Disease

It is evident from this review on pulmonary surfactant and extrinsic allergic alveolitis that the range of exogenous materials challenging the integrity of the lung of the agricultural worker is growing, not only in relation to the capacity of clinicians to establish etiology, but also as the variety of agricultural chemicals and processes is extended. A case of "Goodpasture's syndrome" in a young male pesticide sprayer, which occurred one month after having sprayed a range of 17 pest control chemicals, was recently reported. This arouses speculation concerning the possibility that alveolar damage, by one or more of the chemicals in this list or interference with C'3 (B1C) by the action of one or more of three anticholinesterase agents on the C'1 esterase of the sprayers complement system, was the etiological factor in precipitating the train of immune events now generally accepted as underlying this clinical entity of unknown pathogenesis. This speculation is in some measure supported by experiments this year in which, for the first time, progressive glomerulonephritis with glomerular basement membrane deposition of autoantibodies and complement was induced in sheep by injection of human lung tissue.[169] This glomerulonephritis was, by clinical and immunopathologic criteria, identical to the nephritis induced in sheep by human glomerular basement membrane and closely resembled the nephritis of Goodpasture's syndrome. Furthermore, Markowitz et al.[170] have just disclosed that linear depositions of and B1C globulins have been found on both glomerular and pulmonary basement membranes of three unpublished cases of Goodpasture's syndrome investigated by Dixon and in a further case by the authors. Other points of interest lie in the pulmonary picture in this syndrome which shows variably, necrotizing alveolitis,[171] abnormalities in the membranes of the alveoli including hyaline membranes,[172] degenerative splitting, and partial dissolution of alveolar capillary basement membranes.[173]

The purpose of this review does not permit a recapitulation of the clinical picture in this example of pulmonary-renal immune disease. This subject was fully reviewed by Benoit et al.[172] in 1965 and again this year with particular respect to the immunology by Dixon.[174] The role of circulating nonglomerular antigen-antibody complexes seems secure. Only the initiating insults remain to be exposed.

In 1963 Schmidt et al.[175] suggested that endogenous protein may be altered by inhalation or ingestion of exogenous materials as well as by physical stresses of cold, heat and sunlight. In the report of a clinical conference on a case of Goodpasture's syndrome in a painting contractor this year, Tucker of the Mayo Clinic[171] has conjectured that inhalation of solvent or other chemicals sensitized large areas of the pulmonary surfaces. Furthermore, this clinician observed that the pulmonary-renal syndrome is becoming more common perhaps in view of the number of chemicals to which the individual is now exposed. There is a rationale for possible interference with complement by exogenous anti-esterases. It may also be recalled that Halpern et al.[176] succeeded a few years ago in producing in monkeys and rabbits collagen-like disease by repeated injection of papain which had earlier[177] been shown to deplete cartilage constituents, particularly acid mucopolysaccharides. Disseminated pathological changes were observed in Halpern's animals and included glomerulonephritis, presence of complement binding and precipitating antibodies against autologous and isologous tissue extracts containing glycoproteins.

Glomerulonephritis caused by anti-glomerular basement membrane antibodies has also been reviewed by Dixon.[174] The possibility that exogenous materials may play a role by initiating tissue damage from which autoimmunizing processes may commence in sensitive persons has also been widely considered for this disease. A 1965 review on toxic nephropathy lists a number of nephrotoxic substances occurring in the agricultural milieu.[178] To this may now be added further items. Recent clinical observations by Davies et al.[179] in organophosphate poisoning cases showed disturbances of renal tubular function and amino acid metabolism. Fucik[180] has also reported kidney damage by organic phosphates. Methoxychlor been found to produce chronic nephritis in swine.[181] Pulmonary and renal injury was observed in two human cases of paraquat poisoning in New Zealand in 1966.[182] Clearly, the potentialtities for renal tissue damage by exogenous materials exist. It only remains to be demonstrated whether immune reactions accompany the tissue insult in particular cases, which will depend largely upon the possibilities of producing test antigens as accomplished by Centeno et al.[168] for DDT and malathion — or upon definitive studies concerning the effect on antigen-antibody - complement interactions.

References

1 Kay, K.: Recent Advances in Research on Environmental Toxicology of the Agricultural Occupations. *Amer. J. Public Health* **55**: (No. 7, Pt.2), 1-9, 1965.

2 Kay, K.: Recent Research on Esterase Changes Induced in Mammals by Organic Phosphates, Carbamates and Chlorinated Hydrocarbon Pesticides. *Ind. Med. Surg.* **35**: 1068-1074, 1966.

3 Crevier, M.; et al: Observations on Toxicity of Aldrin. II. Serum Esterase Changes in Rats Following Administration of Aldrin and Other Chlorinated Hydrocarbon Insecticides. *A.M.A. Arch. Industr. Hyg. Occup. Med.* **9**: 306-314, 1954.

4 Ball, W. L.; et al: Modification of Parathion's Toxicity for Rats by Pretreatment with Chlorinated Hydrocarbon Insecticides. *Can. J. Biochem. Physiol.* **32**: 440-445, 1954.

5 Ball, W. L. and Kay K.: Serum Esterase Response in Rats Exposed to Carbon Tetrachloride Vapor. *A.M.A. Arch. Industr. Health* **14**: 450-457, 1956.

6 Frawley, J. P., et al: Marked Potentiation in Mannalian Toxicity from Simultaneous Administration of Two Anticholinesterase Compounds. *J. Pharmacol. Exp. Ther.* **119**: p. 147, 1957.

7 Conney, A. H., et al: Adaptive Increases in Drug Metabolizing Enzymes Induced by Phenobarbital and Other Drugs. *J. Pharmacol. Exp. Ther.* **130**: 1-8, 1960.

8 Kalow, W., and Staron, N.: On Distribution and Inheritance of Atypical Forms of Human Serum Cholinesterase, as Indicated by Dibucaine Numbers. *Can. J. Biochem. Physiol.* **35**: 1305-1317, 1957.

9 Conney, A. H.: Pharmacological Implications of Microsomal Enzyme Induction. *Pharmacol. Rev.* **19**: 317-366, 1967.

10 Cram, R. L., et al: Stimulation by Chlordane of Hepatic Drug Metabolism in the Squirrel Monkey. *J. Lab. Clin. Med.* **66**: 906-911, 1965.

11 Kato, R., et al: Stimulating Effect of Some Inhibitors of the Drug Metabolisms (SKF 525A, Lilly 18947, Lilly 32391 and MG 3062) on Excretion of Ascorbic Acid and Drug Metabolisms. *Med. Exp.* **6**: 254-260, 1962.

Men With Intensive Occupational Exposure to DDT

A Clinical and Chemical Study

Edward R. Laws, Jr., MD; August Curley, MS; and Frank J. Biros, PhD

THE study of workers involved in the manufacture and formulation of any toxic or potentially toxic substance can produce extensive information pertinent to the evaluation of its potential hazard to people exposed to a lesser degree. Such studies are particularly useful in the case of DDT (1,1, 1-trichloro-2,2-*bis* [*p*-chlorophenyl] ethane and other isomers) because long-term, low-level exposure of the general population has existed for some years and will probably continue. The onset of the pharmacological and other effects of chemicals is usually hastened by increased dosage. The present study describes the clinical and analytical findings for men with prolonged, intensive, occupational exposure to DDT.

Since the Montrose Chemical Corporation began its operation in 1947, it has produced DDT continuously and exclusively. Current products are technical DDT chips, technical DDT flakes, 100% technical DDT dust, and 75% technical DDT dust formulated with precipated silica (Hi-Sil). The capacity of the plant is 7.5 million pounds a month; current production averages 6 million pounds a month. The plant employs from 125 to 165 workers divided into three eight-hour shifts. Excluding the 165 persons employed at the time of this study, there have been, during the 19 years of operation, 292 people who worked 6 months or more and 806 who worked for fewer than six months—a total of 1,098 former employees. This annual turnover represents an average of about 58 positions or 35%. According to the California State Employment Office, the turnover rate for the Chemical and Allied Industry for the same metropolitan area in 1965 was 41.8%.

The major criterion in the selection of volunteers was a work history of more than five years of relatively heavy occupational exposure to DDT. The 63 men in this category at the plant were graded subjectively by management, foremen, and themselves according to the average intensity of their contact with DDT. All the workers with high or medium exposure were asked to participate in the study, and all of them did except four who refused for personal reasons (eg, a death in the family) unrelated to their employment. Since it was considered feasible to study only a limited number of sets of samples, only three men with relatively low exposure were chosen for study. It was realized at the time that even these three men with the least opportunity for contact had far greater contact with DDT than people in the general population. After the samples were taken, two sets proved unsatisfactory. Thus, the study was made on a total of 35 men (20, 12, and 3 in the high, medium, and low exposure groups, respectively).

The volunteers were asked to undergo their usual routine annual physical examination, which includes medical history and physical examination, laboratory tests of blood and urine, and a chest x-ray. In addition, they agreed to donate samples of blood, urine, and fat for chemical analysis.

Methods

Age distribution and number of years on the job are presented in Table 1. Various job categories represented are given in Table 2. Of the

Table 1.—*Statistics on DDT Workers*

	No. of Men per Exposure Group	Age			Years of Exposure		
		Range	Mean	Median	Range	Mean	Median
Total	35	30-63	43	43	11-19	15	15
High	20	30-60	42	43	11-19	14	15
Medium	12	33-52	41	49	13-19	16	15
Low	3	48-63	54	51	17-19	18	18

35 men eighteen were of Mexican origin; none were Negro.

Medical histories were compiled, and physical examination and chest x-ray examinations were performed in the clinic of the industrial physician for the plant. The industrial physician and the senior author both examined each volunteer. All of the chest x-ray films were sent to a qualified radiologist for final interpretation. Samples of blood and urine from each man were sent to a reliable commercial laboratory. Blood tests were performed for hemoglobin, hematocrit, white blood cell (WBC) count and differential, and venereal disease research laboratory test for syphilis (VDRL). Routine characterization, pH, specific gravity, determination of glucose, protein and acetone, and microscopic examination were done on urine. Previous records of routine annual physical examinations given the men were available, and these were reviewed carefully.

Table 2.—*Job Categories of 35 DDT Plant Workers Represented With Exposure Rating*

Job Title	No. of Men
Exposure Rating 1 (Most Heavily Exposed)	
Production supervisor	2
Dust plant foreman	2
Grinder operator leadman	1
Grinder operator	1
Flaker operator	3
Dust plant operator	2
Chemical operator	3
Chemical operator helper	3
Laborer	3
Total, exposure rating 1	20
Exposure Rating 2 (Less Heavily Exposed)	
Chemical laboratory supervisor	1
Workshop foreman	1
Laboratory research technician	1
Chemical operator (precursors)	1
Oiler	1
Welder	1
Mechanic's helper	1
Lift-truck operator	2
Laborer	3
Total, exposure rating 2	12
Exposure Rating 3 (Least Exposed)	
Chemical operator (acid recovery)	1
Chief pipefitter	1
Machinist	1
Total, exposure rating 3	3

Additional blood and urine samples were obtained for analysis in our laboratory. The blood samples (8 to 10 ml) were taken by venipuncture after an eight-hour fast. No anticoagulant was added. After clotting, the samples were centrifuged; the serum was removed and refrigerated until analyzed.

Voided urine samples totaling 250 ml for each man were collected in glass bottles with 1 ml of 4% formaldehyde added to each bottle as a preservative. Most of these samples were taken after an eight-hour fast. The urine was not refrigerated.

Fat samples were taken by needle biopsy from the subcutaneous fat of the buttock. Disposable needles (18-gauge) and disposable syringes were used. The small amounts of aspirated fat were placed in tared, glass-stoppered centrifuge tubes and refrigerated until weighed. Two of the original 37 fat samples were unsuitable for analysis, one because it was too small, the other because of gross contamination with blood. The 35 satisfactory samples weighed 5.4 to 95.8 mg and averaged 38.3 mg.

The serum samples were analyzed for chlorinated insecticides and related materials by electron-capture gas-liquid chromatography according to the method of Dale et al.[1] Fat samples were weighed and extracted with hexane. An appropriate aliquot was subjected to gas chromatographic analysis for the determination of chlorinated insecticides and related materials. The details of the method are given by Hayes et al.[2]

The urine samples were analyzed for the complete range of chlorinated insecticides and related neutral materials by gas liquid chromatography.[3]

The metabolic product DDA (2,2-*bis* [*p*-chlorophenyl] acetic acid and other isomers) was determined by conversion of the acid to a methyl ester and subsequent analysis by electron-capture gas-liquid chromatography, as follows: 20 ml of urine were pipetted into a 250-ml capacity Erlenmeyer flask equipped with a ground glass joint. Four ml of 25% sulfuric acid were then added and the solution was maintained at the boiling point for one hour. The solution was cooled to room temperature, and a 4-ml aliquot was transferred to a centrifuge tube. The aliquot was then extracted with two 4-ml portions of chloroform. The chloroform extracts were combined and dried over 0.1 gm of sodium sulfate. A 5-ml aliquot of this extract was transferred to a clean centrifuge tube and concentrated to dryness in a water bath at 40 C with a gentle stream of clean, dry air. In order to methylate DDA in the residue, 4-ml of anhydrous methanol and 0.4 ml of concentrated sulfuric acid were added to it. The solution was shaken well and heated in a water bath maintained at 50 C for 10 minutes. We added 10 ml of water (previously extracted with hexane) were added to the methanol solution and the entire mixture then shaken with four 2-ml aliquots of n-hexane. The n-hexane extracts were combined, dried over 0.2 gm of anhydrous sodium sulfate, and a 6-ml aliquot was evaporated to dryness in a water bath at 40 C with a gentle stream of clean, dry air. The residue (representing 1.8 ml urine) was then taken up in 100μl of n-hexane, and a 10μl aliquot was analyzed.

For comparison, DDA was also analyzed by the colorimetric method of Cueto et al.[4]

All statistical analyses were made by methods described by Ezekiel.[5]

Results

Medical History and Physical Examination.—Tables 3 and 4 present the positive findings on history and physical examination. It should be emphasized that more than one finding was present in most of the men who had any findings at all.

Aside from respiratory tract infections, the most frequent finding was that of hypertension, with five men claiming a history of hypertension, and three men actually having it on physical examination. Two of the five men with a history of hypertension had the original diagnosis made prior to employment at the DDT plant. Three of the 35 men had a history of peptic ulcer.

Diabetes mellitus was found in three of the 35 men, and the diagnosis was made for the first time in one of them at this examination.

There were no cases of either cancer of blood dyscrasia among the 35 men, and neither of these illnesses was discovered in a review of the medical records of the 63 men at the plant who had more than five years of exposure to DDT. The plant management is confident that no employee in the 19 years of operation has ever developed either of these conditions while employed there. It was known, however, that two men had been hired in whom cancer had been diagnosed and treated prior to employment.

One individual case of some interest was that of a 39-year-old man who has been working at the plant since 1951. In 1954, his head was struck by a blunt object. The resulting fracture of the skull and bilateral subdural hematomas were treated by two craniotomies on successive days. Before returning to work, he developed epileptiform seizures occurring two to three times a week. He was placed on anti-convulsant medication, which he took erratically. He returned to work at a job with maximal exposure to DDT. Although his seizures persisted for another six months, they gradually became less frequent, and since one year after his injury he has been free of seizures in spite of total avoidance of medication, continued intake of alcohol, and continued exposure to DDT.

Clinical Laboratory Examinations and Chest X-ray.—The positive findings of the

Table 3.—*Findings From Medical History of 35 DDT Plant Workers (Many Individuals Had More Than One Finding)*

Condition	No. of Cases
Acute subdural hematoma followed by post-traumatic epilepsy	1
Allergic cutaneous sensitivity to monochlorbenzene	1
Acute labyrinthitis	1
Chronic otitis	1
Chronic sinusitis	2
Pterygia*	3
Thyroidectomy for toxic goiter*	1
Pneumonia	2
Pulmonary abscess secondary to pneumonia, treated by segmental resection of lung	1
Gunshot wound of chest requiring thoracotomy	1
Influenza	4
Acute bronchitis	3
Asthma with allergic bronchitis	1
Acute myocardial infarction	1
Hypertension†	5
Acute multiple embolic episode of undetermined etiology, accompanied by transient hematuria and transient cerebral ischemia in a man with hypertension antedating employment	1
Peptic ulcer	3
Acute urinary tract infection with hematuria	1
Non-diabetic glycosuria with proteinuria	1
Diabetes mellitus	3
Iron deficiency anemia secondary to bleeding peptic ulcer	1

*Condition existed prior to employment.
†Two cases of hypertension were manifest prior to employment.

Table 4.—*Positive Findings on Physical Examination of 35 DDT Plant Workers*

	No. of Cases
Obesity	3
Congenital deformity, hand and leg	1
Arcus senilis	2
Bilateral pterygia	2
Poor vision, corrected to 20/20 by glasses	6
Poor vision, refusal to wear glasses	1
Minimal hypertensive retinopathy	1
Grade II hypertensive retinopathy	1
Bilateral chronic otitis with bilateral hearing loss	1
Wheezes in lung fields	1
Irregularly irregular pulse (post-myocardial infarction)	1
Blood pressure greater than 160/95*	3
Soft systolic cardiac murmur	1
Bilateral varicose veins	1
Slack inguinal ring	1
Pilonidal sinus	1

*World Health Organization criterion for hypertension.[25]

clinical laboratory and chest x-ray examinations are presented in Table 5.

A lymphocyte to granulocyte ratio greater than 1.0 (expressed as a quotient) was found in five (14%) of the men, and they and 16 others in the group of 35 had shown such a ratio one or more times in earlier examinations. The lowest total WBC count among the five men showing a ratio greater than 1.0 in the current examination was 5,800; the lowest neutrophil count was 2,495; and the lowest lymphocyte count was 2,770.

One man had a positive VDRL serologic test for syphilis; appropriate treatment was started promptly. Glycosuria was found in all three diabetics. Another man had significant proteinuria, which had been noted at previous examinations, along with nondiabetic glycosuria which was not present this time.

Table 5.—*Positive Findings on Laboratory and X-Ray Examination of 35 DDT Plant Workers*

Laboratory finding	No. of Cases
Lymphocyte/granulocyte ratio >1	5
Positive STS	1
4+ glycosuria (one known diabetic, one probable new diabetic)	2
Trace glycosuria (known treated diabetic)	1
4+ proteinuria (history of glycosuria + proteinuria)	1
Trace protein (no known illness)	1
X-ray	
Old, healed tuberculosis	1
Arteriosclerosis of the aorta	1
Old fibrotic chest nodule	1
Bullous changes and pleural effusion secondary to old gunshot wound of the chest	1
Left ventricular hypertrophy	1
Cardiomegaly	1

The results of routine chest x-ray examination did not produce any unexpected findings.

Analysis of Blood, Fat, and Urine for Pesticides and Related Materials.—The range and means of analytical results for DDT, other chlorinated insecticides, and related materials in blood, fat, and urine are presented in Tables 6, 7, and 8. The mean values for *p,p'*-DDT in fat of men in the different exposure groups were from 39 to 128 times those found in the general population by Hayes et al[2] using the same analytical method, but the corresponding mean values for total DDT-related material were only 12 to 32 times greater than those for the general population. The mean values for total DDT-related materials individually in serum were three to ten times the mean for the general population as determined by Dale et al.[1] The mean values for total DDT-related materials in urine of the three groups were 28 to 78 times the corresponding mean for the general population.

DDE expressed as DDT constituted only 25% to 63% of the total DDT-related material found in the fat of the workers (Table 9); this range is in contrast to that of 72% to 92% found recently in the general population. The mean values were 41% and 81%, respectively, and the difference is significant ($P < 0.001$). DDE also constituted a smaller proportion of total DDT-related material in the urine of workers than in the urine of the general population (Table 9), but the difference for urine described in the next paragraph was more extreme.

There was a striking difference between the concentration of *p,p'*-DDA in the urine of the combined exposure groups as compared to the general population; in fact, no *p,p'*-DDA was found in the particular sampling of urine from the general population. The findings just recorded were reflected in the total excretion of DDT-related material expressed as *p,p'*-DDT; there was some difference in the concentrations found in the different exposure groups but a striking difference in the concentration found in the combined exposure groups as compared to that in the general population. In summary, DDA was the major excretory form of DDT in the occupationally exposed men but was less important in the general population (actually undetected in that set of samples) and *p,p'*-DDE, which is of major importance as an excretory product in the general population, was found at only slightly higher concentration in the urine of workers and showed no increase whatever corresponding to the increasing exposure of the different groups. Thus, there was a striking change in the relative importance of different metabolites of DDT in men with occupational expo-

Table 6.—*Concentration and Distribution of Chlorinated Insecticides and Related Materials in the Men of the General Population) as Determined by Electron-Capture Gas-Liquid*

No. of Men per Exposure Group		Statistical Value	p,p'—DDT	o,p'—DDT	p,p'—DDE	o,p'—DDE	p,p'—DDD	Total as p,p'—DDT
		Range	0.0490— 0.9955	0.0034— 0.1001	0.0443— 0.8546	<0.0027— 0.0053	0.0252— 0.2076	0.1387— 2.2017
High	20	Mean	0.2730	0.0290	0.277	0.0024	0.1128	0.7371
		SE*	±0.0570	±0.0056	±0.042	±0.0002	±0.013	±0.107
		High vs med	$P > 0.05$	$P > 0.40$	$P < 0.05$	$P > 0.50$	$P < 0.005$	$P < 0.025$
		Range	0.0271— 0.3029	0.0053— 0.0708	0.0315— 0.3539	0.0015— 0.003	0.0149— 0.1418	0.1089— 0.7985
Medium	12	Mean	0.1202	0.0218	0.1432	0.0022	0.0503	0.3584
		SE	±0.02	±0.0057	±0.0260	±0.0001	±0.0100	±0.0660
		Med vs low	$P > 0.10$	$P > 0.50$	$P > 0.40$	$P < 0.01$	$P > 0.20$	$P > 0.20$
		Range	0.0447— 0.2914	0.0037 0.0201	0.1387— 0.2232	0.0023— 0.0042	0.0748— 0.1472	0.2898— 0.7279
Low	3	Mean	0.1978	0.0135	0.1934	0.0034	0.0993	0.5412
		SE	±0.0770	±0.0050	±0.0270	±0.0006	±0.0240	±0.1300
All Exposure Groups	35	Range	0.0271— 0.9955	0.0034— 0.1001	0.0315— 0.8546	<0.0015— 0.0053	0.0149— 0.2076	0.1089— 2.2017
		Mean	0.2142	0.0253	0.2240	0.0024	0.0903	0.5905
		SE	±0.0359	±0.0038	±0.0274	±0.0002	±0.0095	±0.0719
Exposed vs General Population			$P < 0.010$	$P < 0.025$	$P < 0.001$	$P > 0.05$	$P < 0.001$	$P < 0.001$
		Range	0.0114— 0.0490	0.0014— 0.0129	0.0101— 0.0416	<0.0006— 0.0096	0.0029— 0.0199	0.0429— 0.1081
General Population§	10	Mean	0.0253	0.0060	0.0257	0.0036	0.0083	0.0732
		SE	±0.0038	±0.0015	±0.0040	±0.0011	±0.0018	±0.0013

*SE indicates standard error; Med, medium.
†Note that value is actually less than that for general population.
‡General population (no + values).
§From Dale et al.[1]

Table 7.—*Concentration and Distribution of Chlorinated Insecticides and Related Materials in the men of the General Population) as Determined by Electron-Capture Gas-Liquid*

No. of Men per Exposure Group		Statistical Value	p,p'—DDT	o,p'—DDT	p,p'—DDE	o,p'—DDE	p,p'—DDD
		Range	35.58— 283.83	3.79— 38.59	37.57— 319.32	<0.05— 1.99	<0.3
High	20	Mean	147.13	14.39	91.19	0.28	
		SE	±17.90	±2.24	±15.10	±0.14	
		High vs Med	$P < 0.05$	$P < 0.005$	$P > 0.05$	$P > 0.50$	
		Range	18.12— 114.15	1.96— 9.37	16.48— 96.89	<0.05— 1.13	<0.3
Medium	12	Mean	69.63	5.11	49.79	0.28	
		SE	±7.50	±0.65	±7.50	±0.13	
		Med vs Low	$P > 0.10$	$P > 0.50$	$P > 0.50$		
Low	3	Range	23.03— 62.51	2.59— 8.84	28.90— 53.20	<0.05— 0.68	<0.30
		Mean	45.20	5.96	44.24	0.22	
		SE	±11.9C	±1.82	±7.70	1 + value	
All Exposure Groups	35	Range	18.12— 283.83	1.96— 38.59	16.48— 319.32	<0.05— 1.99	<0.30
		Mean	111.85	10.49	73	0.28	
		SE	±12.61	±1.51	±9.59	±0.09	
			$P < 0.001$	$P < 0.001$	$P < 0.001$	$P > 0.5$	
		Range	0.15— 2.78	<0.02— 0.23	0.56— 9.91	<0.05— 0.53	<0.09— 0.83
General Population*	13	Mean	1.15	0.04	5.69	0.26	0.32
		SE	±0.22	±0.02	±0.82	±0.05	±0.07

*From Hayes et al.[2]

sure as compared to the general population (Tables 8, 9, 10).

In addition to the determination of *p,p'*-DDA as the methyl ester by electron-capture, gas-liquid chromatography (Table 8) without prior attempts at special cleanup, this acidic component in the urine was also

Serum of 35 Occupationally Exposed Men (and 10
Chromatography Expressed in ppm

γ—BHC	β—BHC	Heptachlor Epoxide	Dieldrin
<0.0013—	<0.0018—	<0.0014—	<0.0015—
0.0126	0.0071	0.0038	0.016
0.0045	0.0038	0.0012	0.007
±0.0012	±0.0003	±0.0003	±0.0009
$P > 0.50$	$P > 0.10$	$P > 0.20$	$P > 0.20$
<0.0015—	0.0020—	<0.0015—	<0.0015—
0.0170	0.0089	0.0034	0.0249
0.0055	0.0047	0.0017	0.0048
±0.0013	±0.0005	±0.0003	±0.0019
$P > 0.20$	$P > 0.50$	$P > 0.50$	$P > 0.50$
0.002—	0.0033—	<0.0015	<0.0015—
0.0039	0.0055	0.0016	0.0046
0.003	0.0046	0.0005	0.0025
±0.0006	±0.0007	...	±0.0013
<0.0013—	<0.0018—	<0.0014—	<0.0015
0.0170	0.0089	0.0038	0.0249
0.0047	0.0042†	0.0013	0.0059
±0.0008	±0.0003	±0.0002	±0.0009
$P > 0.50$	$P < 0.025$†	‡	$P > 0.05$
<0.0005—	0.0024—		<0.0012—
0.0132	0.0184		0.0063
0.0057	0.0065		0.0028
±0.0019	±0.0015		±0.0007

Fat of 35 Occupationally Exposed Men (and 10
Chromatography and Expressed in ppm

Total as p,p'—DDT	β—BHC	Heptachlor Epoxide	HEOD
89.97—	<0.5	<0.1	<0.1
646.63	...	...	...
263.48	...	...	...
±34.0	...	...	...
$P < 0.010$	...	...	...
38.45—	<0.5	<0.1	<0.1
215.02	...	...	...
130.50	...	...	...
±15.50	...	...	...
$P > 0.20$	...	...	...
51.13—	<0.50	<0.10	<0.10
126.11	...	...	...
98.47	...	...	...
±24.00	...	...	...
38.45—	<0.50	<0.10	<0.10
646.63	...	...	...
203.74	...	...	...
±23.27	...	...	...
$P < 0.001$	...	...	...
1.14—	0.03—	0.03—	0.03—
13.24	0.88	0.33	1.15
8.17	0.42	0.17	0.25
±1.14	±0.07	±0.03	±0.08

determined colorimetrically. The mean results and standard error by the colorimetric method expressed as parts per million were: 1.35 ± 0.19 for the high exposure group, 0.61 ± 0.07 for the medium exposure group, and 0.45 ± 0.16 for the low exposure group. While there was a statistically significant difference ($P < 0.01$) between the excretion of this compound by the high exposure group as compared to the medium exposure group, there was no statistical significance ($P > 0.2$) between the mean concentration of *p,p'*-DDA in the urine of the medium versus the low exposure groups. These results correspond very well with those obtained by the gas chromatographic method. In the interest of formal consistency, the results for DDA in Table 8 are those obtained by gas chromatography. However, in its present state of development, the gas chromatographic method for DDA offers no advantage over the colorimetric one. Because of technical factors, including a high baseline and some uncertain peaks, the gas chromatographic method used was just as unsatisfactory as the colorimetric one at low concentrations such as those found in the general population. The basic difficulty is the relative insensitivity of the electron-capture detector to the methyl ester of DDA. Thus, the major drawback of both the gas chromatographic and colorimetric methods for DDA is insensitivity. It will be recalled that Durham et al[6] reported that the colorimetric method failed to detect DDA in 70% of samples from the general population.

As shown on Tables 6, 7, and 8, the significance of the analytical values for each exposure group as compared with the next lower was tested. There were few instances of statistically significant difference, and it is probably best to view all the exposed individuals as constituting a single group with some internal variation but with consistently high exposure to DDT. When the analytical values for all the exposed men were grouped and compared with the values for the general population, the differences were statistically significant for *p,p'*-DDT, *o,p'*-DDT, and total DDT-derived material in serum, fat, and urine. There was also a statistically significant difference for *p,p'*-DDE in serum and fat and for *p,p'*-DDD in serum.

Statistical analysis of the concentrations

Table 8.—*Concentration and Distribution of Chlorinated Insecticides and Related Materials in the 10 men of the General Population) as Determined by Electron-Capture Gas-Liquid*

No. of Men per Exposure Group		Statistical Value	p,p'—DDT	o,p'—DDT	p,p'—DDE	o,p'—DDE	p,p'—DDD	p,p'—DDA as Methyl Ester
		Range	0.0050—	0.0012—	0.0140—	<0.0002—	<0.0017—	0.44—
			0.0238	0.0119	0.0371	0.0015	0.0094	2.67
High	20	Mean	0.0113	0.0032	0.0201	0.0002	0.0049	1.27
		SE	±0.0010	±0.0005	±0.0014	±0.0001	±0.0005	±0.43
		High vs Med	$P > 0.50$	$P > 0.50$	$P > 0.50$	$P > 0.05$	$P > 0.10$	$P < 0.005$
		Range	0.0023—	0.0018—	0.0133—	<0.0002—	0.0033—	<0.01—
			0.0175	0.0065	0.0302	0.0019	0.0132	1.45
Medium	12	Mean	0.0109	0.0036	0.0216	0.0006	0.0065	0.6
		SE	±0.0039	±0.0004	±0.0015	±0.0002	±0.001	±0.12
		Med vs Low	$P > 0.50$	$P > 0.50$	$P > 0.10$	1 + value	$P > 0.40$	$P > 0.40$
		Range	0.0082—	0.002—	0.0209—	<0.0002—	0.0037—	<0.01—
			0.0150	0.0082	0.0330	0.0013	0.0070	0.86
Low	3	Mean	0.0126	0.0039	0.0267	0.0004	0.0048	0.41
		SE	±0.0022	±0.0012	±0.0035	1 + value	±0.0011	±0.24
All Exposure		Range	0.0023—	0.0012—	0.0133—	<0.0002	0.0017—	<0.01—
Groups	35		0.0238	0.0119	0.0371	0.0019	0.0132	2.67
Exposed vs		Mean	0.0113	0.0034	0.0212	0.0004	0.0055	0.97
General		SE	±0.0007	±0.0003	±0.0010	±0.0001	±0.0005	±0.11
Population (GP)			$P < 0.05$	$P < 0.025$	$P > 0.05$	$P > 0.50$	$P > 0.50$	...
		Range	0.0018—	0.0007—	0.0114—	<0.0001—	0.0008—	
General			0.0138	0.0011	0.0209	0.0008	0.0067	
Population*	5	Mean	0.0070	0.0010	0.0156	0.0004	0.0031	
		SE	±0.0023	±0.0001	±0.0015	±0.0001	±0.0008	

*Cueto and Biros.[3]

Table 9.—*Comparison of DDE Storage and Excretion in 35 Workers and in the General Population*

Storage in fat	Exposed Workers	General Population
Range (ppm)	18.36—355.72	0.85—11.33
Mean ± SE	81.62±10.7	6.62±0.94
Range (%)*	25—63	72—92
Mean ± SE (%)*	41±1.6	81±1.9
Excretion in urine		
Range (ppm)	0.0131—0.0413	0.0127—0.0236
Mean ± SE	0.024 ±0.0012	0.0178±0.0018
Range (%)*	0.6—49.4	36—86
Mean ± SE (%)*	5.95±2.1	63±8.7

*Sum of p,p'-DDE plus o,p'-DDE (both expressed as DDT) as a percentage of total DDT-related material (expressed as DDT) in the same sample.

Table 10.—*Estimated Average Daily Intake of DDT by Workers With Different Degrees of Exposure in a DDT Plant*

No. of Men per Exposure Group		Intake of DDT (mg/man/day) Based on: DDT in Fat	DDA in Urine
High	20	18	17.5
Medium	12	6.2	8.4
Low	3	3.6	6.3

for total DDT-related materials in serum and in fat revealed a coefficient of correlation (r) of +0.64. This coefficient indicates a reasonably high positive relationship between the blood serum levels and fat storage levels of DDT, indicating that one value could be predicted from the other but with an average error of 23% (100% minus the coefficient of alienation (k), where $k = \sqrt{1-r^2}$) and, therefore, a larger error in many instances. The average concentration in fat was 344 times greater than the average concentration in serum.

The observed values were plotted (graph not shown) and a line of regression was calculated. This line proved to be statistically different from a line drawn through the origin and a point of which the coordinates were the mean values for fat and serum, respectively. However, no meaning can be attached to the difference because there is no logical reason to take either the values for fat or those for serum as a reference for the others.

For chlorinated pesticides unrelated to DDT, the values in blood and urine were generally the same order of magnitude as those found in the general population, indicating the heavy exposure to DDT did not significantly alter the storage of other chlorinated insecticides. Limitations of the method excluded the possibility of making such a comparison for fat. The small size of the fat samples combined with the limits of the detector and the large amounts of DDT-

Urine of 35 Occupationally Exposed Men (and Chromatography and Expressed in ppm

Total as p,p′—DDT	α—BHC	γ—BHC	β—BHC	α—BHC	Heptachlor Epoxide	Dieldrin
0.5942— 3.3727 1.6296 ±0.1697 $P < 0.005$	<0.0002— 0.0046 0.0016 ±0.0004 $P < 0.05$	<0.0002— 0.0015 0.0005 ±0.0001 $P > 0.50$	<0.0004— 0.0054 0.0009 ±0.0003 $P > 0.20$	<0.0001	<0.0002— 0.0015 0.0001	0.0008— 0.0043 0.0019 ±0.0002 $P > 0.05$
0.0505— 1.8438 0.7911 ±0.157 $P > 0.50$	<0.0002— 0.001 0.0003 ±0.0001 $P > 0.05$	<0.0002— 0.0012 0.0006 ±0.0001 $P < 0.025$	<0.0004— 0.0017 0.0003 ±0.0002	<0.0001	<0.0002— 0.0017 0.0004 ±0.0002	0.0011— 0.0083 0.0028 ±0.0005 $P > 0.5$
0.0521— 1.1292 0.5629 ±0.3120	<0.0002— 0.0046 0.0016 ±0.0015	0.0004— 0.0036 0.0015 ±0.0010	<0.0004	<0.0001	<0.0002— 0.0015 0.0005 1 + value	0.0009— 0.0041 0.0021 ±0.0009
0.0505— 3.3727 1.2507 ±0.1348 . . .	<0.0002— 0.0046 0.0012 ±0.0003 $P > 0.20$	<0.0002— 0.0036 0.0006 ±0.0001 GP only 2 + values	<0.0004— 0.0054 0.0006 ±0.0002 GP no + values	<0.0001	<0.0002— 0.0017 0.0002 ±0.0001 GP 1 + value	0.0008— 0.0083 0.0023 ±0.0002 $P < 0.05$
0.0254— 0.0067 0.0291 ±0.0018	<0.0001— 0.0005 0.0003 ±0.0001	<0.0001— 0.0002 2 + values . . .		<0.0001— 0.0009 1 + value . . .	<0.0001— 0.0005 1 + value . . .	0.0005— 0.0014 0.0008 ±0.0002

related compounds in the samples prevented the accurate determination in the fat of DDD and of all chlorinated pesticides not related to DDT. Since the values were below the limit of detection of the method both for the samples under study and for those from the general population, the concentrations for α- and δ-BHC (< 0.0015 ppm) in serum and for α-, γ- and δ-BHC (< 0.01 ppm) in fat were omitted from the tables.

Comment

Medical History, Physical Examination, and Clinical Laboratory Findings.—In screening examination of insurance company employees (men and women) over 45 years of age, McCombs and Finn[7] found 20% with hypertension. In a group of executives at a naval shipyard, Guild and Mansfield[8] reported five cases (13%) of hypertension in the 43 men they studied for "physical fitness." In 500 executives with a mean age of 48.6 years receiving a periodic health examination, Bolt et al[9] found 50 (10%) with hypertension. D'Alonzo and Rogers[10] in physical examinations on some 27,000 DuPont workers with an age range from 18 to 65 found hypertension in 6%. None of the studies cited is adequate as a standard for judging the results of this study, and it is practically impossible to find another study suitably matched for age, socioeconomic status, or psychosocial factors. However, in other studies the percentages of persons found to be hypertensive are similar to the value of 8.6% found in this study. This is of particular interest because the hiring practices at the plant under study definitely selected anxious, driving people who may, in general, be more prone to have essential hypertension or peptic ulcer.

Of the 35 men in the present study, 8.6% did have a history of peptic ulcer. McCombs and Finn[7] found peptic ulcer in 6% of their group, as did Bolt et al.[9] Guild and Mansfield[8] did not report peptic ulcer, while D'Alonzo recorded this diagnosis in 2% of his group. A Public Health Service publication[11] claims that "over 7% of the population now have or have had an ulcer of the stomach or duodenum."

The studies cited above revealed an incidence of 2% to 3% for diabetes mellitus as compared to 8.6% in the present study. The difference is probably explained by the small size of the present sample.

No cancer or blood dyscrasia was found in the 35 men studied. This is particularly interesting in view of Birk's[12] finding of 19 proven or probable malignancies in 400 industrial workers he studied. It should be emphasized that differences between the two investigations exist both in the nature of the groups studied and the methods of study.

The occurrence of a lymphocyte/granulocyte ratio greater than 1:0 has been observed before in DDT formulators.[13] The percentage of men showing this finding was 5% in Ortelee's study and 14% in the present one. However, there is no reason to suppose that DDT was the cause of the substantial incidence of high lymphocyte/granulocyte ratio. Apparently, the highest incidence ever reported was 22%, and it was found before the introduction of DDT in an exceptionally careful study of 269 normal adults.[14] It is possible that the relatively high incidence of a ratio greater than 1:0 is the only truly normal condition, for many common infections cause granulocytosis and a shift to the left and, hence, a ratio less than 1:0. Certainly, the incidence of a ratio greater than 1:0 is highly variable in healthy populations. Booth and Hancock[15] found this condition in 14% of the 104 men they studied, while D'Alonzo and Rogers[10] found it in only 2.5% of the men they studied in a survey of a large chemical industry.

The absolute number of WBCs, polymorphonuclear granulocytes, and lymphocytes in the five men in this study who had a lymphocyte/granulocyte ratio greater than 1:0 were within the recognized limits of normal[14,16,17] or even within the 95% confidence limits of normal.[18]

No cases of clinical poisoning attributable to occupational exposure to DDT have ever occurred at the plant under study. A few cases of systemic, respiratory, dermal, or other toxicity occurring in the state of California listed under the category "DDT, chlordane, lindane, and kelthane" have been tabulated.[19-21] The nature of the cases or whether any of them really involved DDT is obscure.

The overall clinical findings for this group of men with heavy exposure to DDT for 9 to 19 years do not differ significantly from those one might expect from a group of similar age and socioeconomic status with no occupational exposure to DDT.

Analysis of Blood, Fat, and Urine for Pesticides and Related Materials.—The fact that both the storage and the excretion of DDE were relatively small in workers as compared to the general population is paradoxical. Decreased excretion should lead to increased storage. The observed facts suggest that metabolism of DDT to DDE may be of relatively less importance in workers than in the general population. That the difference is related chiefly to intensity of exposure rather than duration of exposure is indicated by the following: (1) for the workers under study, no significant correlation ($r = +0.18$) was found between years of exposure and storage of DDE (expressed as a percentage of total DDT), (2) as shown in Table 9, the storage of DDE (expressed as a percentage of total DDT) in the workers exposed for 9 to 21 years to DDT was not only small (41%) in comparison to that found recently in the general population (81%) but was also small in comparison to the 62% found in the general population in 1954-1956[22] only ten years or less after the commercial introduction of DDT.

The concentration of DDA found in the urine of the workers was similar to that found by Ortelee[13] in an earlier study of men in DDT formulating plants.

A graph prepared by Durham et al[23] based on data from numerous studies permits estimation of daily intake of DDT in people from the concentration of DDT in their fat. A similar estimate is possible from the concentration of DDA in urine and based on data from one of the studies used in the graph just mentioned in which men were given known daily doses of DDT as an emulsion.[24] Table 10 shows such estimates of daily average intake of DDT for the three exposure groups under study. The correspondence of the estimates based on fat and those based on urine is excellent, especially for the groups with more than ten samples available for study. The estimates for the different groups ranging from 3.6 to 18.0 mg per man per day may be compared to a value of 0.04 mg per man per day for the general population.[23]

Summary

A study was made of 35 men with 11 to 19 years of exposure in a plant that has produced DDT continuously and exclusively since 1947 and now produces an average of 6 million pounds/month. Findings from medical history, physical examination, routine clinical laboratory tests, and chest x-ray film did not reveal any ill effects attributable to exposure to DDT. The overall range of storage of the sum of isomers and metabolites of DDT in the men's fat was 38 to 647 ppm as compared to an average of 8 ppm for the general population. Based on their storage of DDT in fat and excretion of DDA in urine it was estimated that the average daily intake of DDT by the 20 men with high occupational exposure was 17.5 to 18 mg per man per day as compared to an average of 0.04 mg per man per day for the general population. There was significant correlation ($r = +0.64$) between the concentration of total DDT-related material in the fat and serum of the workers. The concentration in fat averaged 338 times greater than that in serum. Workers store a smaller proportion of DDT-related material in the form of DDE, and the difference is related chiefly to intensity rather than duration of exposure. DDE is relatively much less important and DDA much more important as excretory products in occupationally exposed men than in men of the general population.

The employees and management of the Montrose Chemical Corporation, Torrance, Calif, cooperated in this study. Lyle A. Kaliser and Virlyn W. Burse assisted with gas chromatographic analysis. Mrs. Barbara Adam assisted with the colorimetric determinations. The statistical analysis was done by Mrs. Estelle Gray. Wayland J. Hayes, Jr., MD, C. Cueto, PhD, and W. E. Dale gave advice and guidance.

References

1. Dale, W.E.; Curley, A; and Hayes, W.J., Jr.: Determination of Chlorinated Insecticides in Human Blood, *Industr Med Surg,* **36**:275-280, 1967.

2. Hayes, W.J., Jr.; Dale, W.E.; and Burse, V.W.: Chlorinated Hydrocarbon Pesticides in the Fat of People in New Orleans, *Life Sci* **4**:1611-1615, 1965.

3. Cueto C., Jr., and Biros, F.J.: Chlorinated Insecticides and Related Materials in Human Urine, *Toxicol Appl Pharmacol,* **10**:261-269, 1967.

4. Cueto, C.; Barnes, A.G.; and Mattson, A.M.: Determination of DDA in Urine Using an Ion Exchange Resin, *J Agr Food Chem* **4**:943-945, 1956.

5. Ezekiel, M.: *Methods of Correlation Analysis,* ed 2, New York: John Wiley & Sons, Inc., 1947.

6. Durham, W.F.; Armstrong, J.F.; and Quinby, G.E.: DDA Excretion Levels: Studies in Persons With Different Degrees of Exposure to DDT, *Arch Environ Health* **11**:76-79, 1965.

7. McCombs, R.P., and Finn, J.J., Jr.: Group Health Surveys in a Diagnostic Center, *New Eng J Med* **248**:165-170, 1953.

8. Guild, W.R., and Mansfield, R.J.: Physical Fitness of Executives at a US Naval Shipyard, *Arch Industr Health* **11**:121-122, 1955.

9. Bolt, R.J.; Tupper, C.J.; and Mallery, O.T., Jr.: An Appraisal of Periodic Health Examinations, *Arch Industr Health* **12**:420-426, 1955.

10. D'Alonzo, C.A., and Rogers, S.M.: Health Examinations in Industry, *Industr Med Surg* **24**:75, 1955.

11. US Department of Health, Education, and Welfare, Public Health Service: *Peptic Ulcer,* Public Health Service publication No: 280, Health Information Series No 71, 1965.

12. Birk, R.E.: A Clinical Survey of a Cross-Section of an Industrial Worker Group, *Arch Environ Health* **1**:291-296, 1960.

13. Ortelee, M.F.: Study of Men With Prolonged Intensive Occupational Exposure to DDT, *AMA Arch Industr Health* **18**:433-440, 1958.

14. Osgood, E.E., et al: Total Differential and Absolute Leukocyte Counts and Sedimentation Rates, *Arch Intern Med* **64**:105-120, 1939.

15. Booth, K., and Hancock, P.E.T.: A Study of the Total and Differential Leucocyte Counts and Haemoglobin Levels in a Group of Normal Adults Over a Period of Two Years, *Brit J Haematol* **7**:9-20, 1961.

16. Blackburn, C.R.B.: Normal Leucocyte Count, *Med J Australia* **1**:525-528, 1947.

17. Jeffrey, H.C.: Normal Variations in Leucocyte Counts, *J Roy Army Med Corps* **107**:93-98, 1961.

18. Chamberlain, A.C.; Turner, F.M.; and Williams, E.K.: The Evaluation of White-Cell Counting in Radiation Protection, *Brit J Radiol* **25**:169-176 1952.

19. Kleinman, G.A.; West, I.; and Augustine, M.S.: Occupational Disease in California Attributed to Pesticides and Agricultural Chemicals, *Arch Environ Health* **1**:118, 1960.

20. California State Department of Public Health: Occupational Disease in California Attributed to Pesticides and Other Agricultural Chemicals, 1963.

21. California State Department of Public Health: Occupational Disease in California Attributed to Pesticides and Other Agricultural Chemicals, 1964.

22. Hayes, W.J., Jr., et al: Storage of DDT and DDE in People With Different Degrees of Exposure to DDT, *AMA Arch Industr Health* **18**:398-406, 1958.

23. Durham, W.F.; Armstrong, J.F.; and Quinby, G.E.: DDT and DDE Content of Complete Prepared Meals, *Arch Environ Health* **11**:641-647, 1965.

24. Hayes, W.J., Jr.; Dale, W.E.; and Pirkle, C.I.: The Effect of Known Repeated Oral Doses of DDT in Man, to be published.

25. World Health Organization: Hypertension and Coronary Heart Disease: Classification and Criteria for Epidemiological Studies, *WHO Tech Rep Ser No. 168,* 1959.

Toxicology of Abate in Volunteers

Edward R. Laws, Jr., MD; Francisco Ramos Morales, MD;
Wayland J. Hayes, Jr., MD, PhD; and
Charles Romney Joseph, MS

CONTROL of the breeding of *Aedes aegypti* mosquitoes in containers used for the collection and storage of drinking water is essential for the eradication of this species from many areas of the world. Biological evaluation in the branch laboratories of the Communicable Disease Center at Savannah, Georgia, has demonstrated that Abate (0,0,0′,0′-tetramethyl 0,0′-thiodi-*p*-phenylene phosphorothioate) produces effective larvicidal action against *A aegypti* for at least five weeks when added to barrels of drinking water at the rate of 1.0 ppm.[1] The insecticide dissolves gradually but also hydrolyzes gradually so that after addition of the compound, the concentration of Abate in the stored water increases rapidly to about 0.5 ppm and then decreases slowly. Repetition of the treatment at intervals of a month or even three weeks does not lead to a progressive increase in the concentration of Abate in the water.

Because of its excellent activity as a larvicide and initial reports of its low mammalian toxicity, the toxicity of Abate was evaluated in laboratory animals.[2] Acute and subacute studies were conducted using rats, mice, rabbits, guinea pigs, dogs, chickens, and ducks. All the results indicated that Abate has a low toxicity. The values for acute oral 50% lethal dose (LD_{50}) in white rats and mice are 4,000 mg/kg of body weight or greater. All mammalian species tested tolerated 10 mg/kg without clinical effect and 1 mg/kg without effect on cholinesterase. These latter dosages were given daily, and the tests were carried out for a minimum of a month.

Animals made ill by Abate showed typical signs and symptoms of organic phosphorus poisoning, and marked depression of blood cholinesterase activity always preceded clinical symptoms. Abate consistently produced a larger and more rapid depression in red blood cell (RBC) cholinesterase than in plasma cholinesterase.

On the basis of the results of the studies in animals, it was felt that a toxicological evaluation of Abate in humans could be conducted with safety. A study in volunteers was designed (1) to discover the acute oral dosage of Abate necessary to produce cholinesterase depression in humans, and (2) to test a relatively high dosage which would not produce cholinesterase depression when given daily for four weeks.

Materials and Methods

Twenty-eight volunteers were selected from the inmate population of the Penitenciaría Estatal in Río Piedras, Puerto Rico. The project was explained to candidates in detail, describing both the potential benefits to mankind and the possible hazards involved before any of them actually volunteered. The men were told that there was no guarantee that participation in the project would be entirely safe and that the remote possibility of illness or even death did, in fact, exist. Each man signed an agreement, in Spanish, stating these facts before he was accepted. No coercion was involved either on the part of the Public Health Service personnel or the prison officials, and the only rewards accruing to the volunteers were a small sum of money and a certificate presented at the end of the study. Each man was free at any time to decline further participation with or without reason.

The men agreed to take a daily dose of milk which might contain Abate and to give periodic samples of blood by venipuncture. It was also stipulated that the physician who conducted the study be available at all times during the course of the experiment to handle any medical problems that might arise among the volunteers.

Reprint requests to Communicable Disease Center, Atlanta 30333 (Dr. Hayes).

All the participants were men between the ages of 22 and 44, and in excellent health as determined by comprehensive medical histories and physical examinations. Two candidates with histories of hepatitis were rejected, but four men included in the volunteer group had histories of treated schistosomiasis. All other findings on history and physical examination were unremarkable.

The daily doses given to the men in the experimental groups were prepared using technical Abate of batch A, as designated in animal studies.[2] The formulation was as follows: 20 ml ethyl alcohol, 30 ml cod liver oil, 10 gm gum arabic, technical Abate in calculated volume for desired concentration, and water to make up 250 ml. All the ingredients were homogenized in a high-speed mixer. The formulation for the control group was identical except for the omission of Abate. In the preparation of each dose, 5 ml of the appropriate formulation was added to 30 ml of milk and stirred. The volunteers were required to drink the preparation in the presence of the investigator. Fresh formulations were prepared at least every three days, and were rehomogenized daily prior to use. At no time during the experiment were the men told which doses contained Abate.

During the first week of the experiment, the entire group of volunteers received daily doses of formulated milk which contained no Abate, and RBC and plasma cholinesterase determinations were made on three separate blood samples from each man, providing a set of pretreatment control values. It was expected that psychosomatic complaints might become manifest during this "sham" dosage period, but none appeared.

At the end of the first week, 20 men were selected and divided at random into two groups of 10, one to receive Abate and one to serve as controls. The experimental group started with a dosage of Abate at the rate of 2 mg/man/day, and the dosage was doubled every three to four days for a period of four weeks. Levels of RBC and plasma cholinesterase were determined on blood samples taken three times a week from each of the volunteers. Voided urine samples were collected from all the men while the experimental group was receiving the highest dosage (256 mg/man/day) and at weekly intervals for three weeks following cessation of dosage.

In the subsequent part of the study, 18 volunteers were divided into two groups of nine, one experimental and one control. None of the men in the experimental group had previously received Abate. They were given Abate at a daily dosage of 64 mg/man for four weeks. Red blood cell and plasma cholinesterase determinations were made on blood samples taken twice a week from each volunteer, and an additional blood sample was taken five days after the last dose. Urine samples were collected during the first, third, and fourth weeks of dosage and at weekly intervals for three weeks following the last dose.

Throughout the study, blood and urine samples were collected from control and experimental groups simultaneously.

Red blood cell and plasma cholinesterase were determined using the method of Michel.[3] The laboratory technicians were not told whether samples were from experimental or control subjects.

Urine samples were analyzed in the laboratories of the Toxicology Section of the Communicable Disease Center using a method for determination of ether-extractable organic phosphorus.[4] The results are expressed as Abate equivalents based on the molecular weight and phosphorus content of Abate.

Results

At no time during the entire experiment was there a significant difference between the RBC or plasma cholinesterase values of the volunteers receiving Abate and those who were controls, and at no time were there clinical symptoms or adverse reactions resulting from the administration of the compound.

In the first phase of the study, the daily dose of Abate was gradually increased to a level of 256 mg/man. At this point the taste of the doses containing Abate was powerful enough to allow the volunteers to distinguish which of them were in the experimental group and obnoxious enough to cause transitory nausea in one man. The concentration of Abate in the formulation offered was 6,400 ppm. Therefore, it was necessary to terminate this phase of the experiment before reaching the goal of determining a dosage level which would cause cholinesterase depression.

In the second phase of the experiment, a dosage of Abate at the rate of 64 mg/man/day was tolerated without complaints, clinical symptoms, or reactions related to Abate, and without any detectable effect on cholinesterase.

During the course of the study transitory fever, coryza, and myalgia which disappeared in three days developed in one man,

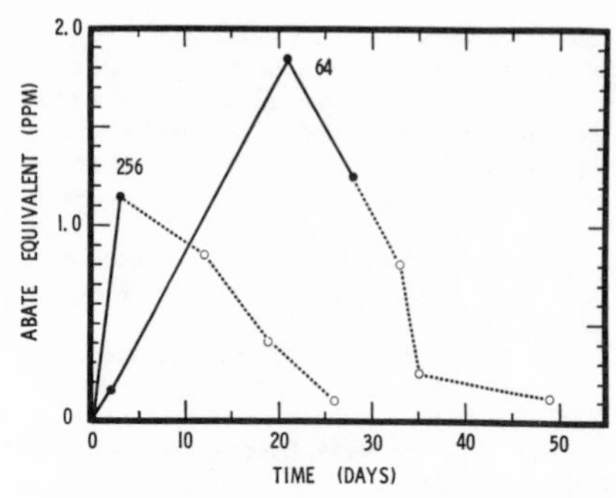

Urinary excretion of organic phosphorus derived from Abate by groups of men while receiving the compound (solid symbols and lines) and after dosage was stopped (open symbols and dotted lines). Groups received Abate at the rate of 256 and 64 mg/kg/day, as indicated, with ten men in the former group and nine in the latter.

and there were four other brief cases of headache and coryza. The five cases were randomly distributed between the men who were taking Abate and those who were not.

The levels of urinary excretion of Abate are given in the Figure. The method used accounted for only a small percentage of the total amount of Abate given to the volunteers; however, the recovery was always proportional to dosage. No conclusion can be drawn from the available data as to whether a steady state of excretion was achieved within the period of dosage. Excretion of organic phosphorus decreased slowly after dosage was stopped but was still detectable three weeks later.

Comment

This study corroborates the low toxicity of Abate and contributes to the establishment of the safety of the use of Abate as a larvicide in treating potable water.

The nominal concentration of Abate proposed for practical larvicidal treatment under field conditions for containers of drinking water is 1.0 ppm. The maximal concentration actually achieved is about 0.5 ppm, and the concentration gradually decreases over a period of weeks. Thus, if a man drinks 2 liters of treated water each day, he will receive Abate at the rate of about 1 mg/day when the water is first treated and at a decreasing rate for the remainder of one month. Then the water is retreated and the cycle repeated. The results of this study in volunteers indicate that an intake of 256 mg/day for five days or 64 mg/day for four weeks is unlikely to produce either symptoms of poisoning or even detectable effects on blood cholinesterase, thus providing a great margin of safety over the expected intake under conditions in the field.

It should be emphasized that, using the formulation presently employed in the field, Abate in water is tasteless even to an informed observer at all concentrations up to 256 ppm.

A limited study of Abate in use under actual field conditions involving people more representative of the general population is under way and, if successful, should pave the way for the widespread use of the compound wherever it is indicated.

Summary

Volunteers tolerated a dosage of 256 mg/man/day for five days or 64 mg/man/day for four weeks without clinical symptoms or side effects attributable to Abate and without detectable effect on red blood cell or plasma cholinesterase.

Technical assistance was provided by Nilda Monroig de Havner and Gladys Sánchez Ruiz who did the cholinesterase analyses. Vincent Sedlak did the urinary analyses.

The Puerto Rico Departments of Health and Justice, the Puerto Rico section of the *Aedes aegypti* Eradication Branch of the Communicable Disease Center, United States Public Health Service, and the officials of the Penitenciaría Estatal cooperated in this investigation.

Technical Abate was supplied by the American Cyanamid Co., Danbury, Conn.

References

1. Brooks, G.D.; Schoof, H.F.; and Smith, E.A.: Effectiveness of Various Insecticides against *Aedes aegypti* Infestations in Water Storage Drums in U.S. Virgin Islands, *Mosquito News* **25**:423, 1965.

2. Gaines, T.B.; Kimbrough, R.; and Laws, E.R., Jr.: Toxicology of Abate in Laboratory Animals, *Arch Environ Health,* to be published.

3. Michel, H.O.: An Electrometric Method for the Determination of Red Blood Cell and Plasma Cholinesterase Activity, *J Lab Clin Med* **34**:1564, 1949.

4. Mattson, A.M., and Sedlak, V.: Measurement of Insecticide Exposure: Ether-Extractable Urinary Phosphates in Man and Rats Derived From Malathion and Similar Compounds, *J Agr Food Chem* **8**:107, 1960.

Parathion Residue Poisoning Among Orchard Workers

Thomas H. Milby, MD, Fred Ottoboni, ChE, MPH, and Howard W. Mitchell, MD, MPH

HEALTH HAZARDS associated with the manufacture, formulation, distribution, and application of the organic phosphate pesticide, parathion, have been frequently described and will not be reviewed here. It has not been so clearly recognized, however, that for many days or weeks after application of parathion spray formulations to argicultural field crops, resulting residues may constitute an important health hazard to argricultural workers.

In 1958, Quinby and Lemmon [1] summarized 11 episodes of poisoning from contact with parathion residues involving a total of more than 70 workers who were involved in harvesting, thinning, cultivating, and irrigating such crops as apples, grapes, citrus, and hops. Although six of the outbreaks occurred within two days of pesticide application, in the remainder of the episodes, the residues had been from 8 to 33 days old. In general, the illnesses were characterized by a gradual onset and a relatively benign clinical course. Percutaneous absorption was thought to be the primary route of entry of the toxicant.

Although we are aware of no other published reports describing poisoning by organic phosphate residues, cases have occurred on a sporadic basis in California over the past several years and, in 1959, more than 275 cases of parathion residue poisoning were reported among workers harvesting citrus crops throughout the state.[2]

In early August, 1963, the California Department of Public Health was notified of an outbreak of parathion poisoning among orchard workers who were harvesting peaches in the northern part of California's San Joaquin Valley.

From the Bureau of Occupational Health, California State Department of Public Health.

Description of Area and Workers

The epidemic was centered around the town of Hughson in Stanislaus County. Several hundred peach orchards with a total of about 24,000 acres under cultivation are located in this major peach-growing area. For the annual harvest, during August and September, these orchards employ 7,500 to 8,500 agricultural workers, most of whom are migrants. The orchards in this area grow a number of varieties of peaches and because each variety of peach becomes ready for picking at a slightly different time, the harvest season extends over a six to eight week period. As a result of this prolonged harvest period, a grower in the area is able to employ a small crew, usually 15 to 25 workers, for the entire season. The pickers move from one variety to another and often from one grower to another as the fruit becomes ready for harvest.

The harvesting of peaches has not been mechanized. Each piece of fruit must be picked from the tree by workers using ladders and chest-slung baskets or canvas bags. In this process there is manual contact with all of the fruit and a great deal of contact between the upper half of the body and tree foliage. The picker's breathing zone is often closely surrounded by branches thick with leaves, affording maximum opportunity for inhalation of pesticide residues rendered airborne by the picking process. The San Joaquin Valley summer heat, the constant use of ladders, and the pace induced by piecework combine to make the job hot and uncomfortable. As a result, clothing is light and often sweat impregnated. Shirts are open at the collar, and often shirt sleeves are short or rolled above the elbow.

General Study Plan

The general study plan consisted of relating the worker's health to his occupational environment. During the course of the field work, three groups consisting of 186 peach orchard workers exposed to parathion residues were identified and studied. The first group was made up entirely of cases of poisoning reported by local physicans. These cases were selected to provide information on the clinical manifestations of the toxicant involved and also served to identify orchards with unsafe residue levels. A second group was comprised of workers employed in a sample of these unsafe orchards. Some of these workers had been poisoned and sought medical attention and some had not. This group was selected to provide information on the prevalence of clinical illness in these orchards as well as an estimate of the prevalence of subclinical illness as reflected by depression of blood cholinesterase levels. This group also provided subjects for environmental studies from which maximum daily assimilation of residue could be estimated. A third group was selected at random in order to estimate the prevalence of cholinesterase depression in the universe of orchard workers employed in the Hughson peach-growing area.

To study the relationship between pesticide application and the occurrence of reported clinical illness, pesticide spraying schedules were collected from all growers in whose orchards illness had been reported. These schedules were then compared to a second group of schedules selected from orchards in which no illness had been reported. The second group of schedules was obtained from two large canneries and represented all of the fruit purchased in the epidemic area by these two firms. These schedules were readily available because a copy of the grower's pesticide application schedule is required by all fruit processors in California as a condition of purchase. Except for several schedules which were excluded for technical reasons such as incompleteness or illegibility, all were used for comparison purposes.

Finally, leaf and fruit specimens were collected from both orchard groups, those with associated illness and those with no associated illness. These specimens were analyzed for residues in an attempt to relate residue levels to presence of illness.

For the purpose of this paper, cholinesterase depression is defined as depletion of *either* red blood cell (true) cholinesterase *or* plasma (pseudo) cholinesterase, or *both* to a level below the range of normal variation reported by Wolfsie and Winter [3] as:

RBC: 0.53–1.21 pH units per hour.
Plasma: 0.44–1.38 pH units per hour.

All cholinesterase values were determined by the electrometric method of Michel [4] as modified by Hamblin and Marchard.[5] All parathion analyses were carried out using the method of Averill and Norris.[6] However, because this method does not differentiate between parathion and its S-phenyl isomer, its S-ethyl isomer, or its oxygen analog (paraoxon), any value reported as parathion may reflect the presence of these other forms.

Results

Group 1.—The first group was made up of 94 orchard workers who became clinically ill during the period between Aug 4 and Sept 15 and who sought medical care from local physicians because of the severity of their complaints. The figure shows these 94 cases as they appear when converted to a weekly attack rate based on the total picker work force.[7] The spread of cases with time generally coincides with the period of peak peach harvest, which extended from July 28 to Sept 15, 1963.

At the time of illness, cholinesterase levels were determined on blood collected from 68 of these cases and found to be depressed in 66 of them. The 26 cases in which no cholinesterase levels were determined were considered by the attending physicians to be so typical of organic phosphate poisoning that no laboratory verification of the clinical diagnosis was necessary. The most consistent complaints described by these clinically ill workers were nausea, vomiting, occipital headache, profound weakness, and extreme malaise. Other manifestations of parasympathetic stimulation including miosis, blurred vision, dizziness, excessive sweating, salivation, diarrhea, and abdominal cramping were reported, but not consistently so. It is noteworthy that several clear-cut cases failed to demonstrate miosis at any time during the course of illness. Although a number of patients were hospitalized for 24 to 48 hours, symptomatic treatment with large, parenteral doses of atropine (1.2 to 2.4 mg) repeated as necessary appeared to give satisfactory relief in every case.

One death during the epidemic was attributed to parathion poisoning by a local pathologist. Although it was determined that the deceased had worked for no more than 1½ days in an orchard from which one other case of clinical illness had been reported, his activities and exposure during several of the days immediately prior to his hospitalization could not be traced. The clinical course of this fatal illness is obscure, but it was reported that upon hospitalization seven days after known work exposure, both red blood cell and plasma cholinesterase levels were depressed. Sixteen days after exposure, following nine days of hospitalization, the patient died. Cholinesterase levels at the time of death were reported to be in the low-normal range. Postmortem examination led to a final diagnosis of "bilateral terminal bacterial pneumonia" and "organic phosphate poisoning."

Group 2.—The second group consisted of 68 volunteers from a total work force of about 100 workers who were employed in six orchards from which clinically recognizable cases of organic phosphate poisoning were being reported. Among these 68 volunteers were 62 pickers, 2 fruit graders, 2 orchard owners, 1 labor contractor, and 1 cook. All 68 were interviewed, and a blood specimen for cholinesterase determination was obtained from each of them. In addition, 14 volunteers from these six orchards were examined for skin parathion contamination. In the course of this procedure, various skin surfaces of measured area were scrubbed with alcohol-moistened cotton swabs which were then sent to the laboratory for analysis. To further evaluate skin exposure, a shirt was acquired from one worker who stated that it had not been laundered for eight days. The condition of the shirt bore out his claim. To estimate respiratory exposure to airborne residues, breathing zone dust samples were collected on lapel-mounted filter paper air samplers and analyzed for parathion. The results of the analysis of these environmental samples will be described later in this paper.

From Table 1, it can be seen that 84% of the 68 volunteers in this group showed evidence of significant organic phosphate absorption reflected by depression of blood cholinesterase levels. Moreover, from this table it is clear that once an individual becomes sufficiently ill to seek medical care, his cholinesterase levels are very likely to be depressed. However, it is equally as clear that cholinesterase depression, in itself, is not always accompanied by symptomatic illness, even of a mild degree.

It is of interest to note that of the six volunteers who were not employed as pickers, none had sought medical care, and five were asymptomatic with normal cholinesterase levels. Only one, the labor contractor, was placed in a category indicating both complaints and cholinesterase depression.

Group 3.—In order to arrive at some estimate of the prevalence of cholinesterase depression among peach pickers in the epidemic area, 45 workers residing in seven groups of living units were studied. The living units consisted of trailer parks, labor camps, and motels and were selected only on the basis of proximity to the harvest area. The workers were interviewed and blood samples were obtained from each of them. Only individuals who were actively engaged in peach picking were included in the study group. None reported working in orchards with which clinical illness had been associated, nor had any of them sought medical attention during the peach harvesting period. Sixteen, however, complained of minor signs and symptoms, including nausea, vomiting, headache, dizziness, weakness, insomnia, and anorexia, which several thought referable to their work. As shown in Table 2, 35% of these 45 orchard workers had absorbed a significant amount of organic phosphate pesticide as reflected by depression of blood cholinesterase levels. Table 2 also relates cholinesterase levels to presence and extent of clinical illness. In this group, as in the second group, it is clear that a worker may be asymptomatic even though his blood cholinesterase is significantly depleted.

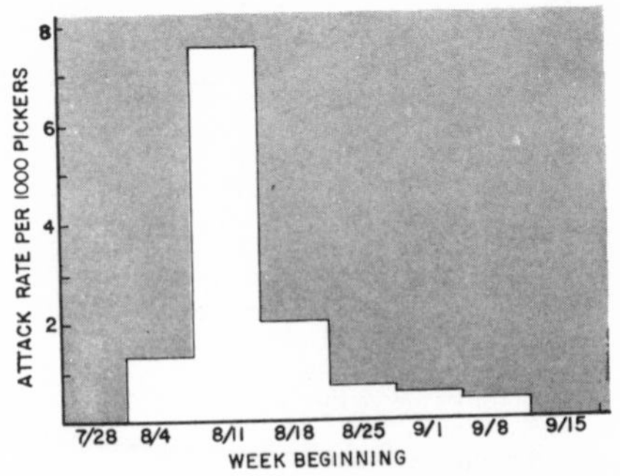

Clinical attack rate per 1000 pickers by week.

The status of the cholinesterase levels of the entire population of several thousand pickers cannot be realistically extrapolated from these 45 cases. However, the prevalence of cholinesterase depression in this small sample suggests that the problem of significant parathion residue absorption extended beyond the few score cases reported by physicians or discovered by study of a highly selected group of pickers working in orchards from which illness had been reported.

Pesticide Spraying Schedules.—In all orchards studied, parathion was applied in the form of 25% wettable powder in water suspension. Essentially all spraying was done by tractor-drawn spray blowers. Schedules varied from one to seven applications over the growing seasons at rates of one to two pounds of parathion per acre at each application. The frequency and rate of pesticide application depended largely upon the degree to which the orchard owner felt his crop was threatened by insect predators.

Pesticide spraying schedules from 16 illness-producing orchards were compared to pesticide spraying schedules from 43 no-illness orchards. This com-

Table 1.—Cholinesterase Levels by Extent of Illness Among 68 Workers From Six Illness-Producing Orchards

Extent of Illness	No. of Workers Sampled	% With Depressed Cholinesterase
Clinical poisoning by physician diagnosis	21	95
Complaints but no medical care sought	14	86
Asymptomatic	33	76
Total Cases	68	84

parison indicated that the only cholinesterase inhibiting compound used in every illness-producing orchard was parathion. However, it was apparent that application of parathion, in itself, was not the determining factor in the causation of illness because 40 of the 43 orchards without associated illness also applied it. Organic phosphate pesticides other than parathion were used only irregularly, and therefore, their effect in the causation of the epidemic could not be assessed.

The important differences between the orchards with associated clinical illness and those with no associated illness appeared to be related to total parathion application and to the time interval between the last application and the start of harvest. The 40 parathion-using orchards with no associated clinical illness applied a mean parathion dosage of 4.38 pounds per acre for the season and waited a mean interval of 45 days between the last application of parathion and the start of harvest. Comparable means from the 16 orchards with associated clinical illness were 7.14 pounds and 23 days. Thus, the orchards without illness applied less total parathion per acre and waited longer between the last application and harvest. While this information is of interest, it is not really useful as time and dosage have an uncertain effect on residue levels at harvest and both vary in this particular comparison.

The interval between the final parathion application and the first day of harvest (the spraying-picking interval) was eliminated as a variable by matching the group of illness-producing orchards and a group of no-illness orchards with comparable mean intervals. This was accomplished by eliminating all of the orchards in the no-illness group with a spraying-picking interval of more than 40 days. The final matched groups consisted of 16 illness-producing orchards with a mean interval of 22.9 (standard deviation 7.7) days and 26 no-illness orchards with a mean interval of 22.4 (standard deviation 6-1) days.

Table 2.—Cholinesterase Levels by Extent of Illness Among 45 Peach Pickers From Miscellaneous Camps, Motels, and Trailer Parks

Extent of Illness	No. of Workers Sampled	% With Depressed Cholinesterase
Clinical poisoning by physician diagnosis	0	0
Complaints but no medical care sought	16	25
Asymptomatic	29	41
Total Cases	45	35

Table 3 compares the mean parathion dosages for these two matched groups of orchards. Note that the total dose for the season after Jan 1 was significantly higher in the orchards from which clinical illness had been reported. The dosage applied after July 1, 1963, however, does not show a significant difference between the two groups of orchards. It was not possible to make a similar comparison (holding dosage constant) to determine the effect of the spraying-picking interval on illness because matching groups could not be constructed from the data available.

Parathion Residues and Total Worker Dose.—Foliage specimens were collected from 11 orchards, six with associated illness and five with no associated illness. Leaves from the former group of orchards were obtained within three days of the onset of clinical illness among employees and leaves from the latter group of orchards were collected at random during the harvest period. Residues were removed by surface stripping with benzene and analyzed for parathion. The mean parathion content of the foliage from the six illness-producing

Table 3.—Comparison of Matched Groups of Orchards by Parathion Dosage

Time Interval and Orchard Group	Mean Dose, Pounds/Acre	Standard Deviation	Significance of Difference by "t" Test
After July 1, 1963			
Orchards with illness	2.24	1.10	Not significent
Orchards without illness	1.78	0.81	
After Jan 1, 1963			
Orchards with illness	7.14	2.60	0.01
Orchards without illness	4.99	1.95	

orchards was 4.4 parts per million by weight (PPM) with a range of 1.0 to 7.2 PPM. The mean parathion content of the foliage from the five no-illness orchards was 1.9 PPM with a range of 0.9 to 3.6 PPM. While these figures suggest that illness is more likely to be associated with higher parathion residues levels, the considerable overlap between the two orchard groups rendered impractical any attempt to use residue levels as an indicator of worker risk.

Fruit samples were collected from five of the above orchards, three of which had produced clinical illness and two of which had not. Results of the analysis of these samples for parathion content ranged from 0.02 PPM to 0.5 PPM parathion by weight. No levels were found which exceeded the permissible tolerance for peaches of 1.0 PPM. (US Department of Health, Education, and Welfare, Food and Drug Administration, Federal Insecticide, Fungicide and Rodenticide Act [1947] and Miller Pesticide Residue Amendment [1954] to the Federal Food Drug and Cosmetic Act of 1938.)

Variations in weather are known to exert a complex effect on parathion degradation rates and resulting residue levels. However, in this study

Table 4.—Parathion Contamination of Pickers From Orchards That Produced Clinical Illness

Description of Sample	No. of Observations	Parathion, μg per per square inch of skin surface	
		Range	Mean
Palm of hand, sorters	3	2.0-4.7	3.4
Palm of hand, pickers	5	0.5-7.0	2.8
Forearm, pickers	3	0.4-4.7	2.0
Upper arm, pickers	2	0.2-1.4	0.8
Back of neck, pickers	3	0-1.4	0.5

One shirt, worn eight days: total parathion—960μg

weather was not a variable because all of the orchards in the small peach-growing area were necessarily subjected to essentially identical weather conditions.

Table 4 lists residues found on the shirt of 1 and on the skin of 14 pickers working in orchards that produced maximum rates of clinical illness. These data confirm the presence of parathion on the arms and trunk as well as on the palms of hands and suggest that contact with leaves and tree surfaces contribute to total exposure. Because of limitations of the alcohol swab sampling technique, the values for skin contamination are probably about 10% low.[8] Breathing zone air samples from these same orchards were collected on filter paper and analyzed for parathion. The highest value obtained by this method was 35 μg of parathion per cubic meter of air. According to Durham and Wolfe [8] these values may be 10% to 15% low due to evaporation of parathion into the airstream passing through the filter. Airborne parathion vapor exposure was not measured but estimated as zero, based on studies by others.[9]

Table 5 details an estimate of the maximum fruit, skin, and air exposure to parathion which could be encountered by a picker. These are maximum values based on measurements obtained from the two orchards which had produced the highest rates of clinical illness.

Although the total daily dose of parathion absorbed through the skin cannot be precisely determined, work by Durham and Wolfe [8] suggests that parathion is slowly and inefficiently absorbed and that residues found on the skin constitute many times the quantity which will be absorbed in eight hours. Likewise, the quantity found on a shirt is more than would be absorbed in a single day. Thus, the maximum quantity of parathion absorbed through the skin of the individuals studied was probably less than 3,000μg per day and the maximum total dose by all routes was less than about 4,000μg per day.

Comment

From the very beginning of the epidemic, local physicians had implicated parathion as the causative agent. Their reasoning was based on observation of the clinical syndrome and on some knowledge of the spraying practices prevalent in

Table 5.—Daily Maximum Exposure Which Could be Encountered by a Picker by Sources of Agent and Routes of Entry

Route of Entry and Source of Exposure	Parathion, μg
Ingestion	
4 peachers per day with residue level of 0.5 PPM by weight	500
Inhalation	
Airborne dust, based on highest breathing zone value of 35 μg/cu m and breathing rate of 10 cu m/day	350
Airborne vapor, estimated negligible	0
Dermal*	
Palms of two hands about 63 square inches total and 7μg per square inch maximum	440
Backs of hands, forearms, and face about 351 square inches total and 4.7μg per square inch maximum	1650
Back of neck and "V" of neck about 40 square inches and 1.4μg per square inch maximum	56
Upper arms and remainder of trunk based on shirt	960
Lower limbs assumed negligible because contact with foliage minimal	0
Total dermal exposure	**<3000**
Total exposure all routes	**<4000**

*Surface areas determined using Berkow's method as quoted by Durham et al.[8]

surrounding orchards. Our evaluation of spraying schedules indicated that parathion was the only organic phosphate applied in every illness-producing orchard and supported their contention. The slight variance of the clinical picture of poisoning described here from the classical syndrome of intense parasympathetic stimulation seen in cases of poisoning among workers exposed to sprays or concentrates is almost certainly related to the insidious manner in which the poison was assimilated. Thus, the slow rate of absorption produced a gradual but progressive depletion of blood cholinesterase elements until, at a critical level, mild to moderately severe clinical illness became apparent.

Careful environmental sampling revealed that although parathion could be recovered and identified without much difficulty, the maximum daily dose with which a worker could come into contact was not in excess of 4 mg. Because this quantity of parathion, even if completely absorbed, constitutes only about one-half of the daily dose reported to be capable of producing progressive cholinesterase depletion,[10] the presence in the orchard environment of one or more compounds, derived from parathion, but considerably more toxic is hypothesized. The presence of such a compound could be the result of contamination of the original spray material (unlikely) or the product of parathion aging, weathering, plant alteration, or other form of degradation. The most likely suspect is the oxygen analog of parathion, paraoxon, a cholinesterase inhibiting compound with cutaneous toxicity ten times that of parathion.[11] (Routine analytical capacity for the determination of paraoxon was not available at the time of the study. One leaf residue sample analyzed for paraoxon indicated the presence of 3.0 PPM paraoxon and 2.8 PPM parathion.) Also to be considered are the S-phenyl and S-ethyl isomers of parathion. Both of these isomers and the analog are potent, direct, in vitro inhibitors of cholinesterase. (Extrapolation from parathion toxicity suggests that absorption of as little as 2 mg per day of paraoxon would be sufficient to produce progressive cholinesterase depletion.) Cook and Pugh [12] reported the presence of these three compounds plus an unidentified "light product" in a sample of parathion irradiated with ultraviolet light under laboratory conditions. Although they did not identify this "light product," they reasoned that its cholinesterase inhibiting properties were at least as potent as paraoxon.

Available information [13-15] has indicated that parathion disappears in a rapid and continuous manner and has suggested that multiple sprayings applied over a six-month period should not result in residue accumulation. This concept would tend to implicate the final spray application closest to harvest in the causation of residue poisoning. However, our data cannot support this concept for the following reasons: (1) In no orchard studied, regardless of illness experience, did the final parathion application exceed the recommended rate of 2.5 pounds per acre followed by a 14-day interval between spraying and harvest.[16] (2) Analysis of data previously presented here can detect no significant difference between matched orchard groups when compared by amount of parathion applied during the five weeks preceding the outbreak described. Such differences appear only when earlier applications are included in the analyses. These findings suggest that the observed illnesses were the result of residue accumulation related to total amount of parathion applied during the entire growing season. It should be noted that the relative contribution to the total leaf residue levels of the winter, spring, or early summer parathion applications is not known. Likewise, while the last spraying was not the deciding factor in the causation of illness, it certainly contributed by adding to the already present foliage residues.

References

1. Quinby, G.E., and Lemmon, A.B.: Parathion Residues as Cause of Poisoning in Crop Workers, *JAMA* **166**:740, 1958.
2. *Doctor's First Report of Work Injury,* State of California, Division of Labor Statistics and Research, 1959.
3. Wolfsie, J.H., and Winter, G.D.: Statistical Analysis of Normal Human Red Blood Cell and Plasma Cholinesterase Activity Values, *Arch Environ Health (Chicago)* **6**:43, 1952.
4. Michel, H.O.: Electrometric Method for Determination of Red Blood Cell and Plasma Cholinesterase Activity, *J Lab Clin Med* **34**:1564, 1949.
5. Hamblin, D.O., and Marchard, H.F.: *Cholinesterase Tests and Their Applicability in Field,* New York: American Cyanamid Co, 1951.
6. Averell, P.R., and Norris, M.V.: Estimation of Small Amounts of 0,0-Diethyl 0,0-Nitrophenyl Thiophosphate, *Analyt Chem* **20**:753, 1948.
7. *Weekly Farm Labor Report,* No. 922-930, State of California, Department of Employment, 1963.
8. Durham, W.F., and Wolfe, H.R.: Measurement of Exposure of Workers to Pesticides, *Bull WHO* **26**:75, 1962.
9. *Parathion Vapor Concentrations in Atmosphere of California Groves During and After Application,* New York: American Cyanamid Company, 1951.
10. Hayes, W.J., Jr.: *Clinical Handbook on Economic Poisons, Emergency Information for Treating Poisoning,* US Dept of Health, Education and Welfare, 1963.
11. Metcalf, R.L.: *Organic Insecticides. Their Chemistry and Mode of Action,* New York: Interscience Publishers, Inc, 1955.
12. Cook, J.W., and Pugh, N.D.: Quantitative Study of Cholinesterase Inhibiting Decomposition Products of Parathion Formed by Ultraviolet Light, *J Assoc Agric Chem* **40**:277, 1957.
13. Culver, D., et al: Study of Exposure to Parathion in Greenhouse, *Arch Environ Health (Chicago)* **18**:235, 1958.
14. Fahey, J.E.; Hamilton, D.W.; and Rings, R.W.: Longevity of Parathion and Related Insecticides in Spray Residues, *J Econ Entomol* **45**:700, 1952.
15. Brunson, M.H.; Koblitsky, L.; and Chisholm, R.D.: Effectiveness and Persistence of Insecticides Applied During Summer Months to Control Oriental Fruit Moth on Peach, *J Econ Entomol* **55**:728, 1962.
16. *Manual of Manufacturing Information for Thiophos Parathion,* New York: American Cyanamid Company, 1958.

PROTECTING THE AG PILOT

So long as certain insects have a greater capacity per unit time than do humans to create population explosions, we'll find ourselves in competition with these tiny rapacious fellow earthlings for available agricultural products.

Additionally, man will wage widespread winged warfare against specific chitinous culprits, so long as aerial applicator techniques remain superior in efficiency and effectiveness to ground applicator techniques.

Man's weapons in this warfare are chemicals, the most potent falling within the chlorinated hydrocarbon and organic phosphate pesticide categories.

Sometimes, just as is the case of conventional warfare, we find that an enthusiastic warrior wounds himself rather than his foe. Let us look at a case in point.

Near Paris, Texas, in August of 1963, a very experienced aerial applicator pilot crashed while at work. A thorough investigation revealed that the pilot was not feeling well on the morning of the flight, and had a headache. Four days earlier, he had accidentally spilled parathion on his clothes while pouring the concentrate from a 55 gallon drum. He neither changed clothes nor washed the skin where it was touched by the chemicals, but continued to work the rest of the day (although a fellow worker encouraged him to take these precautionary steps).

In the next two to three days after getting the chemical on his skin, the pilot, who usually was good-natured and outgoing, became irritable and introverted. It was under these circumstances that he undertook flight on the fourth day. The accident occurred when the plane stalled out of a turn following completion of five swath runs.

"Impaired physical efficiency of the pilot caused by exposure to toxic material" is listed in the Probable Cause ledger for this accident by the Civil Aeronautics Board (Docket 2–2515).

What is the magnitude of the role played by the impairment of pilot proficiency under the circumstances of aerial applicator activities? We note that more than half of the aerial applicator operations last year were concerned with pesticides. Also, from the best figures we can glean concerning total active aerial applicator aircraft in the world, it appears that there are 16,000 airplanes involved (6,000 are estimated to be in Russia and its allied countries, with a similar number in the U.S.).

It is clear, then, that a little carelessness across-the-board, in the handling of the chemicals mentioned above, can have a general adverse impact upon the safety statistics. All of us would prefer to avoid a bad safety record.

Let's see what the accident statistics in aerial applicator operations show us. A detailed review of 113 aerial applicator accidents which were pulled in no particular sequence from our 1963 files containing reports of 304 accidents (of which 24 involved fatalities) has been accomplished.

Each of these 113 accidents has been placed in one of three categories, depending upon where the major responsibility was felt to rest with respect to arranging the circumstances leading to the respective accident.

Major Responsibility for Accident	*Number*	*Percent*
Pilot Factors	75	66%
Aircraft	33	29
Outside Force	5	5
	113	100%

We recognize that for various reasons certain contributory factors in given accidents may never come to light. In general, though, it appears that the investigators accomplished a creditable job, especially in the cases where the pilots were able to give detailed accounts. Incidentally, one gets the distinct impression that aerial applicator pilots, as a group quite seasoned, are very cooperative and candid in reconstructing the events surrounding their accidents. Certainly they are as concerned as anyone in preventing recurrences.

Pilot Factors

1. Stalls out of Turn:

Nine accidents involved stalls out of turns. Most of the pilots who fell victim to this pitfall were experienced, with several thousand hours flight time, much of which was in aerial application.

Attempting to hurry the turn to start another swath appears to be a significant factor here, complicated by fatigue in several cases (producing sloppy flying).

2. Collisions with Objects:

In 22 cases, wires, trees, and other objects were struck. Interviews with surviving pilots indicate that in a number of cases fatigue played a key role in the impairment of their flight proficiency.

In one case, occuring in June of 1963 near Olathe, Colorado, where the pilot flew into a tree during a pull-up, the pilot had been flagging earlier and by accident was exposed to drifting parathion. He experienced certain symptoms, in cluding headaches, and was ill enough to require atropine therapy. This accident points up the fact that these chemicals can adversely affect the proficiency of a pilot, a circumstance which can become clearly apparent under the close-quarters flight conditions of aerial application.

The combined effects on pilot proficiency of fatigue and certain potent chemicals, can be quite detrimental to flight proficiency. Interviews appear to indicate that these two factors, alone or together, continued to affect the aerial applicator safety record during 1963.

3. Exercise of Judgment in Take-off and Landing Areas:

Nine pilots in the group failed to become airborne, or ran off of their landing area, because their selected ground sites were marginal or inadequate under the circumstances of the attempted flights. In the cases of the take-off accidents, the gross-weights were usually too high for the field lengths, ground texture, or air density. Rising air temperature during the day was a factor in several cases. In landing accidents, the available width or length of the terrain was too limited in several cases (occasionally complicated by cross-winds or tail-winds). Once again, in tight operational situations, where peak pilot proficiency is required, fatigue, accidental toxic effects, and certain other factors, seem to make the difference between a successful ground/air or air/ground transition and an unsuccessful transition. Additionally, these factors adversely affect pilot judgment in these matters.

4. Proficiency:

Four accidents were defintely attributable to a lack of proficiency in flight techniques.

Most aerial applicators appear to have obtained between 2,000 and 6,000 flight hours, and to have learned through various combinations of training and experience how to perform adequately their low level maneuvers.

Some, however, especially the low-time pilots and the pilots who have perpetuated bad flight habits, represent "accidents waiting to happen."

Mr. Joe Fallin of the FAA Academy at Oklahoma City, has pointed out that a large number of pilots are mistaught concerning the proper means of accomplishing a given ground track in turns under wind-drift conditions (a key maneuver in aerial application). Accidents continue to occur where the plane drifts into objects while on a down-wind turn.

The last page of this paper presents this example which we at CARI have administered to many pilot groups consisting of all categories of pilots.

Ninety percent of the pilots in all groups except professional pylon race and stunt pilots place the X at the 6:00 point on the circle. This includes most instructors. The result is that many pilots are deliberately waiting to place their craft in its steepest banked turn until the 6:00 point is reached. This means that they are drifted away from the pylon (the pylon may be real or imaginary) prior to reaching the 6:00 point (due to the fast ground speed at the 9:00 downwind point), necessitating an even steeper bank to get back on their intended ground track.

Actually, the steepest bank should occur at the 9:00 point, and as the plane swings around on its proper ground track, wind correction angle is fed in to maintain the track. The 6:00 point should encompass the maximum wind correction angle. An identical WCA and bank is at 12:00. There is no wind correction at 9:00 or 3:00.

We call this "making the mental image match the maneuver", and find that in many cases the instructors will actually perform the maneuver correctly (steepest bank at 9:00), but explain it to the student incorrectly (steepest bank at 6:00). The result is that the student puts forth his best effort to do the maneuver according to the explanation. Either the individual never masters the maneuver and gets ultimately into trouble, or he gradually unconsciously discovers the proper means of accomplishing the maneuver.

The situation is analogous to the old-time teaching by many that the rudder was the control that turned the plane (although in practice most of those teaching this performed coordinated turns, zealous students would at times be found ruddering the planes around—especially under tight circumstances).

The solution rests in good initial flight training, plus periodic proficiency training. Even seasoned old hands should cross-check each other from time to time.

5. Other

The remaining pilot factors were concerned with such things as allowing the plane to run out of fuel through carelessness, attempting to take-off heavily loaded with the carburetor heat on resulting in inadequate take-off power for the length of the strip, carelessness in taxiing resulting in collision with some object, propping an unattended improperly tied-down craft (which on starting runs amuck), attempting take-off under adverse load and field conditions with the propeller in high pitch, attempting take-off's and landings in the dark, attempting exuberant low level aerobatic maneuver following completion of spray job, attempting "spray formation" flying and becoming caught in the vortex of a colleague's craft, and attempting to fly too slow.

In many of these cases the pilots stated that they had been in a rush and simply made mistakes. The National Pilots Association slogan "If you are in a hurry you are in danger", continues to be validated.

In some cases, preoccupation with financial worries was felt to have produced forgetfulness.

In certain instances, a certain amount of celebration the night before the day of flight, resulted in fatigue, complicated by hangover effects, certainly an adverse circumstances to safe flight. An aerial applicator pilot at work must be quick and alert, and should avoid celebrations on the eve of flights. In this respect, the occupation of these pilots differs from desk occupations. The desk worker can physically appear at work after a night of partying (although he may be very uncomfortable internally) and somehow get through the day. The ag pilot can't coast through the day—he's got to deliver—with a sharp eye and with sharp reflexes. It will pay many times over not to fly if one doesn't feel well—regardless of the reason for the indisposition.

We should note here that several instances of collisions between two taxiing aircraft, or landing aircraft, or one landing and one departing aircraft, occurred during 1963. Vigilance during these operations is of the utmost importance, especially in planes with poor visibility features.

Interestingly, aerial applicator activities encompass the spread of many kinds of chemicals and objects, including, along the Gulf Coast, irradiated screw worm larvae. This work is vital to the Gulf Coast economy. Some cases of susceptibility of certain pilots to become allergic to certain dusts generated by these insects has come to our attention. Dr Robert Dille and I of CARI, together with Dr. Harry Gibbons of the Southwest Region, Dr. Peter Siegel of the Aeromedical Certification Division, and certain Aviation Medical Examiners, have been studying certain of these cases which have taken on the clinical picture of acute asthmatic attacks (some occurring during flight necessitating a forced landing). We are seeking means of testing for individual sensitivity to screw worm fly dust and of preventing in-flight exposure to this substance.

The aerial applicators' world is diverse, complex, and challenging for all concerned.

Aircraft:

Due to the hard-hitting, rapid-paced, nature of aerial applicator work, and to the fact that certain flights must be made on very short notice, we note that at times corners are prone to be cut which can very likely result in accidents.

Not infrequently marauding insects determine the moment in time of undertaking to spray in what might be thought of as defensive warfare. This is the worst type of situation, since such short notice for preparation prevails, and so many farmers want immediate action within a short span of time.

Anticipation and a continued state of preparedness are wise characteristics of the aerial applicator operator. The accident factors attributable to the aircraft appear as follows.

Aerial applicator aircraft are, in general, designed especially for maneuverability and load carrying capabilities. The new generation agricultural aircraft, with their crashworthy structure, their considerably delethalized cockpits, their shoulder harnesses, and their slow-flight characteristics, are a tribute to the engineering accomplishments of today's manufacturers. These aircraft are paving the way for similar safety achievements in the future light aircraft not intended for aerial applicator use. The record shows that these new generation aircraft are enabling the pilots to survive (frequently with no or only minor injuries) some severe impacts which would surely be fatal in non-aerial applicator aircraft.

1. Engine or Propeller Failure:

Twenty eight cases of engine failure necessitated forced landings. This group does not include improper fuel management which we have classified under "pilot factors".

The successful accomplishment of many of the forced landings after low altitude engine failures is a tribute to the skill of the pilots in general engaged in aerial applicator work.

Operator "human factor" maintenance errors included such items as an improperly installed propeller which lost a blade in Arizona and the failure to tighten certain spark plugs resulting in a power loss in South Carolina. Other causes of power failure included several cases of water contamination of the fuel (the fuel was improperly stored by the operator and the pilots neglected to drain the sumps), some instances of protracted general neglect of the engines, an instance of maintenance inattention to a deteriorating fuel selector valve, and improper maintenance of failing magnetos.

Four cases of broken crank shafts, three cases of "swallowed" exhaust valves, and several instances of bearing failure or gear fracture (especially in helicopters), point up the demanding nature of aerial applicator activities upon the machinery.

Other failures included an instance of a stuck carburetor float and some instances of detonation and backfiring with engine stoppage. Sometimes the real reason for engine failure remains obscure.

2. Airframe Failure:

In an interesting accident in California, the main wing spar failed during flight and the wings folded over the fuselage. The impact was quite hard but the pilot survived. Many modifications had been made to the craft (within the then approved techniques) including a marked increase in horsepower and hopper capacity, and apparently under the rugged conditions of aerial applicator work, the spar reached its limits.

Two instances of horizontal stabilizer failure (both survived) occurred, the failures apparently assisted by the corrosive nature of certain chemicals.

3. Chemical Ducting Failure:

In July of 1963 near Caldwell, Idaho, a connector hose came loose shortly after take-off and sprayed a mixture containing parathion in the pilot's face. His body become saturated and he lost control of the plane which crashed. His hard hat made a deep dent in the instrument panel. He suffered no fractures and was fortunate in crashing near a water canal. He immediately jumped in, disrobed, and washed off the chemicals. His quick action resulted in no serious poison effects.

Sometimes the ducting failure occurs after an accident, and in September of 1963 near Lambert, Mississippi, a pilot was splashed with a defoliant during and after an accident. The chemical effects of the defoliant almost resulted in his death, for his blood pressure fell to 50 mm mercury and it took six hours treatment in a hospital to get him stabilized and on his way to recovery.

Outside Force:

Under this heading we place those things which were not reasonably under the pilots control, the operators control, or due to the aircraft.

In July of 1963, a sudden high velocity wind caught an aircraft on landing and caused an accident. At least two, and possibly more, whirlwinds caused accidents. A sudden downdraft caused a heavily loaded plane on take-off to have an accident. In September near Sunflower, Mississippi, a cow walked in front of an aircraft attempting to take off and caused an accident.

In previous years, children have thrown fruit in the path of aerial applicator aircraft causing accidents, and hard-shelled beetles have broken the goggles of pilots in two cases necessitating forced landings.

General Statement:

Aerial applicator activities are a safe or as dangerous from the aeromedical standpoint as those engaged in them care to determine.

It is fortunate that most of these activities take place in the spring, summer and early fall months, and are conducted in relatively calm winds. Almost never does an aerial applicator pilot get caught in instrument flight conditions.

Some aerial applicators, for example the U.S. Forest Service's fire-fighters, must undertake sudden activities under very adverse circumstances. Fortunately, their chemicals (for example, borate compounds) are relatively nontoxic. Recently, a CARI team of scientists has worked with the U.S. Forest Service pilots in Montana, in an effort to pin-point the specific effects of fatigue itself on flight proficiency. Details will be reported later, but one observation is the increased roughness on the controls and sloppiness of maneuvers which accompanies the development of marked fatigue.

For each accident which occured in 1963, there were probably more than a dozen near-accidents, which by the merest margin escaped becoming statistics. Herein lies our fertile ground for safety improvement. Prevent the recurrence of near-accidents! When one has a close call due to pilot factors—hold off flying until the factor is remedied. If rest is the problem, then get some rest.

If an aircraft is developing a mechanical deficiency hazardous to flight—stop—and rectify the deficiency. Actually, an operator should reward an alert pilot who calls a potential defect to the attention of the mechanic. The psychology of this approach will benefit all concerned in the long run.

The operator and the pilot (and if possible the ground personnel) should know the nature of the specific chemicals they use, the preventive techniques in working with the chemicals (gloves, respirator, etc.) the symptoms produced by body absorption, and the emergency treatment in case of contamination.

For example, poisoning by the organic phosphates produces a tendency to sweat more than one should, an increased flow of saliva, an upset stomach with nauseous feelings and vomiting, increased tear formation in the eyes, intestinal cramps, diarrhea, difficulty in breathing (a later stage of the poisoning), difficulty in focusing the eyes on objects and seeing clearly (especially in the distance), irritability, headaches, muscle tremors and a generally bad feeling.

One may experience any or all of the above symptoms, and death can follow. Atropine can be given for emergency, and later, symptomatic treatment, and a "new drug" (actually known and used in other countries, notably Japan, for many years), named Protopam, marketed by Campbell Pharmaceuticals, 121 East 24th Street, New York, New York, can be used as primary therapy. A doctor should administer the atropine and Protopam treatments.

Some pilots have felt that they have built up a resistance to the organic phosphates (parathion, etc.). Possibly, a slight resistance can be developed, but no real resistance can occur.

If one feels the symptoms of organic phosphate poisoning developing, one should knock off for several days and clear his system of the compond. A low blood cholinesterase is clear evidence that the body contains too much of the organic phosphates, since these inactivate the blood cholinesterase (as well as affect other parts of the body) and, thus, tell us of their presence.

Each operator should touch base with a nearby doctor (possibly an Aviation Medical Examiner) who can be aware ahead of time that he may be called upon to treat a possible poison case. This is good and responsible planning.

Among the other chemicals are the chlorinated hydrocarbons (which can cause nervous tensions, anxieties, nausea, dizziness, headache, giddiness and muscular tremors), and the dinitrophenolic compounds (cause sweating, thirst, euphoria, and later fatigue—note: if alcohol is consumed the combined effects can speed the onset of symptoms of blushing, a feeling of heat, rapid breathing and a dropping blood pressure).

In all cases of skin contamination, the first principle of treatment is removal of the chemicals with water. Plenty of soap should be used, and denatured alcohol should be swabbed on an rubbed off of the area. Keep denatured alcohol (or isopropyl alcohol) handy. Use rubber gloves, boots, and aprons when mixing and loading chemicals. Wear a respirator when dealing with the more toxic chemicals.

It appears that aerial application is here to stay. Let's strive through planning, preparation, preventive medicine and proficiency, to achieve the top safety record commensurate with our potentialities and wherewithal.

Note to Physicians and First-Aid Personnel

As a final note, we can observe that a number of the victims of aerial applicator accidents succumb to what is termed "shock". Shock is a a condition which can develop some hours after an individual has survived the initial accident. The initial accident may have involved any one of the following: the absorption by the body of toxic substances, the crushing of muscles or organs by impact forces, internal or external hemorrhage, or severe burns secondary to fire.

Following conventional approaches to accident victim therapy, certain individuals appear to respond, but, after a period of a few hours, begin to manifest the following symptoms of the shock syndrome:

1. Pallor of the Skin (this is due to collapse of the veins);

2. Tachycardia (fast heart rate is due to low pulse pressure—the margin between systolic and diastolic blood pressure—and represents an attempt to move more blood through the circulation per unit time);

3. Cold Skin (this is due to the slow filling of the capillary bed due to sympathetic nervous system activity);

4. Sweating Skin (excessive sweating results from increased activity in the sympathetic nervous system);

5. Oliguria (the low blood pressure results in less urine formation).

There is some strong indication at present that the treatment of shock with noradrenalin as is now generally accomplished, is perhaps unwise. This is because the noradrenalin has the capability of constructing the venous side of the capillary bed, as well as the arteriolar side, producing still greater losses of fluid into the tissues from the blood.

The Civil Aeromedical Research Institute and the University of Oklahoma School of Medicine, in a cooperative FAA-University undertaking, have just completed a five day "Shock Seminar" (November 16–21), where forty scientists conducted experiments at CARI on the specific therapy of shock. Each scientist was able to observe the other's technique and many points were resolved which otherwise would have remained disputed. Some of the most noted physicians, surgeons, physiologists and pharmacologists in the U.S. and Canada participated in this experimental program which is aimed at enabling physicians ultimately to provide more effective treatment to injured survivors of aerial applicator accidents (and other types of accidents).

The results of the symposium will be made universally available to physicians. The utility of using adrenergic blocking agents (opposite in effect to noradrenalin) together with blood, plasma, dextran, or other plasma expanders, will be reported. The new drug, dibenzyline, is a blocking agent specifically studied. Its promising beneficial effects are that it increases the total blood flow to the organs where the blood is needed, it shifts fluid from the pulmonary circulation (which otherwise is in danger of being overloaded with water which causes lung congestion) to the systemic circulation, it results in the reestablishment of urine formation, and it effects a sensitive blood pressure response which gives the physician a good index of the adequacy of the circulating blood volume. Noradrenalin masks this latter effect, and the physician is often hard-pressed to know whether his intravenous fluid infusions are too little, enough, or too much. CARI personnel can provide further medical information on these new advances in improving accident survival. The full report is available: *J. Okla State Medical Assoc.* 59:8, August 1966, pp. 407-485.

Stanley R. Mohler, M.D.
Charles R. Harper, M.D.

Allergic Purpura Induced by Exposure to *p*-Dichlorobenzene

Confirmation by Indirect Basophil Degranulation Test

Robert M. Nalbandian, MD, and James F. Pearce, MD

FOR MANY YEARS an extensively used moth repellent, *p*-dichlorobenzene (Di-chloricide) has only rarely been implicated in the production of adverse reactions.[1-3] We have recorded here the first case to our knowledge of allergic (anaphylactoid) purpura induced by exposure to *p*-dichlorobenzene.

Since acute glomerulonephritis developed in this patient as a complication in the course of the allergic purpura (a relationship previously reported[4]), reexposure to the probable offending agent *p*-dichlorobenzene, to establish etiology, was considered but rejected.

The etiologic relationship between *p*-dichlorobenzene and the provoked allergic purpura was later unequivocally established by positive results of a basophil degranulation test.[5] Additional injury to the patient in the search for the etiologic factor was thus avoided. A brief report of the significant features of the patient's hospitalization follows.

Report of a Case

Present Illness.—This 69-year-old white retired man was in apparent good health until three weeks prior to admission, when he noted stiffness in his neck and "tightness" in his chest, with "gas pains" in the abdomen. There was an asymptomatic interval until two days prior to admission, when he had a gradual onset of petechiae and purpura in a symmetrical pattern over the upper and lower extremities, associated with swelling and discomfort.

Subsequent history indicated that the patient had been in a chair which earlier in the day had been treated with *p*-dichlorobenzene crystals by his wife. While seated in the chair, the patient had a bout of dyspnea followed in 24 to 48 hours by the previously described skin eruptions.

There were no other significant features in the history except that he had taken tablets of aspirin, phenacetin, and caffeine (A.P.C.) frequently for years without adverse reactions.

Physical Examination.—The patient was an elderly white man in minimal distress, alert, cooperative, and oriented. His blood pressure was 150/90 mm Hg; pulse rate, 120 beats per minute; temperature, 98 F (36.7 C); and respirations, 16 per minute.

Skin.—There were numerous petechial and purpuric lesions over the skin of hands and forearms (Figure), feet and legs, in an essentially symmetrical pattern. Also, there was swelling of the hands and feet.

Laboratory Data.—SECOND HOSPITAL DAY.—Examination of the blood disclosed the following values: hemoglobin, 15 gm/100 cc; hematocrit, 48%; red blood cell count, 5,100,000/cu mm; white blood cell count, 13,000/cu mm, with 35% band cells, 52% neutrophils, 4% monocytes, and 9% lymphocytes; reticulocyte count, 0.7%; sedimentation rate, 26 mm/hr; color index, 1; mean corpuscular volume, 94/cu μ; mean corpuscular hemoglobin, 30.2/$\mu\mu$g; mean corpuscular hemoglobin concentration, 32.1%; bleeding time, 4 minutes; clotting time, 24 minutes; and platelets, 418,000/cu mm.

THIRD HOSPITAL DAY.—A stool examination for occult blood gave positive results. Urinalysis showed a pH of 5.5; specific gravity of 1.028; reaction for albumin, negative; 0 to 1 white blood cells (WBC) per high-power field; and 6 to 8 red blood cells (RBC) per high-power field. The glucose value was 124 mg/100 cc; the blood-urea-nitrogen (BUN) value was 28 mg/100 cc; and findings from the venereal disease research laboratory test for syphilis were negative.

FOURTH HOSPITAL DAY.—The BUN value rose to 57 mg/100 cc.

SEVENTEENTH HOSPITAL DAY.—An Addis count gave the following values: RBCs, 536 million/12-hr specimen; WBCs, 198 million/12-hr specimen; casts, 720,000/12-hour specimen. The total volume was 720 cc/12 hr; specific gravity, 1.011; and quantitative albumin, 225 mg/12 hr.

EIGHTEENTH HOSPITAL DAY.—BUN fell to 15 mg/100 cc.

FIVE MONTHS AFTER ADMISSION.—The findings from an indirect basophil degranulation test were strongly positive (62% of basophils showing degenerative changes when tested with *p*-dichlorobenzene). The control (*p*-dichlorobenzene with normal serum) was 6% positive. (This test "requires one drop each of the patient's serum, the test drug solution, and rabbit leukocytes [buffy coat]. A positive reaction is a progressive degranulation observed in the basophil cells."[5])

EIGHT MONTHS AFTER ADMISSION.—An indirect basophil degranulation test was done on the patient's serum to rule out the possibility of sensitivity to aspirin, phenacetin, and caffeine. The antigens tested were acetylsalicylic acid, phenacetin, and caffeine. The results in percentage of basophil degranulation were 14%, 10%, and 8%, respectively. The control reaction was 6%.

Thus, there was *no* demonstrable sensitivity to the drug components of aspirin, phenacetin, and caffeine in this patient, when tested by the indirect basophil degranulation test technique. The absence of sensitivity to salicylates was confirmed also by an immunologic method.[6]

Treatment.—Treatment was supportive. Prednisolone, 10 mg four times a day, was given from the 7th to the 31st hospital day, at which time the patient had improved and was discharged.

Summary

Allergic (anaphylactoid) purpura was induced by *p*-dichlorobenzene, an apparently innocuous and widely used moth repellent. Acute glomerulonephritis presented as a complication of the allergic purpura.

The offending agent was *p*-dichlorobenzene, proved by the indirect basophil degranulation test. Serum antibodies for *p*-dichlorobenzene were still demonstrable in this patient five months after the initial exposure.

The basophil degranulation test has obvious advantages over clinical methods in establishing a suspected drug or chemical as the offending agent in a clinical reaction. It undoubtedly deserves wider application, which will result in a higher order of accurate identification between offending drugs and chemicals and alleged clinical manifestations.

Particularly attractive features of the indirect basophil degranulation test are the accuracy of identification of an offending chemical or drug achieved as a laboratory procedure, without subjecting a patient to the hazards and jeopardy of a cautious clinical trial or reexposure to the offending agent.

References

1. Hollingsworth, R.L., et al: Toxicity of Paradichlorobenzene *Arch Indust Health* **14:**138-147 (Aug) 1956.

2. Weller, R.W., and Crellin, A.J.: Pulmonary Granulomatosis Following Extensive Use of Paradichlorobenzene, *Arch Intern Med* **91:**408-413 (March) 1953.

3. Perrin, M.: Nocivité possible du paradichlorobènzene employé comme anti-mites, *Bull Acad Med* **125:**302-304 (Nov 25) 1941.

4. Levitt, L.M., and Burbank, B.: Glomerulonephritis as a Complication of the Schönlein-Henoch Syndrome, *New Eng J Med* **248:**530-536 (March 26) 1953.

5. Shelley, W.B.: Indirect Basophil Degranulation Test for Allergy to Penicillin and Other Drugs, *JAMA* **184:**171-178 (April 20) 1963.

6. Weiner, L.M.; Rosenblatt, M.; and Howes, H.A.: The Detection of Humoral Antibodies Directed Against Salicylates in Hypertensive States, *J Immun* **90:**788-792 (May) 1963.

Feasibility of Prophylaxis by Oral Pralidoxime

Cholinesterase Inactivation by Organophosphorus Pesticides

Griffith E. Quinby, MD, MPH

IN March 1964, one of medicine's few specific antidotes became available for prescription in organic phosphorus pesticide poisoning.

There have been numerous reports on the prophylactic value in experimental animals of salts of pralidoxime (2-pyridine aldoxime methyl chloride [2-PAM chloride], praloxidime iodide [2-PAM iodide], or pralidoxime mesylate [P_2S]) with[1] and without atropine by all routes of administration.[2-4] The prophylactic use of the drug has been limited in the manufacturer's literature to "not longer than one working day and will not be repeated within less than a week." By present label directions, patients are required to remain within reach of the prescribing or another physician, experienced in the treatment of intoxication due to phosphate ester anticholinesterases, for 48 hours after exposure. These recommendations and consideration of more prolonged prophylactic use have been subject to controversy.

Pralidoxime chloride has been shown to be relatively nontoxic when administered in doses of from 1 to 4-gm orally daily for two to eight months in 29 women and eight men.[5]

Pralidoxime mesylate administered once to 78 normal human subjects in 1-gm doses has produced no toxic effects.[6] A report completed since our study states that 2-PAM chloride, administered orally to dogs at much higher dosages (75 to 100 mg/kg/day), produced epithelial defects in gastric mucosa and symptoms of systemic poisoning.[7]

Prophylaxis by drugs of other industrial diseases, usually of chronic nature, has been criticized[8] because of the possibility that management or labor may rely upon the drug instead of maintaining or improving standards of plant hygiene. However, the present study was conducted to determine the value of a cholinesterase reactivator administered orally as a supplement to the very best environmental safety measures that were practical under existing operating

---

Reprint requests to PO Box 1313, Wenatchee, Wash 98801

conditions in four plants formulating organic phosphorus and other pesticides. Dispensing of the drug by management and periodic history-taking afforded additional opportunities to remind workers about minimizing exposures and detecting or correcting intermittent carelessness.

The seasonal pattern of cholinesterase inactivation in formulators of pesticides was noted in 1951.[9] Red cell cholinesterase levels of formulators were gradually reduced during the summer until the average level for the occupational group reached as low as 15% of "normal," using the Michel method. At such low levels (Fig 1), there were occasional cases of overt poisoning. In more recent years periodic testing of cholinesterase levels has resulted in requiring workers to discontinue exposure for most of the remaining season when their cholinesterase activity levels declined below an arbitrary value, usually 50% in Washington state. If such a worker was not a permanent employee, this usually resulted in his discharge without compensation with financial loss to the employee, the pesticide industry, and the agricultural economy.

Materials and Methods

This study was conducted mainly during the 1964 pesticide formulating season. A few observations from the 1965 season were fortuitously added. The experiment was designed as a double-blind test of the efficacy of oral pralidoxime chloride as a periodic prophylactic reactivator of cholinesterase in four formulating plants in central Washington state at Wenatchee and Yakima.

The 24 volunteers were white men aged 18 to 65. They were classified as to predominant job activity in decreasing order of estimated intensity of exposure (Table 1). The blinder divided volunteers into treated and placebo groups. He assigned individuals from each exposure classification alternately to achieve maximal comparability of exposure. However, during the study, some men shifted from one job to another so it was necessary to reclassify them to reflect their heaviest occupational exposure. This caused the unequal distribution of workers in exposure groups. Only the final classification is shown. Variable exposures occurred almost daily in the work week of five to seven days. Of the volunteers, 13 received 2-PAM chloride and 11 received placebo.

The four plants of the three formulating companies produced a wide variety of pesticid al mixtures in liquid and dust forms. Becaus the agricultural needs varied in the area served, types and conditions of exposure wer hardly uniform. All of the plants, howeve handled considerable volumes of organic phos phorus insecticides, mainly parathion, tetra ethylpyrophosphate (TEPP), malathion, an azinphosmethyl (Guthion). Lesser quantities o demeton, ethion, mevinphos (Phosdrin), an dimethioate were processed. The market de mand for organic phosphorus pesticides in cen tral Washington results in fairly frequent for mulation of organic phosphorus insecticides o longer half-life (parathion, malathion, and az inphosmethyl) during May and June followe by heavier demand for insecticides with rela tively short half-lives (TEPP and mevinphos in July and August. Carbaryl, a carbamat rather than an organic phosphorus compound was the only other cholinesterase inactivato formulated sporadically at the plants. Expo sures to carbaryl were sporadic and considere inconsequential.

Pralidoxime chloride was administered orally in 500 mg capsules by management personnel placebos appeared identical. The dosage rat was 1 gm at the end of every other exposur day (Monday, Wednesday, Friday) of the usu al five day work week. Administration of th drug started in mid-April and continue through the first week of September for fiv months or 22 weeks at most. In one plant th men usually worked Saturdays and took a dos at the end of Satuday also. Thus 12 men re ceived 3 gm per week, but one man received gm a week.

The ineffectiveness of 2-PAM chloride as a cholinesterase reactivator, when administere too long after inactivation, has been recognize in animals and man.[10,11] This is believed du to chemical transformation or "aging" of th phosphorylated enzyme. For this reason, th design of this study provided that workers wer prophylactically treated so that cholinesteras inactivation gradually occurring during th work week might be subjected to reactivatio every 48 hours at the longest. It was estimate that most phosporylation occurred less than 3 hours earlier and that over half occurred withi eight hours. It was planned to compare th anticipated seasonal decline of cholinesteras activities of men with comparable exposures or the reactivator with those on placebo.

Upon entrance into and departure from th study, a medical history was taken, and a phys ical examination and routine laboratory work were performed. Chest x-ray films were not taken. Blood cholinesterase tests (modified Mi-

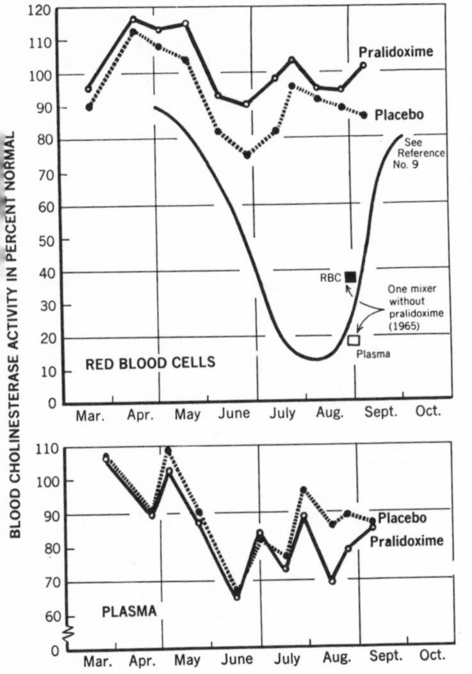

Fig 1.—Influence of 2-PAM chloride on blood cholinesterase levels in pesticide formulators (mixers only).

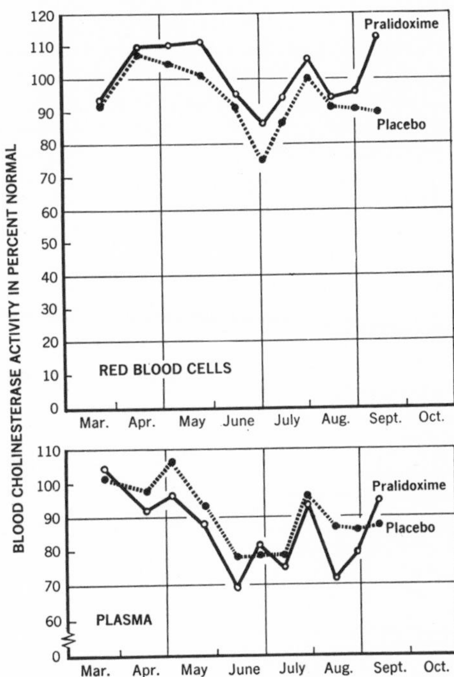

Fig 2.—Influence of 2-PAM chloride on blood cholinesterase levels in pesticide formulators (all types).

chel method) were performed before significant exposure except for three late entrants whose "normal" levels were determined as postexposure values four months after they ceased to handle organic phosphorus compounds.

Medical histories, exposure histories, and blood cholinesterase tests were repeated every two to three weeks until formulation and exposure practically ceased the first week in September. Re-examinations were carried out whenever intercurrent complaints and illnesses occurred. All illnesses occurring during the study, including those treated by the patients' attending physicians, were investigated by the author.

The formulating workers in this industry are seasonally employed for the most part. As a result, they were somewhat unstable with a variety of physical defects and chronic illnesses (Tables 1 and 2). One young man was recovering from rubella contracted while he had a hernia repaired. Another had had many partially successful plastic repairs for hypospadias. A third worker's recurrent duodenal ulcer was not being helped by a successful paternity suit. Only one pre-dose physical finding altered the protocol of the study. One patient had a repeated trace to 1+ albuminuria and a history of intractable chronic urethritis with a penile discharge since military service. His physician was unable to offer a satisfactory explanation for the two-tube positive albuminuria but did not perform renal function tests. Because the pralidoximes are excreted primarily through the kidneys and reduced renal function was once considered a contraindication to the nonemergency administration of the antidote, this man's dosage status was not blinded to be sure he did not receive more than the three doses he had already had by the time his clinical studies had been completed. Since he was found to be on the active agent, the record of a man on the placebo with comparable exposure was also revealed and the two volunteers exchanged in the protocol. They continued on the test for the full 22 weeks without anyone being informed of the change except the blinder and the investigator. Analysis and interpretation of the data comparing the treated and placebo groups were first completed without these two volunteers and their data were added after the dosage status of the other volunteers was revealed.

Table 1.—*Group Placebo*

Case No.	Age	History, Physical Findings, and Impressions	Weeks on Dosage	Intercurrent or Postdosage Clinical Findings	% Change From Pre-treatment WBC	% Change From Pre-treatment Systolic BP
Mixers						
12	54	Nephrolithiasis, 1936; parathion poisoning, 1952	20	Diarrhea, cramps, weight loss	+23	+ 6
23	52	Ricketic chest; ruptured disc, 1952; rhonchi right chest; unable to understand toxicity hazards	19	O.P. Poisoning?; discharged carelessness; finer rales right chest	−20	−18
24	18	Son of pesticide chemist; dermatophytosis; infectious mononucleosis, 1963; early deafness	7	Allergic (?); giddiness after first pill; insubordination on job	−12	0
18	25	Chronic urethritis 1959; albuminuria, repeatedly	22	Gastroenteritis, 1 tarry stool; contact dermatitis legs and arms	*	− 1
Packagers						
4	23	Duodenal ulcer; paternity suit	14	Blamed headaches on parathion	+44	+16
6	19	No doctor in 10 years	17	Fell out of a jeep; bruised	− 6	0
20	32	Regional ileitis 1961; adhesions?, 1963	22	Bloody nasal discharge after ziram, also itchy skin	−38	−15
Warehouseman						
9	30	Hypochondriac, "allergy, hematemesis, sinus"	8	Bizarre multiple hypochondriac complaints	+ 4	+14
Foreman						
2	43	Neurodermatitis; nervous gastritis; TB scapula	21	Recurrent gastric distress with tensions; miosis from TEPP; headache; tightness in chest	+26	− 4
Managers						
15	56	Hayfever; submucous resection, 1930; diphenylamine poisoning, 1963	22	Hayfever; gastroenteritis	+34	+17
10	47	Borderline hypertension 1962; "nervous" most of life; June hayfever; nasal polyps; slight exophthalmos; fundic nicking	21†	Blood pressure fell on repeated exams to 140/90; erysipelas of foot; took pills irregularly May and June	+44	− 3
		Total man-week dosage	193	Mean change	+10	+ 1

*Specimen lost.
†Irregular administration; missed five weeks in May and June.

Results

The dosage of 1 gm of 2-PAM chloride three to four times a week for up to 22 weeks was tolerated without any signs or symptoms referable to the antidote. The total man-weeks of dosage for the treated group was 221 and for the placebo was 193. Only one worker with a prior history of frequent "heartburn" believed that his gastric discomfort was more frequent after the capsules than beforehand. When he was encouraged to take the capsules with milk or to cease taking them long enough to test his judgment, he found no association in time sequence with dose and took the capsules for the full 22 weeks of the test. The patient was found after unblinding to have been on placebo.

One 21-year-old employee was influenced by his parents after only three weeks of receiving the test dose to stop taking the capsules when his private physician re-prescribed three other drugs that he had been taking in earlier months for nervousness, pylorospasm, and perennial hayfever. However, he continued in the study for blood testing and other observations realizing he may have been on the placebo. His cholinesterase and other laboratory results have been included in Fig 2 and Table 2 but not in Tables 3 and 4 because he was treated only three weeks.

Table 2.—*Pralidoxime Chloride*

Case No.	Age	History, Physical Findings, and Impressions	Weeks on Dosage	Intercurrent or Postdosage Clinical Findings	% Change From Pre-treatment WBC	% Change From Pre-treatment Systolic BP
Mixers						
3	34	Frequent "indigestion"	21	Tightness in chest from TEPP; rhinitis; nasal septal ulcer	−13	+ 4
13	25	Rubella; infectious hepatitis 1954; recent herniorrhaphy	21	No complaints	+19	− 8
14	21	Hypospadias with many repairs; anxiety neurosis; tensions	3	"Sinus" attack; started three other drugs; stopped participation after 3 weeks	−34	+10
19	21	Perforated eardrum; dermagraphia	18	Gastroenteritis	+10	+ 5
22	20	Migraine; school drop-out; ruptured testis; phosdrin poisoning 2 years ago; hypochondriac	19	Extreme fatigue; furuncles; migraine; insomnia; upper respiratory infection	−29	− 7
25	18	Son of plant owner; emotional antipathy to blood	9	Ultraviolet burns from welding	+27	0
17	45	No doctor in many years	22	Headaches; contact dermatitis wrists; fever unknown origin	−59	− 4
Packagers						
5	36	Penile ulcer; spontaneous pneumothorax 1948; old shrapnel wound in leg	8	Penile lesion and sinus still present; scooter accident with lacerations after release	+125	+14
Warehouseman						
11	36	Marital problems; rheumatic heart disease age 19	21	Gastroenteritis; upper respiratory infection	+30	− 3
Foreman						
16	65	Raynaud's disease; anemia; varicosities, left leg; allergy to weeds; frequent epistaxis; stress headaches; diphenylamine poisoning 1963	22	Headaches continue; gastroenteritis; leg cramps	+80	−10
Manager						
1	44	Long history severe extra systoles; hypertension 1962; nervous stomach; hayfever; asthma	20	No complaints	+78	+ 9
Trucker						
8	44	Allergic rhinitis; perennial stutterer	15	Rubella-like infection without adenopathy quit to study for the ministry	−39	−13
Mechanic						
21	51	Not remarkable	22	No complaints	− 5	− 7
		Total man-week dosage	221	Mean change	+15%	− 1%

The means of the postdose white blood cell counts (WBC) of the two groups (Table 1 and 2) increased 15% over the pre-dose WBC for pralidoxime group 2 and 10% for placebo group 1, indicating that there was a difference of only 5% in WBC between the two groups. Since determinations were done in two commercial laboratories, this small difference cannot be attributed to pralidoxime but probably represents random variation. These results are in accord with those in an earlier, more detailed study.[5]

The brief pressor effect of large intravenous doses of pralidoximes has been reported for experimental animals and man,[12,13] but was not confirmed by this study, nor another recently published,[5] as any long-term action of moderate doses of pralidoxime. Tables 1 and 2 reveal that the mean change of systolic blood pressure after 3 to 22 weeks of antidote was an insignificant 1%

Table 3.—*Minimum Blood Cholinesterase Values*

% of Pre-exposure Control	% of Cases	
	Placebo (8 cases)	Pralidoxime (10 cases)
RBC		
<100	100	100
< 90	87	90
< 80	63	30
< 70	63	10
< 60	25	0
< 50	0	0
Plasma		
<100	100	100
< 90	100	100
< 80	87	100
< 70	63	80
< 60	13	30
< 50	13	10

Table 4.—*Month of Occurence of Minimum Blood Cholinesterase Values*

% of Pre-exposure Control	No. of Cases							
	Placebo				Pralidoxime			
	March	June	July	Aug	May	June	July	Aug
RBC								
91-100	1					2		
81-90		2				2	1	1
71-80						2	1*	
61-70		2	1					1
51-60		1		1				
41-50								
Plasma								
91-100								
81-90		1						
71-80		1		1	1			
61-70		1	1				2	2
51-60		2				3		1
41-50		1				1		

* The same minimum also occurred in August.

deviation in each direction in both groups.

Seasonal trends of cholinesterase activities of pralidoxime and placebo groups are shown in Fig 2 for all types of formulation workers. Seasonal inactivation of red blood cell (RBC) cholinesterase level to 76% of normal during 1964, occurring in placebo group 1, as a whole does not compare closely with earlier studies (Fig 1) in which the reduction was to 15% of normal.[9] One factor considered responsible was inclusion of managers, foremen, and truckers in this study. They sometimes helped with mixing procedures, packaging, and loading, but their exposures were usually much less than that of mixers.

Since mixers had much greater exposure than the other workers, their results were considered separately. Cholinesterase levels of only the seven mixers taking the drug (group 2) and the four mixers on placebo (group 1) are graphed in Fig 1. In both figures, levels for both RBC and plasma declined during mid-June and mid-August, corresponding with the greatest seasonal formulation of parathion (June) and TEPP (August). In comparing the two groups of formulators including all types of exposure (Fig 2), the widest difference was only 10% at the end of June. However, in comparing only mixers (Fig 1), the maximal difference was 20% in mid-July. Possibly more germane to the study was the finding that the level of the RBC cholinesterase of the treated men remained consistently above that of the placebo group 1. As expected for men chronically exposed to TEPP and other organic phosphorus compounds, the plasma cholinesterase levels in percent of normal generally remained lower than the level for RBC until exposures were about over. Plasma levels for both groups showed no significant differences.

The foregoing considerations based upon the means in the group of all formulators and also the group of mixers only partly obscures the time and extent of the lowest inactivations at which cholinesterase levels were measured in each man. This minimum cholinesterase value for each individual worker is probably the best indication of whether clinical poisoning is likely to occur. Therefore the minimum cholinesterase level for each worker was tabulated by month for the eight men on pralidoxime and the ten men on placebo who remained in the experiment throughout its course. The other six men were not included because they were in the study for substantially shorter periods.

Over half the subjects on placebo (Table 3) had drops in RBC cholinesterase levels to the 60% to 70% range. One fourth of the subjects, fell to the 50% to 60% range whereas only one of ten under treatment with pralidoxime had an RBC drop to below 70% of pre-exposure level. None of the subjects on pralidoxime fell below 60%. These differences occurred when exposure was believed to be greatest. Such data are consistent with a prophylactic effect by the drug.

The month of occurrence of lowest RBC cholinesterase level is shown in Table 4. Again, the lowest RBC cholinesterase level occurred during the months of heaviest exposure. The minimal RBC cholinesterase levels in the treated group were higher, suggesting a partial prophylactic effect.

The results on the lowest plasma cholinesterase level in Table 3 show no suggestion of differences in the two groups.

One man in Yakima had rubella when examined before entrance into the study. An extensive outbreak of that disease occurred in Washington state during the spring and summer of 1964. A second man in Wenatchee developed fever and a rubella-like rash after three weeks on the capsules, but he lacked the characteristic postauricular adenopathy. Because of the remote possibility that this might have been a drug-induced rash, this man was taken off exposure until his rash cleared and he was judged well clinically. For this period his blood cholinesterase values were omitted. Under constant observation, he was returned to work and again began receiving a 0.5 gm dose on the 16th and 18th days after onset of his three-day rash. The rash did not recur and the 1 gm dose was continued without untoward effect through the 17th week when the man resigned to enter the ministry. Had this study not been double-blind, most investigators, including the author, might not have resumed administration of the pralidoxime. This man was found much later to have been taking pralidoxime.

Comment

After recognized heavy overexposure of any worker, immediate oral atropine alone has been widely recommended as a preventive or ameliorant of symptoms while medical help was sought. Such a recommendation has changed over the years from rather general advocation to condemnation. This is because atropine is a nonspecific antidote. In many incipient cases of poisoning, the early signs have been masked without correcting the cause of poisoning. Consequently, some serious complications and even deaths occurred because workers and physicians relied upon the antidote's protection within the critical 24-hour period after exposure and the worker no longer received postexposure medical surveillance. Moreover, prophylactic atropine became so popular that its administration sometimes escaped the physician's control and was taken over by management or even the worker. Pesticide workers and even some physicians mistakenly thought that atropine was specific in organic phosphorus poisoning rather than a nonspecific partial pharmacological antagonist. This bias should not be carried over to pralidoxime which is a specific antidote.

Nothing occurred in this study that was inconsistent with the premise that pralidoxime can be safely used as a prophylactic. It was unfortunate that 1964 was the least opportune year for a clinical trial to establish firmly or refute the prophylactic value of oral pralidoxime in Washington. Washington had the coldest spring and early summer in 1964 in many years. The pest problem was very small on tree fruits and row crops. Correspondingly, less organic phosphorus compounds were formulated in 1964 than in many years. Particularly absent were the long weekends and overtime hours of extra heavy exposure entailed in other years when pesticides were usually required over longer periods and in greater frequency and volume. Moreover, the use of parathion and TEPP was considerably supplanted by other less toxic pesticides. The Wenatchee plant in 1964 formulated only 22% of the parathion and 55% of the TEPP of the mean annual production of these compounds for the preceding 14 and 16 years respectively. For the four plants, the parathion dust volume formulated was onethird and the TEPP dust formulation less than onehalf of prior years. For 1965, the year after the prophylactic trial, the TEPP formulation was estimated at four times that for 1964.

No data are yet available for man which would enable determination of the critical percentage of reactivation of phosphorylated cholinesterase required to prevent clinical poisoning in occupational workers with gradual inactivations from repeated exposures. There are concrete data[11] to indicate that administration of pralidoximes sufficient to raise the RBC cholinesterase level 25%, and often less, results in radical improvement or complete disappearance of symptoms in acutely or subacutely poisoned individuals. Therefore it seems plausible that the maintenance by oral prophylaxis of RBC cholinesterase levels of a group of workers at a level 10% to 20% above that of untreated workers may prevent occurrence of even mild poisoning or ameliorate the severity of cases that may occur despite oral prophylaxis. It is recommended that dose be administered at the end of each day's exposure to minimize "aging" of the phosphorylated enzymes.

The conduct of this study made both management and labor more conscious of safety precautions indicated in formulating and handling organic phosphorus compounds. With each employee reporting to take his capsule three times a week, one or more managerial staff members could discuss the need for proper care and protective devices, and for bathing after each day's exposure. At the time of blood testing (usually every two weeks) and taking clinical and exposure histories, the investigator reminded the workers of the continuing need for following the best safety practices practicable. Sometimes this advice was specific for compounds other than organic phosphorus compounds. The clinical toxicologist who conducted this study had other frequent opportunities to emphasize safety precautions to both management and labor. A great deal of conversation was stirred up between the workmen at lunch, at home with relatives, with visiting salesmen, customers, and other business men who happened to observe the drug-testing program. This educational process spread to the volunteers' physicians who called to learn more about the drug under test and its possible impact. Irrespective of the outcome of this test, there was a definite educational value which could not be gauged by the available staff. Under the conditions prevailing in 1964, the results were all beneficial. Prophylaxis was not continued into 1965 either as a clinical investigation or as a company-sponsored new industrial hygiene practice. Formulation was more extensive and exposures were heavier. No blood tests were done until the formulating season was practically over. By then the blood cholinesterase values of the formulating crew had dropped to lower levels than in 1964. The RBC cholinesterase level of one mixer had declined to below 40% of normal as shown in Fig 1 and his plasma level was even lower.

In connection with the use of this drug as a regular periodic prophylactic antidote, no one had yet measured the relative protective effect under conditions of prescription by a physician who may not be able to devote as much time to instructions in good hygienic practices that should be given to plant workers or to field workers.

All the volunteers knew that the study was double-blind and that they could not count on any prophylactic benefit should they be on the placebo. It is possible that the safety practices of the men might have been more lax had they reasonable assurance of protection from the capsules. The duration of action of oral pralidoxime is so short that prophylactic administration is apt to be required frequently to be of greatest benefit, perhaps daily or twice a day. Yet if the dispensing of the drug is kept in managerial hands, and prescription in the plant physician's hands, thus providing reinforced opportunity to advocate protective measures, it is difficult to understand how any harm can come from its intelligent use as a prophylactic. Both management and labor must realize that no tolerable oral dose of pralidoxime can be expected to protect infallibly against massive contamination at one time or prolonged heavy over-exposures. In view of the timing, the mode of action of this drug, and clinical experience with early phosphate ester poisoning, very few clinical breakthroughs of poisoning should be expected from exposures to organic phosphorus pecticides of the types known to be amenable to reactivation therapy. Major merits of ral prophylaxis should rest in maintenance of seasonal levels of cholinesterase activity that would be as high as practical under existing conditions by reactivating periodically the enzymes that have been phosphorylated by the small frequent practically unavoidable exposures. By accompanying the prophylactic antidote with continued education in minimization of exposures, industrial poisoning in formulating plants should decrease in frequency and severity.

If the feasibility of this type of prophylaxis accompanied by constant education in safety can be established under more prolonged conditions of exposure in other plants and in other years, then consideration should be given to similar tests of its prophylactic use in commercial pesticides applicators' firms that employ loaders, "swampers," or others with the highest attack rates of poisoning.

Supported by the Campbell Pharmaceuticals, Inc., and Ayerst Laboratories Blood cholinesterase determinations and double-blinding were performed by the Toxicology Section of the Communicable Disease Center, Public Health Service, Wenatchee, Wash.

Generic and Trade Names of Drug

Pralidoxime chloride—*Protopam*.

References

1. Crook, J.W., et al: Adjunctive Value of Oral Prophylaxis with the Oximes of 2-PAM Lactate and 2-PAM Methanosulphate to Therapeutic Administration of Atropine to Dogs Poisoned by Inhaled Sarin Vapor, *J Pharmacol Exp Therap* **136:**397-399 (June) 1962.

2. Durham, W.F.; and Hayes, W.J., Jr.: Organic Phosphorus Poisoning and Its Therapy, *Arch Environ Health* **15:**21-47 (July) 1962.

3. Ellin, R.I., and Wills, J.H.: Oximes Antagonistic to Inhibitors of Cholinesterase, *J Pharm Sci* **53:**(pt 1) 995-1007 (Sept); (pt 2) 1143-1150 (Oct) 1964.

4. Holmes, J.H.: Unpublished Reports from the University of Colorado Medical Center 1962 Cited in Protopam Chloride (Brochure), Campbell Pharmaceuticals, Inc. (Jan) 1964.

5. Calesnick, B.; Christensen, J.A.; and Richter, M.: Human Toxicity of Various Oximes, *Arch Environ Health* **15:**599-608 (Nov) 1967.

6. Taylor, W.J.R., et al: Effects of a Combination of Atropine, Metaraminol, and Pyridine Aldoxime Methanosulfonate on Normal Human Subjects, *Canad Med Assoc J* **93:**957-961 (Oct) 1965.

7. Albanus, L.; Jarplid, B.; and Sundwall, A.: The Toxicity of Some Cholinesterase Reactivating Oximes, *Brit J Exp Path* **45:**120-127, 1964.

8. Kehoe, R.A.: Misuses of Edathamil Calcium-Disodium for Prophylaxis of Lead Poisoning, *JAMA* **157:**341-342 (Jan 22) 1955.

9. Summerford, W.T.; et al: Cholinesterase Response and Symptomatology from Exposure to Organic Phosphorus Insecticides, *AMA Arch Industr Hyg* **7:**383-398 (May) 1953.

10. Hobbiger, F.: Protection Against the Lethal Effects of Organophosphates by Pyridine-2-aldoxime Methiodide, *Brit J Pharmacol* **12:**438-446 (Dec) 1957.

11. Quinby, G.E.: Further Therapeutic Experience with Pralidoximes in Organic Phosphorus Poisoning, *JAMA* **187:**202-206 (Jan 18) 1964.

12. Kewitz, H., et al: A Specific Antidote Against Lethal Alkyl Phosphate Intoxication: II Antidotal Properties, *Arch Biochem Biophys* **64:**456-465 (Oct) 1956.

13. Calesnick, B., et al: Hypertensive Response Produced by the Intravenous Administration of 1-Methylpyridinium-2-Aldoxime Chloride (2-PAM Cl), abstracted, *J New Drugs* **4:**221-222 (July-Aug) 1964.

Pesticide Concentrations in the Liver, Brain and Adipose Tissue of Terminal Hospital Patients*

J. L. RADOMSKI, W. B. DEICHMANN and E. E. CLIZER,
with the assistance of A. REY

INTRODUCTION

Extensive animal experimentation has shown that chronic exposure to the chlorinated hydrocarbon pesticides produces liver damage. In addition, acute and chronic exposure to these substances, particularly aldrin and dieldrin, have been shown to affect the central nervous system. If humans are developing liver or neurological disease due to excessive exposure to pesticides, then it would seem reasonable to assume that those so affected would bear higher concentrations of pesticides than normal. We therefore set out to compare the pesticide concentrations present in the tissues taken at autopsy following deaths from liver and brain disease at Jackson Memorial Hospital with the samples obtained previously from the general population (Fiserova-Bergerova, Radomski, Davies & Davis, 1967).

While we have found that certain disease conditions appear to be associated with elevated pesticide levels, we were more impressed by the great individual variability of pesticide levels found in this experiment, regardless of the disease category. This has confirmed the impression of marked variability which we observed before in the general population. Now, it is widely held that the diet or the food ingested is the major source of *p,p'*-DDT and presumably other chlorinated hydrocarbons found in the tissues of the general population. Since the presence of pesticides in the food supply of this nation is ubiquitous, it would seem that if food were the primary source of the pesticide body burden, there would be relatively little variability. The existence, for no apparent reason, of widely divergent concentrations of pesticides in individuals, therefore, has caused us to question this premise.

While pesticide concentrations in food and water and the effects of environmental exposure to pesticides have been widely investigated, little attention has been paid to the possible hazardous effects of the home usage of pesticides (other than the danger of accidental poisoning) and to the contribution such usage might make to the pesticide body burden. In order to determine whether this unknown factor could be responsible for the very high levels of pesticides observed in certain individuals, a field investigation of the usage habits of as many cases as possible was conducted. These usage habits were then compared with the pesticide body burden.

EXPERIMENTAL

Selection of Cases. From January 1964 to June 1967, inclusive, all autopsies at Jackson Memorial Hospital with a history of liver disease or brain or neurological disease were selected for this study. In addition, a group selected at random with normal livers and brains was collected. Samples (1–2 g) of liver, adipose tissue and brain were collected for chemical analysis. In addition, sections of all organs showing gross changes were subjected to histopathological examination. Each case was evaluated by consideration of the cause of death and the gross and histopathological tissue changes detected. All cases were classified according to the type of liver or neurological disease present without knowledge of the pesticide levels. In a few instances, cases were listed in more than one category because of the presence of significant multiple pathological changes.

Chemical analysis

Tissue analyses for pesticides were performed by modifications of the method of Radomski & Fiserova-Bergerova (1965).

Extraction procedures. Interior portions of tissue specimens were used to eliminate the possible contamination of the outer surface. These were weighed on aluminium foil to an accuracy of ± 5 mg and placed in a glass tissue-grinder (Duall Size C). Petroleum ether (Burdick and Jackson Laboratories, Muskegon, Michigan) was added, and the samples were homogenized. The homogenate was transferred to a calibrated, glass-stoppered, centrifuge tube. The homogenizer was rinsed, with grinding, with small amounts of petroleum ether which were used to adjust the final volume.

Adipose tissue. Adipose tissue (200 mg) was ground and extracted with 10 ml petroleum ether, and 2 μl of extract was usually injected into the gas chromatograph. As the concentration of *p,p'*-DDE was generally much higher than the concentration of other pesticides, it was usually necessary to make a 1:10 dilution to determine this metabolite.

Liver. A portion of tissue (250 mg) was ground with 3·0 ml petroleum ether. During the grinding, approximately 2 g sodium sulphate was added. The petroleum ether extract was transferred quantitatively to a graduated centrifuge tube using small quantities of petroleum ether. The final volume was adjusted to 5 ml, and 2 μl of extract was usually injected (occasionally 5 μl, when no excessive interferences were present).

Brain. Brain tissue (250 mg) was dried in a desiccator over silica gel for a minimum of 2 days before grinding with a total of 5 ml petroleum ether. The samples were then centrifuged, and 2 μl of extract was injected.

Column preparation. QF-1 (fluorosilicone fluid–Dow Corning FS-1265, 1·25 g) or Dow 200 (Dow Corning silicone fluid 200-12500 CSP, 2·5 g) was dissolved in 150 ml ethyl acetate and added to 25 g Chromosorb W, DMCS-treated, acid-washed, high performance, 80-100 mesh (Johns-Manville) in a fluted round-bottom 1-l flask. After the suspension had been shaken for 20 min, the ethyl acetate was evaporated in a flash evaporator. The columns were packed in 0·25 in. glass tubing, 3 ft long for Dow 200 and 6 ft long for QF-1. After packing, the columns were conditioned for 48 hr at 215°C with a low carrier-gas flow rate (pre-purified nitrogen, Matheson, Coleman and Bell).

Gas chromatography. The Micro Tek G.C. 2000MF with a tritium electron capture detector was used for all experiments. Conditions were as follows:—Temperatures: inlet 210°C, column 185°C, detector 195°C; nitrogen pressure: input 70 psig; flow rate of nitrogen: 60–70 ml/min; polarizing voltage: 15 V d.c. On-column injections with Hamilton 10-μl syringes were utilized. Dilutions were made, however, when the peaks were too high for recording.

Field studies

A former medical student questioned the next of kin, whenever possible, regarding (1) age, sex and race of the deceased, (2) length of sickness and loss of weight before hospitalization, (3) duration of stay in the hospital, (4) loss of weight in the hospital before death, (5) nutrition, (6) occupation, (7) employer, (8) hobbies, and (9) general usage of chemicals, sprays, pesticides or insecticides. On the basis of information secured, each case was classified with regard to extent of pesticide exposure, as a heavy pesticide user (Group 3) one who used pesticides very little or not at all (Group 1) or one whose pesticide usage was considered normal, moderate or average (Group 2). The patients were also categorized according to their loss of body weight. These judgements were made without knowledge of the tissue pesticide levels.

Statistical analyses

Several procedures were applied to these data. The medians of all experimental groups were computed as well as the percentages of cases above the median of control groups. The significance of the percentage of cases above the control median was computed. In addition, rank tests were done to assess the existence of significant differences between the various experimental groups and the control groups, and correlation studies were conducted of the relationship between pesticide levels in the various tissues analysed. Means and standard deviations of the means were also computed.

RESULTS

Originally, it was our intention to collect tissues from three types of case namely those with liver disease, those with brain or neurological disease and a third group of cases made up of a random selection of patients not believed to have suffered from either brain or liver disease. During the examination of these cases prior to final classification, it was discovered that many individuals believed to have had liver disease before autopsy, showed only, for instance, chronic passive congestion secondary to myocardial infarction. A large group of cases in the random group was also found to have diseases such as atherosclerosis and hypertension as well as various malignant diseases. All cases were therefore reclassified taking into account the postmortem findings. When it was discovered that there were groups of cases with atherosclerosis, malignancies, hypertension and leukaemia, it was decided that perhaps additional information could be gleaned from this investigation if these cases were evaluated as individual experimental groups.

Liver diseases

Pesticide concentrations in fat and liver tissue from patients dying with liver disease are listed in Table 1.

Portal cirrhosis. All of the cases in this group had typical alcoholic (nutritional or Laennec's) cirrhosis of the liver. Mean levels of *p,p'*-DDE and dieldrin in the body fat were significantly higher than those of the normal population ($0{\cdot}01 > P > 0{\cdot}001$ and $P < 0{\cdot}001$, respectively). Mean *p,p'*-DDT levels were also approximately double those of the normal population, but because of a high degree of variability, the observation was not significant ($P = 0{\cdot}18$). Levels of *p,p'*-DDE, *p,p'*-DDT, and dieldrin were all significantly elevated in the liver tissue itself, but *p,p'*-DDD levels in the liver were lower than normal. *p,p'*-DDD is a metabolite of *p,p'*-DDT which originates in the liver. Impaired function would presumably lower the rate of this metabolic conversion.

Fatty metamorphosis. Included in the group were cases in which fatty metamorphosis (or infiltration) of the liver was present without necrosis or cirrhosis. Pesticide levels in both fat and liver tissue were essentially normal.

Metastatic liver disease. The most striking elevations in pesticide concentration were observed in the rather large series of cases of metastatic malignancy of the liver. The mean *p,p'*-DDE concentration in the fat was three times as high as normal ($P = 0{\cdot}005$), and seven times as high as normal in the liver tissue itself ($P = 0{\cdot}001$). *p,p'*-DDT concentrations were also elevated in the fat but, surprisingly, not in the liver tissue. Dieldrin and *p,p'*-DDD concentrations, on the other hand, were essentially normal. These would appear to be fairly specific alterations which probably have some real, but at the moment, obscure meaning. It does not seem likely that these altered concentrations were related to increased fat content of the liver, for in this case, consistent increases should have been observed.

Primary malignancy (*hepatoma*). One of our initial primary interests in carrying out this study was a possible relationship between chlorinated hydrocarbon pesticides and tumours of the liver. Unfortunately, only a small number of primary liver-tumour cases were obtained. Three of the four cases had rather normal *p,p'*-DDE, *p,p'*-DDD, *p,p'*-DDT and dieldrin concentrations. The fourth case had 35·04 ppm of *p,p'*-DDE in the fat and 1·25 ppm in the liver. All four cases had abnormally high levels of β-BHC in liver and adipose tissue. however, when compared to the infectious disease control group.

Post-necrotic cirrhosis. This pathological entity was characterized, in contrast to portal cirrhosis, by the occurrence of essentially normal pesticide concentrations, frequently lower than normal.

Other pathological states of the liver. Most noteworthy was one case of ideopathic amyloidosis which had markedly elevated levels of all the pesticides measured. A concentration of 50·8 ppm of *p,p'*-DDE in the fat was observed. This was one of the highest levels of *p,p'*-DDE observed in this study.

Brain and neurological diseases

Pesticide concentrations in fat and brain tissue from patients dying from disease of the brain or nervous system are listed in Table 2.

Table 1. *Pesticide concentrations in the fat and liver tissue of patients terminating with liver disease*

Diagnosis	No. of cases	Tissue	Level of pesticide (ppm±SD)					
			p,p′-DDE	*p,p′*-DDD	*p,p′*-DDT	Dieldrin	β-BHC	Heptachlor epoxide
Normal	42	Fat	6·69± 4·07	0·28±0·38	2·77±1·42	0·21±0·15		
		Liver	0·35± 0·01	0·36±0·03	0·10±0·02	0·03±0·03		
Infectious diseases	20	Fat	8·89± 7·67	0·12±0·27	3·92±5·01	0·38±0·23	0·66±0·23	0·28±0·16
Portal cirrhosis	33	Fat	11·49± 7·46	0·34±0·42	5·19±5·97	0·50±0·36	0·76±0·50	0·33±0·21
		Liver	1·19± 0·90	0·20±0·25	0·53±0·14	0·10±0·19	0·12±0·18	0·05±0·09
Fatty metamorphosis	14	Fat	7·12± 4·61	0·27±0·19	3·53±3·20	0·35±0·15	0·51±0·37	0·22±0·14
		Liver	0·63± 0·45	0·15±0·18	0·05±0·15	0·04±0·09	0·10±0·18	0·03±0·09
Malignancy, metastatic	30	Fat	19·27± 6·35	0·37±0·51	5·87±6·32	0·49±0·37	0·82±0·48	0·27±0·44
		Liver	2·44± 9·16	0·22±0·87	0·08±0·13	0·07±0·07	0·10±0·20	0·03±0·04
Malignancy, primary	4	Fat	13·58±14·44	0·49	7·70±8·00	0·36±0·17	1·06	0·23
		Liver	0·60	0·16	0·02	0·03	0·07	0·02
Acute necrosis	9	Fat	9·51±11·37	0·65±0·15	5·84±1·17	0·36±0·25	0·58±0·48	0·31±0·19
		Liver	1·32± 0·28	0·40±0·10	0·35±0·11	0·03±0·04	0·06±0·10	0·03±0·07
Toxic hepatitis	2	Fat	14·36± 3·90	0·21	4·64±3·87	0·36±0·11	0·86	0·34
		Liver	0·65	0·14	0·44	0·03	0·12	0·04
Adenocarcinoma of bile ducts	2	Fat	10·45± 7·42	0·82	3·68±7·42	0·21±0·03	0·61	0·34
		Liver	0·35	0·14	0·00	0·03	0·07	tr.
Amyloidosis	1	Fat	50·80± 7·42	1·84	—	—	2·28	1·12
		Liver	1·39	0·14	0·50	tr.	0·11	tr.
Post-necrotic cirrhosis	4	Fat	6·90± 4·35	0·07	1·53±1·83	0·21±0·16	0·75	0·27
		Liver	0·29	0·08	tr.	0·03	0·06	0·01

Table 2. *Pesticide concentrations in the fat and brain tissue of patients terminating with brain disease*

Diagnosis	No. of cases	Tissue	Level of pesticide (ppm±SD)					
			p,p′-DDE	*p,p′*-DDD	*p,p′*-DDT	Dieldrin	*β*-BHC	Heptachlor epoxide
Normal	42	Fat	6·69± 4·07	0·28±0·38	2·77±1·42	0·21±0·15		
		Brain	0·12± 0·19	—	0·04±0·01	0·04±0·01		
Infectious diseases	20	Fat	8·89± 7·67	0·12±0·27	3·92±5·01	0·38±0·23	0·66±0·23	0·28±0·16
Ischaemic necrosis	3	Fat	6·22± 5·67	0·00	2·28±2·38	0·67±0·63	0·52	0·28
		Brain	0·14	0·00	0·00	0·02	0·00	0·00
Encephalomalacia	24	Fat	12·63±11·43	0·27±0·19	5·40±8·45	0·50±0·33	0·75±0·34	0·35±0·16
		Brain	0·20± 0·18	0·00	0·00	0·01±0·01	0·00	0·00
Cerebral haemorrhage	16	Fat	12·20± 7·36	0·38±0·38	3·92±2·45	0·59±0·48	0·70±0·26	0·41±0·30
		Brain	0·12± 0·11	0·00	0·00	0·01±0·02	0·00	0·00
Brain tumour	6	Fat	7·33± 8·17	0·27±0·47	3·91±3·93	0·32±0·25	0·46±0·23	0·34±0·29
		Brain	0·08± 0·06	0·01±0·03	0·01±0·02	0·02±0·02	0·00	0·01±0·01
Degenerative changes	10	Fat	10·44± 7·64	0·28±0·70	2·00±1·60	0·30±0·17	0·59±0·35	0·29±0·16
		Brain	0·16± 0·13	0·00	0·01±0·02	0·02±0·02	0·00	0·00
Brain, metastatic tumour	1	Fat	0·00	0·00	0·00	0·00	0·00	0·00
		Brain	0·07	0·00	0·00	tr.	tr.	0·00
Granulomatous encephalitis	1	Fat	6·60± 7·64	tr.	2·10	0·41	0·88	0·42
		Brain	0·14	0·00	0·00	tr.	0·00	0·00
Personality changes	1	Fat	3·66± 7·64	0·32	2·80	0·38	0·37	0·16
		Brain	tr.	0·00	0·00	tr.	tr.	0·00

Encephalomalacia. Significant elevations of the *p,p′*-DDE concentration in the fat ($0{\cdot}01 > P > 0{\cdot}001$) as well as the brain tissue ($0{\cdot}05 > P > 0{\cdot}01$) were observed. Strangely, while there was a highly significant elevation of the dieldrin concentration in the fat ($P < 0{\cdot}01$), brain concentrations were, if anything, lower than normal. *p,p′*-DDT levels were elevated, but not significantly ($P > 0{\cdot}1$), in the fat, but were lower than normal in the brain tissue.

In this category were included all cases of definite encephalomalacia, either old or recent. Patients in this category were, of course, invariably aged, and atherosclerosis and myocardial infarcts were also commonly present.

Cerebral haemorrhage. Cases where there was distinct evidence of haemorrhage, either old or recent, were included in this group. *p,p′*-DDE and dieldrin levels were significantly elevated in the fat ($0{\cdot}015 > P > 0{\cdot}01$, respectively), but were essentially normal in the brain tissue.

Cerebral degeneration. Pesticide concentrations in fat and brain were essentially normal. A small elevation of the mean *p,p′*-DDE level in the fat was observed ($0{\cdot}3 > P > 0{\cdot}2$).

Included in this group were all cases of degeneration of the central nervous system cells, where strokes or haemorrhage had not occurred. Degeneration was frequently of the diffuse or Alzheimer type.

Brain tumour. One of the initial major interests of this investigation was the possibility that chlorinated hydrocarbon pesticides might produce tumours of the brain. Primary brain tumours are rare and only four cases were found in this study. It is, of course, hard to justify the drawing of a conclusion on the basis of four cases, but the fact that normal pesticide levels were observed in both fat and brain tissue would seem to suggest that chlorinated hydrocarbon pesticides are not involved in the development of brain tumours.

Other neurological diseases. Pesticide levels in other neurological diseases observed in the study were not remarkable. In three cases of ischaemic necrosis, pesticide levels were normal with the exception of elevated levels of dieldrin in the fat ($P = 0{\cdot}33$).

Miscellaneous diseases

Results of pesticide analyses carried out on the adipose tissue of patients with various other terminal conditions are given in Table 3.

Table 3. *Pesticide concentrations in the fat tissue of patients with various terminal conditions*

Diagnosis	No. of cases	Level of pesticide (ppm±SD)					
		p,p′-DDE	*p,p′*-DDD	*p,p′*-DDT	Dieldrin	β-BHC	Heptachlor epoxide
Normal	42	6·69± 4·07	0·28±0·38	2·77±1·42	0·21±0·15		
Infectious diseases	20	8·89± 7·67	0·12±0·27	3·94±5·01	0·38±0·23	0·66±0·23	0·28±0·16
Atherosclerosis	54	12·01± 2·53	0·27±0·75	5·10±7·57	0·37±0·23	0·76±0·52	0·33±0·17
Hypertension	8	17·91± 6·28	0·40±0·42	6·54±3·64	0·73±0·59	0·95±0·56	0·48±0·37
Carcinoma	40	15·97± 3·78	0·34±0·51	5·65±5·61	0·55±0·34	0·78±0·43	0·34±0·24
Leukaemia	5	16·10± 5·53	0·58±0·45	4·69±4·28	0·47±0·22	0·77±0·58	0·32±0·23
Chronic renal disease	8	8·11± 4·25	0·21±0·25	2·11±1·10	0·23±0·16	0·42±0·63	0·21±0·11
Pancreatitis	3	11·57±10·81	0·09	1·29±1·05	0·40±0·35	0·56	0·32
Hodgkin's disease	5	10·06± 4·43	0·38±0·27	3·22±1·20	0·51±0·18	0·57±0·22	0·30±0·06

Atherosclerosis. In addition to atherosclerosis, most of these cases had arteriosclerosis; myocardial infarcts were also commonly present and were frequently the cause of termination. Chronic passive congestion of the liver secondary to the cardiovascular failure following myocardial infarction was frequently present, causing many of these to be classified originally in the liver-disease category. Moderately elevated levels of *p,p'*DDE, *p,p'*-DDT and dieldrin were observed in the fat of many of these cases. However, mean pesticide concentrations were not significantly different from those of the normal population.

Hypertension. Although only eight cases of essential hypertension were contained in this study, more consistent elevations of pesticide concentrations in the fat were observed in this group than in any other disease category. Concentrations ranging from two to four times the normal of *p,p'*-DDE, dieldrin, *p,p'*-DDD and *p,p'*-DDT were observed in all eight cases. This seems to be one of the most significant observations of the experiment since it is difficult to postulate depressed liver function as a possible explanation.

Carcinoma. Included in this group were carcinomas of the lungs, stomach, rectum, pancreas, prostate and bladder. Concentrations of all the pesticides were remarkably high, averaging two or three times the normal concentration. Before grouping all of these various types of carcinoma together, each type was analysed individually in an attempt to discover whether the elevated levels were associated with a particular neoplastic disease. However, no such association could be discerned.

Chronic renal disease. Eight cases of chronic renal disease were studied. The usual pathological diagnosis was chronic glomerular nephritis. It is noteworthy that pesticide levels were, if anything, somewhat lower than normal in this group, suggesting (1) that chlorinated hydrocarbon pesticides are probably not involved in the production of this disease condition, and (2) that chronically impaired renal function does not lead to elevated levels of pesticides in the fat.

Correlation of pesticide concentrations in adipose tissue with those of liver tissue and brain tissue

As part of our statistical analysis of the data of this experiment, the correlation of pesticide concentrations in fat with those in liver tissue and with those in brain tissue was studied. Contrary to our expectations, virtually no correlation was observed between the fat and liver concentrations of any of the pesticides measured. On the other hand, fat concentrations of pesticides are correlated with pesticide concentrations in the brain. The results obtained with *p,p'*-DDE are plotted in Fig. 1.

When similar correlation studies were performed on our data from the general population, correlation was observed between liver and fat concentrations of *p,p'*-DDT and its metabolite. It would therefore appear that the lack of correlation between pesticide concentrations in liver and fat was associated with the presence of disease.

The relationship of pesticide levels to the home usage of pesticides

In the examination of these data, the most striking observation is the widely varying difference that exists in pesticide levels in individuals, some having very high and others very low pesticide levels. Hospital records indicated no occupational exposure to pesticides in this study. Therefore, it was considered that perhaps these remarkable discrepancies could be related to the home usage of pesticides.

The data obtained by interviewing the next-of-kin of these cases revealed that there was indeed a relationship (Fig. 2). This was most striking with the insecticide *p,p'*-DDT and its

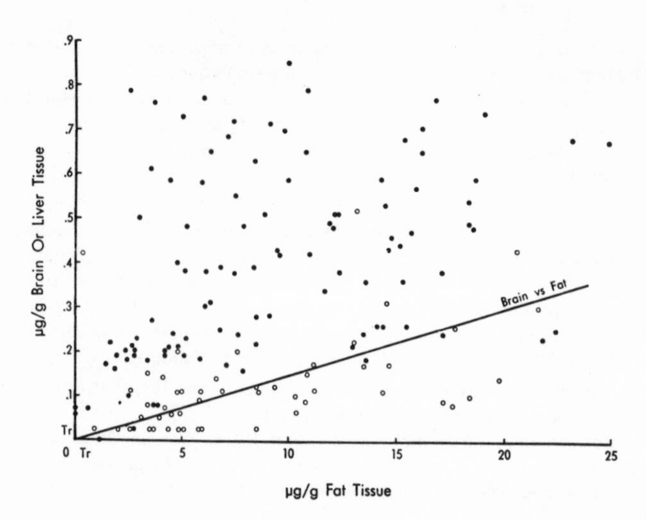

FIG. 1.

Correlation Between p,p'DDE Concentrations in Brain and Fat(o), and Liver and Fat (•).

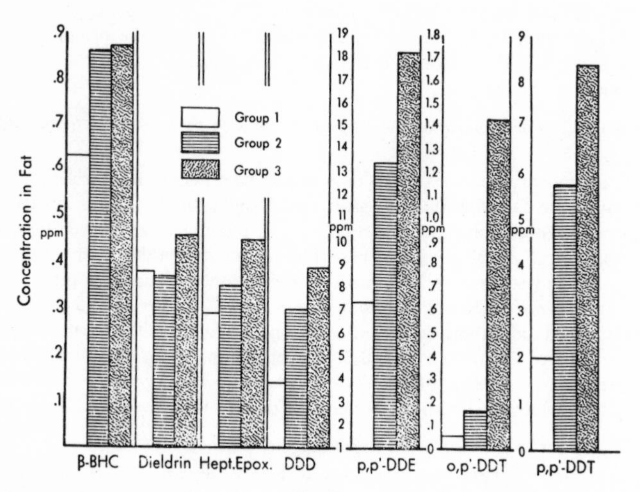

FIG. 2.

Relationship of Fat Concentrations of Organochlorine Pesticides to the Estimated Degree of Exposure in the Home.

Group 1 -- those using little or no pesticides in the home.
Group 2 -- " moderate " or "average"users.
Group 3 -- heavy users.

principal metabolite, *p,p'*-DDE. The mean *p,p'*-DDT concentration in the high exposure group is over four times the mean *p,p'*-DDT concentration in the low exposure group, ($P = 0{\cdot}001$), while the mean *p,p'*-DDE concentration is almost three times greater in the high exposure group than in the low exposure group. In the case of dieldrin and heptachlor epoxide, there were no significant differences in mean pesticide levels between the three exposure groups. β-BHC showed a significant increase ($P = 0{\cdot}01$) from Exposure Group I to Group 2 but not between Groups 2 and 3. These striking results clearly indicate that the home usage of pesticides is one, if not the major, factor in the occurrence of elevated pesticide body levels of *p,p'*-DDT and its metabolites. Similar results with dieldrin and heptachlor epoxide would not be expected since these pesticides are not normally present in pesticide mixtures intended for home usage.

It also occurred to us that all the cases in this study were patients with terminal diseases, and terminal patients frequently become extremely emaciated. Possibly the high pesticide levels observed were related to loss of fat in the adipose tissue, resulting in a higher concentration of pesticides in the remaining fat. However, when the loss of body weight was compared to pesticide concentrations, no such relationship was observed.

DISCUSSION

It is generally recognized that the present attempt to discover whether the current, long-term, low-level exposure to pesticides experienced by the general population in this country is producing harmful effects is beset with monumental difficulties. The ubiquitous distribution of large numbers of compounds and the lack of adequate controls are serious handicaps. Many possible approaches have been suggested and some have been tried; all have serious deficiencies. Certainly, new approaches are needed. One such approach is the evaluation of pesticide levels in the presence of diseases possibly caused by pesticides. The present experiment was conceived as a pilot study designed to test the value of this approach. It was recognized at the outset that the interpretation of the results obtained would be difficult. When a correlation is established between a given disease and a particular pesticide body burden, three possibilities need to be considered. Either the disease is responsible for the pesticide body burden or *vice versa*, or both result from some other factor. There is no evidence in the literature that a diseased liver may cause an elevation in pesticide level in the adipose tissue. However, such a possibility distinctly exists. Although experiments can readily be designed in animals to determine whether liver damage may result in elevated pesticide levels, it is doubtful whether this possibility has yet been investigated.

The most convincing relationship between pesticide concentration and disease revealed in this study was found in the consistently high concentrations present in patients with cirrhosis of the liver, carcinoma, and hypertension. Before any conclusions can be drawn on the role of excessive pesticide exposure in the production of any of these diseased states, confirmation of these preliminary observations is essential. The following step would consist of experiments designed to determine whether the disease caused the elevated pesticide level or *vice versa*.

The question was raised as to whether factors associated with the hospitalization were responsible for these elevated levels. No correlation was found between the elevated pesticide levels and the length of stay in hospital or with inanition.

The statement is commonly made that the major source of pesticide body burden of individuals in the general population is the food intake, quantities taken in via water and air being considered negligible. While it has been repeatedly demonstrated that occupational exposure to pesticides is capable of producing elevated body levels, the role of the non-occupational home usage of pesticides in the production of pesticide storage in the body has not been considered significant. Minute quantities of pesticides are present in all types of foods, however, and it is difficult to imagine one individual ingesting much more pesticide than another. Of course, individual differences do exist in the ability to metabolize and excrete pesticides. However, with regard to the degree of individual use of pesticides, it has been our observation that there seem to be two types of individuals, those who abhor insects and are ignorant and contemptuous of the hazards of chemical intoxication, and those with an instinctive, exaggerated fear of chemical exposure. It is conceivable that there could be a hundred-fold difference in the exposure of these two types of people.

This experiment suggests that the household use of pesticides may be responsible for the accumulation of elevated body levels of these compounds. Actually, this experiment—complicated by the presence of disease and death of the patient—did not provide the optimum conditions for the evaluation of this factor. A similar study of the correlation between home usage of pesticides and body burden in random cases of accidental death would be more suitable. A second approach, perhaps even more valuable, would be the determination of blood levels in individuals who use pesticides in house and garden.

Exposure of rats to *p,p'*-DDT is known to produce elevation of liver microsomal-enzyme activity. Schwabe (1964) and Gerboth & Schwabe (1964) have shown that phenobarbitone-induced sleeping time was reduced by approximately 50% by the feeding of a level of *p,p'*-DDT which produced concentrations of 10–15 ppm total DDT (including metabolite) in the fat. These data applied to this investigation would mean that a high percentage of the patients in this study probably exhibited enhancement of microsomal-enzyme activity which may have modified their response to drugs. Of course, there is as yet no evidence that these rat data are quantitatively applicable to man, nor that significant microsomal stimulation by pesticides occurs in man.

No attempt has been made to correct the concentrations of pesticides in the liver for the fat content of this organ. Many of the diseases encountered are characterized by fatty infiltration or degeneration of the liver. The possibility exists that elevated pesticide concentrations in the liver were wholly or partially a reflection of an increased fat content of this organ. One observation should be emphasized, namely, that elevated pesticide levels were not observed in the presence of fatty metamorphosis of the liver. It should also be noted that the major conclusions of this investigation are based on the concentration of pesticides in adipose tissue: not on the concentrations in liver or brain.

REFERENCES

Fiserova-Bergerova, V., Radomski, J. L., Davies, J. E. & Davis, J. H. (1967). Levels of chlorinated hydrocarbon pesticides in human tissues. *Ind. Med. Surg.* **36,** 65.

Gerboth, G., u. Schwabe, U. (1964). Einfluss von gewebsgespeichertem DDT auf die Wirkung von Pharmaka. *Naunyn-Schmiedebergs Arch. exp. Path. Pharmak.* **246,** 469.

Radomski, J. L. & Fiserova-Bergerova, Vera (1965). The determination of pesticides in tissues with the electron capture detector without prior clean-up. *Ind. Med. Surg.* **34,** 934.

Schwabe, U. (1964). DDT-Speicherung bei der Haltung von Versuchstieren als mögliche Fehlerquelle bei Arzneimittelprüfungen. *Arzneimittel-Forsch.* **14,** 1265.

Pesticides in blood

By

Mary L. Schafer

I. Introduction

Since the introduction of DDT [1,1,1-trichloro-2,2-bis(*p*-chlorophenyl)-ethane] during World War II for the control of malaria, many synthetic organic compounds have been found to be effective pesticides. The commonly used ones, the organochlorine compounds, are not readily degraded under normal conditions. Although extremely small concentrations of these compounds are required for pesticidal control, the effect on public health of their gradual increase in our environment is under close scrutiny. There is considerable interest in evaluating the usefulness of the concentration of these materials in blood to determine the severity of recent exposures and also to estimate the total body burden in the general population.

When ingested in sufficient amount, many of these compounds that have pesticidal activity are clearly poisonous to vertebrates. In the mammalian species, the toxic action is manifested almost entirely through the nervous system. Prominent signs of poisoning include muscle tremor, incoordination, and convulsions. Levels below that required to produce toxic effects cause vague and nonspecific symptoms. Headache, blurred vision, dizziness, and nausea are frequently mentioned symptoms of occupationally exposed individuals.

The concentration of DDT in the fat depots of the general human population has been monitored since 1951 (HOFFMAN *et al.* 1964). Experimental evidence indicates that the general world population now carries a body burden of it. The level in any given individual depends upon his history of exposure. DURHAM (1965) reported a concentration of DDT plus DDE in excess of 1,000 parts per million (p.p.m.) in one occupationally exposed individual. QUINBY *et al.* (1965) reported a mean storage level in the fat of 12.6 p.p.m. of DDT and DDE [2,2-bis(*p*-chlorophenyl)-1,1-dichloroethylene] for the population of the United States in 1961–62. The storage levels observed in this study were not significantly different from those observed for the population of the United States in 1954–56. A significant variation in residue levels has been observed in the fat of subjects from various countries (EGAN *et al.* 1965, WASSERMAN *et al.* 1967). The storage levels of DDT-related compounds in subjects in Alaska, Canada, and most of the European countries is slightly less than it is in the United States. People in Hungary and Israel store at least as much as Americans while subjects in India have significantly greater storage levels than comparable persons in the United States (DALE *et al.* 1965). With the refinement of assay techniques, other commonly used pesticides, their metabolites, and related compounds have been observed in the fat of human beings with no known occupational exposure. The following compounds have been reported: dieldrin (1,2,3,4,10,10-hexachloro-6,7-epoxy-1,4,4a,5,6,7,8,8a-octahydro-1,4-*endo,exo*-5,8-dimethanonaphthalene), heptachlor epoxide (1,4,5,6,7,8,8-heptachloro-2,3-epoxy-2,3,3a,4,7,7a-hexahydro-4,7-methanoindene), *p,p'*-DDT, *o,p'*-DDT, *p,p'*-DDE, *o,p'*-DDE, *p,p'*-DDD, lindane (gamma isomer of 1,2,3,4,5,6-hexachlorocyclohexane), and the three isomers associated with lindane, α-, β-, and δ-BHC (alpha-, beta-, and delta-hexachlorocyclohexane).

The rationale for using blood levels of these insecticidal compounds to reflect levels of exposure has been strengthened by studies on the solubility of these compounds in blood, the mechanism of their absorption from the gastrointestinal tract, their redistribution between the brain and fat stores during recovery from exposure, and the increased susceptibility to toxic reactions from these compounds during mobilization of stored fat.

Most of the organochlorine pesticides have low solubilities in water.[1] In water supplies they are concentrated usually in the silt, sand, plankton, or algae. In biological material, they are found in the fat-containing tissues. MOSS

---

[1]Reviewed by GUNTHER, F. A., W. E. WESTLAKE, and P. S. JAGLAN: Reported solubilities of 738 pesticide chemicals in water. Residue Reviews **20**, 1 (1968).

and HATHAWAY (1964) reported that the serum constituents of the blood exert a powerful solubilizing effect on dieldrin: they reported a solubility in rabbit serum of around 200 micrograms per milliliter (μg./ml.) compared with a solubility in water 0.05 μg./ml.

In vivo and *in vitro* studies have demonstrated that dieldrin is absorbed from the gastrointestinal tract via the hepatic portal system while DDT is absorbed via the lymphatic system. When ^{36}Cl-dieldrin was fed to rats in arachis oil, HEATH and VANDEKAR (1964) observed that it was absorbed from the gastrointestinal tract via the portal vein and was distributed throughout the body within a few hours. After oral administration, maximum concentrations were reached in organs of the body in one to two days, the highest concentrations appearing in the fat. Some redistribution took place in favor of fat within the next four days. These results indicate that blood levels of dieldrin might reflect both the initial absorption and the redistribution phenomenon. ROTHE *et al.* (1957) found 47 to 65 percent of the intestinally absorbed radioactive DDT in the chyle when DDT was administered orally to rats with their thoracic lymph ducts cannulated. This pesticide also undergoes *in vivo* redistribution in favor of fat.

DALE *et al.* (1963) observed a direct correlation between the severity of symptoms when rats were exposed to a single dose of DDT and the concentration in the nervous system as measured by the concentration of the brain. Rats showing severe tremors had brain concentrations ranging from 386 to 433 p.p.m.; those with convulsions, 289 to 606 p.p.m.; and those with convulsions and death had DDT concentrations ranging from 524 to 848 p.p.m. in brain tissue. The animals that recovered from the exposure to DDT showed decreasing concentrations of DDT in the brain within 26 hours (ranging from 138 to 213 p.p.m.) accompanied by significantly increased levels in the fat (from an average of 58 p.p.m. two hours after exposure to 598 p.p.m.).

Episodes have been reported in which the toxic symptoms appeared weeks after the exposure. This phenomenon of delayed reaction was observed by KAZANTZIS *et al.* (1964) in human beings working in an aldrin (1,2,3,4,10,10-hexachloro-1,4,4a,5,8,8a-hexahydro-1,4-*endo,exo*-5,8-dimethanonaphthalene) formulating plant. It has been produced also in animals. While studying factors affecting the LD_{50} dose of dieldrin for rats, BARNES and HEATH (1964) observed that rats surviving toxic doses lost weight progressively for the next few days. During this period in which the body fat was being metabolized, they were more susceptible to a second dose. HEATH and VANDEKAR (1964) reported a marked increase in excretion of ^{36}Cl-dieldrin if the diet of the animals was restricted sufficiently to cause loss in body fat.

In the past few years there has been considerable controversy concerning the causative agents in massive fish kills. These episodes occur frequently in areas of concentrated population and/or industries. Thus it is possible to observe a variety of toxicants in the fat storage depots of the fish. If the storage of these toxicants occurred from exposure to low levels over a long period of time, large fat concentrations might be observed for toxicants that are not causative agents in the death of the fish. In contrast, blood levels should reflect recent exposure and thus aid in identifying the toxin or toxicants that caused the kill.

II. Distribution of pesticides among the components of blood

Data are accumulating in the literature relating the concentration of organochlorine compounds in blood to symptoms of toxicity and to the total body burden. Some of the earlier work was done with plasma and serum. Whole blood was used in most of the recent studies.

MOSS and HATHAWAY (1964) reported studies on the distribution of ^{14}C-dieldrin; most of this insecticide was found in erythrocyte contents and plasma and not in leucocytes, platelets, or stroma. The distribution of ^{14}C-dieldrin among soluble proteins in blood was the same *in vivo* as *in vitro,* and the ratio between cells and plasma was about 1:2 for dieldrin. These authors found dieldrin associated with the hemoglobin and an unknown consistuent in the erythrocyte contents. Their studies give strong evidence for the binding of dieldrin in the serum with albumin, a constituent of the globulin fraction, and a third fraction. The authors suggest that the combination of dieldrin with the soluble proteins of the blood somewhat resembles the complex binding of corticosteroids and thyroid hormones but the actual pattern for distribution of the insecticides among the serum proteins appears to be specific for these substances. They were able to remove the insecticides from whole blood with repeated ether extractions. Data reported in two subsequent studies verify a distribution of organochlorine compounds between the plasma and the cells.

DALE *et al.* (1966) reported that 61 percent of the total dieldrin in whole blood was present in the serum. A range of 45 to 90 percent was observed in the six organochlorine compounds with which they worked. KADIS and JONASSON (1965) found that 8 to 20 percent of insecticides added to citrated blood remained with the cells.

III. Methods of analysis

Before 1960 the assay methods used in studies on the factors affecting storage of DDT in living organisms lacked sufficient sensitivity to detect DDT in blood even after the ingestion of sufficient concentrations to produce toxic symptoms. SMITH and STOHLMAN, however, in 1944 and STIFF and CASTILLO in 1945 reported blood levels of organically bound chlorine equivalent to DDT in rabbits after exposure to toxic levels of DDT. In 1946, LAUG used a bioassay technique to observe levels greater than one p.p.m. of DDT in the blood of rabbits after oral administration of DDT in concentrations of 350 mg./kg. or greater. JUDAH (1949) used a spectrophotometric method of assay to observe 15 p.p.m. of DDT in rat blood after a single oral injection of 500 mg./kg. REYNOLDS (1957) reported a method for the assay of dieldrin in the range of 0.25 p.p.m. in whole blood. In this method, 20-ml. aliquots of whole blood were saponified and extracted with hexane. The hexane extract was subjected to additional cleanup, using column chromatography, and was assayed, using either a colorimetric reduction phenylazide method or a bioassay technique employing *Drosophila melanogaster*. When dieldrin was added to blood at a level of 0.25 p.p.m. the recovery was 100 percent. DALE *et al.* (1963) used the Schechter-Haller colorimetric method to measure blood plasma levels of 119 to 178 p.p.m. of DDT in rats that recovered from a single oral dose of 150 mg./kg.

With the development of the electron-capture detector for the gas chromatograph, it is possible to assay for many organochlorine pesticides in whole blood in the parts per billion (p.p.b.) range. With this increase in minimum detectability the blood levels of these pesticides can be measured after administration of sublethal concentrations. There has been a trend in recent years toward simple, rapid extraction and purification methods followed by assay in gas chromatographs equipped with this selective detector. With the simple extraction methods and electron-capture detectors, it is possible, for example, to assay for organochlorine compounds, such as dieldrin, in blood with a precision of ± five percent at levels of 10 to 50 p.p.b.

The most commonly used procedure for evaluation of the efficiency of any method is the *in vitro* recovery study. With many of the recently developed rapid methods, it is possible to recover 85 percent or more of organochlorine pesticides added to blood just prior to the extraction step. A statement of accuracy based on *in vitro* recovery studies assumes that the pesticides present in blood from *in vivo* experiments are available. Ideally, the accuracy of assay methods is evaluated with collaborative studies in which a given naturally contaminated sample is assayed by alternate methods. Several of the methods used for the assay of pesticides in blood have been evaluated in collaborative studies on other biological fluids such as milk.

In several of the recently developed methods, the organochlorine compounds are extracted from aqueous solutions of the blood after a saponification step. Portions of the solvent extract of the alkaline hydrolysate are injected directly into the gas chromatograph with no further purification. ARCHER and CROSBY (1966) reported a method for the assay of the dehydrohalogenated toxaphene (mixture of isomers of octachlorocamphene) in rat and cow blood, using benzene as the solvent; CROSBY and ARCHER (1966) reported a method for DDT and metabolites in bovine and human blood, using pentane as the solvent; ROBINSON and HUNTER (1966), MOUNT *et al.* (1966), SCHAFER and CAMPBELL (1966), and RICHARDSON *et al.* (1967) reported methods for endrin (1,2,3,4,10,10-hexachloro-6,7-epoxy-1,4,4a,5,6,7,8,8a-octahydro-1,4-*endo,endo*-5,8-dimethanonaphthalene), DDE, and dieldrin in blood of human beings, dogs, and fish, using hexane as the solvent. KADIS and JONASSON (1965) assayed for eight insecticides in whole blood of human beings and animals by placing a blood-acetone mixture on a partially activated florisil column and eluting with methylene chloride; the methylene chloride was evaporated and the residue was dissolved in hexane for gas chromatographic assay. JAIN *et al.* (1965) assayed for 12 organochlorine compounds, ten organophosphorus compounds, and one nitro compound (2-sec-butyl-4,6-dinitrophenyl-3,3-dimethylacrylate, or Morocide) in blood using an extraction mixture consisting of equal volumes of acetone and ether; the solvent was then evaporated and the residue was dissolved in hexane. SASCHENBRECKER and ECOBICHON (1967) assayed for seven commonly used organochlorine compounds in plasma and animal tissues using an initial acetonitrile-acetone extraction step. After the fats and waxes were removed on a polyethylene-coated alumina column, the organochlorine compounds were partitioned into *n*-heptane and suitable volumes of the heptane solutions were assayed by gas chromatography. GOODWIN *et al.* (1962) used an acetone-hexane partition method to separate the organochlorine compounds from fat. DALE *et al.* (1966) reported a method in which only the hexane extractable organochlorine compounds were measured. Recoveries of 45 to 59 percent were attained with this method. The working hypothesis for the method, however, was that the bound insecticides (those not extracted by hexane) did not contribute to the toxic symptoms. These same workers (DALE *et al.* 1967) recently reported a new method based on volatilization of organochlorine compounds. The whole blood or plasma was heated to 125°C., and the volatilized insecticides were trapped on silanized glass wool. The range of recovery from water was 80 to 100 percent, from whole blood, 14 to 62 percent, and from plasma, 23 to 60 percent.

IV. Studies relating blood levels to exposure

Even with the nonsensitive methods available at the time that the early studies were carried out on the storage of organochlorine compounds in the mammalian species, enhanced blood levels were observed after known exposures. The early studies usually emphasized tissue buildup after a single exposure at a level that caused extreme toxic symptoms and death. Studies in the last few years have been designed to measure tissue storage from low-level exposures. Results from the studies in which a single oral dose of insecticide was used are summarized in Table I. The more recent studies include data

for the metabolites of *p,p'*-DDT, *p,p'*-DDE, and DDD, the storage level of dieldrin after ingestion of aldrin, and the storage level of heptachlor epoxide after ingestion of heptachlor (1,4,5,6,7,8,8-heptachloro-3a,4,7,7a-tetrahydro-4,7-methanoindene).

a) Animals

SMITH and STOHLMAN (1944) reported that two days after a rabbit was given a single dose of 550 mg. of DDT/kg., which was sufficient of this pesticide to cause hyperexcitability and mild tremors, organically bound chlorine equivalent to 107 p.p.m. of DDT was found in the blood of the animal. STIFF and CASTILLO (1945) reported that after a single dose of 550 mg./kg. of DDT was given to each of four rabbits, organically bound chlorine equivalent to 68 p.p.m. of DDT was found in the blood within two hours and 98 p.p.m. within 48 hours. They did not directly detect DDT. The samples were alkaline hydrolyzed, however, before analysis by a colorimetric procedure specific for DDT-ethylene; thus, any DDT present was dehydrohalogenated to DDE. With a biological assay method, LAUG (1946) observed levels of 1.5 to 8 p.p.m. of DDT in the blood of rabbits exposed to a single dose of 350 to 400 mg. of DDT/kg. Using a spectrophotometric assay method, JUDAH (1949) reported a mean concentration of 15 p.p.m. (range 8 to 20) within about four hours in the blood of rats receiving an oral dose of 500 mg./kg.; fat levels ranged from 60 to 360 p.p.m. (mean 214). DALE *et al.* (1963) observed the concentration of DDT and DDE in blood plasma of rats before, during, and after recovery from signs of illness following a single oral dose of 150 mg. of DDT/kg. The concentration of DDT increased significantly in the plasma from values ranging from 178 to 417 p.p.m. for asymptomatic animals two hours after the administration of DDT to a maximum level of 1,348 p.p.m. in the animals that had convulsions 6 to 11 hours after ingestion. Those animals that developed convulsions after longer periods of time had lower levels of DDT (417 to 893 p.p.m.) and those that recovered had levels of 119 to 178 p.p.m. within 26 hours. The level of DDE in the blood plasma increased from < 0.5 p.p.m. to a maximum of 476 p.p.m. in the 26-hour period. Four hours after administering 200 mg. of DDT/kg. to fasting rats, JAIN *et al.* (1965) observed 50 p.p.m. of DDE, six to seven p.p.m. of *p,p'*-DDT, and seven p.p.m. of *o,p'*-DDT in the blood. They found 2.2 p.p.m. of dieldrin in whole blood nine hours after giving fasting rats a single dose of 75 mg./kg. and one p.p.m. of endrin after a single dose of 50 mg./kg. With a single dose of 150 mg. of lindane/kg., a concentration of 10 p.p.m. was observed in the blood within four hours. KADIS and JONASSON (1965) fed a single dose of DDT and aldrin (15 mg./kg. of each) to a four-month-old female calf and measured blood levels after three hours and after 1, 7, 21, and 42 days. The blood level for DDT was 0.70 p.p.m. after 24 hours, decreased to 0.06 p.p.m. in seven days, and remained at a level of 0.04 p.p.m. for 21 to 42 days. DDE was not observed in the blood of animals fed DDT. However, DDD was present in concentrations of less than 0.03 p.p.m. The maximum aldrin concentration, 0.09 p.p.m., was detected in the 24-hour sample; the seven-day sample contained only 0.005 p.p.m. A maximum dieldrin level of 0.20 p.p.m. was observed one day after the aldrin was fed. Forty-two days later the dieldrin concentration was 0.09 p.p.m., and no aldrin could be detected. When dieldrin was fed along with lindane and heptachlor in a single dose of 17 mg./kg. each to a four-month-old female calf, a maximum level of 0.14 p.p.m. of dieldrin was observed on the seventh day, and the concentration after 42 days was 0.05 p.p.m. Only traces of heptachlor were found, but a maximum level of 0.42 p.p.m. of heptachlor

Table I. *Blood levels after a single oral dose of insecticide*

Insecticide	Animal		Exposure (mg./kg.)	Interval between exposure and sampling	Assay[a] method	Level in blood (p.p.m.)		Reference
	No.	Species						
DDT						*p,p'-DDT*	*Metabolites*	
	1	Rabbit	550	2 days	1	107	—	SMITH and STOHLMAN (1944)
	4	Rabbit	550	2 hours	1	68	—	STIFF and CASTILLO (1945)
				48 hours				
	2	Rabbit	350	24 hours	3	1.1-1.5	—	LAUG (1946)
	1	Rabbit	400	24 hours		8	—	
	5	Rat	500	4-5 hours	2	8-20	—	JUDAH (1949)
	12	Rat	150	2 hours	2	178-1190	<0.5 DDE	DALE *et al.* (1963)
	13	Rat	150	6-11 hours		215-1348	<0.5-476 DDE	
	5	Rat	150	26 hours		119-178	298-476 DDE	
	3	Rat	200	4 hours	4	6-7	50 DDE	JAIN *et al.* (1965)
							7 *o,p'-DDT*	
	1	Calf	15	3 hours	4	Trace	—	KADIS and JONASSON (1965)
				24 hours		0.700	Trace DDD	
				7 days		0.064	0.030 DDD	
				21 days		0.040	0.010 DDD	
				42 days		0.042	—	
Aldrin						Aldrin	Dieldrin	
	1	Calf	15	3 hours	4	0.04	Trace	KADIS and JONASSON (1965)
				24 hours		0.09	0.200	
				7 days		0.005	0.072	
				21 days		0.008	0.090	
				42 days		None	0.092	

Heptachlor						Heptachlor	Heptachlor epoxide	
	1	Calf	17	3 hours	4	Trace	0.03	KADIS and JONASSON (1965)
				24 hours		Trace	0.100	
				7 days		None	0.420	
				21 days		None	0.04	
				42 days		None	0.005	
Dieldrin						Dieldrin	—	
	1	Rat	50	4 hours	4	0.67	—	JAIN *et al.* (1965)
	1	Rat	75	9 hours		2.2	—	
	1	Calf	17	3 hours	4	0.006	—	KADIS and JONASSON (1965)
				24 hours		0.095	—	
				7 days		0.140	—	
				21 days		0.040	—	
				42 days		0.046	—	
Endrin						Endrin	—	
	1	Rat	50	9 hours	4	1	—	JAIN *et al.* (1965)
	1	Rat	20	4 hours		Trace	—	
	1	Rat	10	4 hours		Trace	—	
	1	Rat	5	4 hours		None	—	
Lindane						Lindane	—	
	2	Rat	150	4 hours	4	10	—	JAIN *et al.* (1965)
	1	Calf	17	3 hours	4	0.62	—	KADIS and JONASSON (1965)
				1 day		2.000	—	
				7 days		0.124	—	
				21 days		0.003	—	
				42 days		0.009	—	

[a] 1 = Measured as organically bound chlorine.
2 = Spectrophotometric assay method using plasma.
3 = Fly bioassay.
4 = Gas chromatographic methods using whole blood.

epoxide was observed on the seventh day; this concentration had decreased to 0.005 p.p.m. after 42 days. The maximum level for lindane, two p.p.m., was observed on the first day with a decrease in concentration to 0.003 p.p.m. after 21 days. The persistence of dieldrin and DDT in the blood after 42 days indicates the usefulness of blood levels for ascertaining the exposure history in calves.

Studies in which animals were exposed to low levels of insecticides over long periods of time are summarized in Table II. Most of these studies were

Table II. *Blood levels after exposure to oral low levels of insecticides*

Insecticide	Animal		Exposure	Exposure time	Level in blood (p.p.m. × 10^{-3})	Reference
	No.	Species				
DDT						
	4	Cow	30 p.p.m.	60 days[a]	1-4[b], 4-7[c]	LABEN *et al.* (1965)
	4	Cow	300 p.p.m.	60 days[a]	15-34[b], 8-41[c]	
	4	Cow	600 p.p.m.	60 days[a]	29-69[b], 36-68[c]	
Dieldrin						
	2	Dog	0.2 mg./kg./day	8 mth[d]	110-220	BROWN and HUNTER (1964)
	4	Dog	0.4-0.8 mg./kg./day	8 mth[d]	270-1,270	
	3	Dog	0.1 mg./kg./day	121 days	97-190	RICHARDSON *et al.* (1967)
Endrin						
	3	Dog	0.1 mg./kg./day	65 days	5-8	RICHARDSON *et al.* (1967)

[a] From 90th to the 30th day before expected parturition.
[b] 19 to 28 weeks postpartum.
[c] 31 to 40 weeks postpartum.
[d] Eight episodes of intoxication.

so designed that the relationship between blood levels and the total storage of the chemical could be ascertained. LABEN *et al.* (1965) dissolved DDT in acetone and included it at levels of 30, 300, and 600 p.p.m. for 60 days in the feed of dry dairy cows and heifers approaching their first lactation. This highest level was equivalent to 18 mg./kg. and produced no recognizable toxic symptoms. The rate of decline in concentration of DDT and metabolites was observed in the body fat and milk during 40 weeks of lactation. Whole-blood levels were measured once during the 19th to 26th week postpartum and once during the 31st to 40th week postpartum. The relationships among milk fat, blood, and body fat varied widely among individual animals. The authors reported a correlation, however, between body fat and blood DDT levels of 0.95 at each sampling date and a correlation of 0.73 to 0.80 between milk fat and blood. They suggest that blood levels should be useful in detecting animals with significant DDT storage.

The animal exposure studies of RICHARDSON *et al.* (1967) indicate different modes in the metabolism of the stereoisomers dieldrin and endrin. When ethanol solutions of either dieldrin or endrin were added to the diet of dogs at a level of 0.1 mg./kg./day, an increase in the concentration of dieldrin in the blood from < one p.p.b. to a maximum of 190 p.p.b. was observed in 121 days. The maximum level for the three animals, however, ranged from 97 to 190 p.p.b. In contrast, endrin levels in the blood remained between three and eight p.p.b. for the entire study. These workers observed correspondingly low levels of endrin in the body fat and tissue but reported a correlation index, R^2, of 0.86 for dieldrin between blood and fat.

BROWN *et al.* (1964) administered 0.2 to 0.8 mg. of dieldrin/kg./day by capsule to fasted dogs. When intoxication episodes occurred with the higher levels, dieldrin was omitted until the animal recovered. Then dieldrin was again administered at the same level. A maximum level of 1.27 p.p.m. was observed in the blood. Animals receiving low levels of dieldrin and showing no signs of intoxication had maximum blood levels of 0.11 to 0.22 p.p.m.

The object of many of the studies summarized above was simply the detection of insecticides in the blood after known exposure. Generally, observed conclusions from these studies may be summarized as follows. Insecticides such as DDT, dieldrin, and lindane were observed in the blood of all animals that were tested. In addition, two metabolites of DDT (DDE and DDD) were observed in some studies. When the insecticides heptachlor and aldrin were administered, their metabolites (heptachlor epoxide and dieldrin) were observed in the blood. With a single injection, the level of each apparently reached a maximum in the blood and then decreased. The time required to produce this maximum concentration was apparently dependent on the insecticide and the animal species. When feeding studies were conducted at different levels of intake, the blood levels at any given time for either dieldrin or DDT seemed to reflect both the length of time between exposure and testing and the exposure level. On a given exposure level, the blood concentration seemed to plateau after a finite period of time.

WITT *et al.* (1966) measured the rate of decline of DDT from the blood of two lactating Holstein cows after a single intravenous injection of DDT (one to two mg./kg.). They reported a half-life of 60 to 80 minutes followed by an eight-hour half-life, with equilibrium nearly reestablished 24 hours after cessation of insecticide absorption into the blood.

b) Fish

Two studies have been reported in which blood levels of endrin in fish have been correlated with known concentrations in water. Both of these studies were designed to establish the concentration of insecticide in the blood of fish killed by the insecticide. During the 1963–64 investigations of large fish kills in the lower Mississippi River, a relationship was observed between endrin concentrations in the blood and intoxication of fish, such as channel catfish and gizzard shad. Because of this observed relationship, MOUNT *et al.* (1966) exposed 95 catfish to 15 different lethal and sublethal concentrations of endrin in water and observed larger concentrations of endrin in the blood of the fish killed by endrin poisoning than in the fish that were exposed but not killed. They concluded that the concentration of endrin in the blood could be used to determine the occurrence of endrin poisoning. Their data indicated a threshold concentration of 0.30 p.p.m. for this species of fish. BRUNGS and MOUNT (1967) exposed gizzard shad to known concentrations of endrin in water and reported a critical level (the level above which few shad can survive) of 0.10 p.p.m. for endrin in this species. These critical values were further strengthened by the observation that fish of each species that were killed in the 1963–64 fish kill had blood levels above the threshold levels established by the laboratory experiments.

c) Human beings

Because of the toxicity of aldrin and dieldrin to human beings, clinical signs of intoxication may be observed in the employees of plants where these insecticides are manufactured and the products formulated. Clinical symptoms of intoxication from aldrin and dieldrin include neurophysiological disturbances, such as tremors, with epileptiform convulsions in the more severe cases. Dieldrin has been found in fat biopsies at levels in excess of 100 p.p.m. in individuals showing symptoms of intoxication. The need of the medical investigators for a simple test to anticipate the symptoms of intoxication stimulated the development of a method for detecting dieldrin in blood. In 1963 ROBINSON reported the development of such a method with sufficient simplicity, precision, and minimum detectability to be of use in assessing the severity of occupational exposure and supporting the diagnosis in suspected intoxication. He reported dieldrin levels of 20 to 280 p.p.b. in the blood of nine occupationally exposed workers and 0.5 to 4.0 p.p.b. in 12 persons with no known occupational exposure. On the basis of these data, he suggested the blood level of 15 p.p.b. as indicative of industrial exposure to aldrin and/or dieldrin.

Using this same assay procedure, KAZANTZIS *et al.* (1964) reported a dieldrin level of 100 p.p.b. in the blood of a worker four months after he had had three convulsive attacks. Three workers with electroencephalogram abnormalities associated with intoxication had dieldrin levels in the blood of 40 to 280 p.p.b. while other workers with no clinical symptoms had blood levels of 30 to 120 p.p.b. Two unexposed individuals had blood levels of 0.6 and 20 p.p.b. Thus, these data substantiated the level suggested by ROBINSON (1963) to differentiate between occupationally exposed workers and the general population. BROWN *et al.* (1964) reported dieldrin blood levels of 130 to 368 p.p.b. in the blood of five workers two to five weeks after they showed clinical signs of intoxication, and an overall range of 2 to 220 p.p.b. for 89 occupationally exposed workers with no clinical signs of intoxication. The range in dieldrin levels in the blood of 20 unexposed individuals was 0.5 to 10 p.p.b. These data, along with those reported for dogs which were described in the previous section, were the basis for a proposed threshold concentration of 150 to 200 p.p.b. According to the authors cited, if a given blood level is greater than this threshold concentration, the man or dog is in critical condition, requiring some as yet unidentified factor of factors to precipitate intoxication.

In 1966, ROBINSON and HUNTER reported a range of one to 260 p.p.b. for dieldrin levels in the blood of 123 workers from plants in the United Kingdom, Holland, and Venezuela. Blood levels of 20 to 530 p.p.b. were observed in 12 workers who were tested one to two weeks after showing clinical signs of intoxication. One worker with an epileptic convulsion on the day his blood was tested had a dieldrin level in the blood of 320 p.p.b. A range of 0.06 to 5.0 p.p.b. of dieldrin was reported for the blood of 44 human beings with no known occupational exposure.

DALE *et al.* (1966) reported dieldrin levels of 279 p.p.b. and aldrin levels of 36 p.p.b. in the plasma of one person eighteen hours after ingestion of aldrin. These workers reported dieldrin levels of nine to 27 p.p.b. and aldrin levels of 0.7 to 2.3 p.p.b. in the plasma of 24 occupationally exposed human beings with no clinical signs of intoxication.

The data for dieldrin and DDT blood levels in occupationally exposed human beings with no clinical signs of intoxication are summarized in Table III, and data for blood levels of male human beings showing clinical signs of intoxication from pesticides are given in Table IV. Data for the general population are summarized in Table V. None of the persons in the general population group had dieldrin levels in blood above 15 p.p.b. All persons with

Table III. *Blood levels for occupationally exposed male human beings with no clinical signs of intoxication*

Insecticide	No.	Whole blood (p.p.b.)	Plasma (p.p.b.)	Type of exposure	Reference
Dieldrin	9	20-280	—	—	ROBINSON (1963)
	4	30-120	—	—	KAZANTZIS *et al.* (1964)
	32	2-50	—	Refinery operative	BROWN *et al.* (1964)
	7	6-120	—	Form. plant	
	50	3-220	—	Manuf. plant	
	52	4-260	—	Manuf. and form. plant	ROBINSON and HUNTER (1966)
	57	4-75	—	Form. plant	
	14	1-11	—	Applicators	
	24	—	9-27[a]	—	DALE *et al.* (1966)
DDT	35	—	395-737[b]	—	DALE *et al.* (1967)

[a] Also reported aldrin levels of 0.7 to 2.3 p.p.b.
[b] Total DDT metabolites.

Table IV. *Blood levels for male human beings showing clinical signs of intoxication from insecticides*

Insecticide	No.	Whole blood (p.p.b.)	Plasma (p.p.b.)	Time between clinical symptoms and blood test	Reference
Dieldrin	1	100	—	4 months	KAZANTZIS *et al.* (1964)
	3	40-280	—	—	KAZANTZIS *et al.* (1964)
	5	130-368	—	2-5 weeks	BROWN *et al.* (1964)
	1	320	—	—	ROBINSON and HUNTER (1966)
	12	20-530	—	1-2 weeks	
	1	—	279[a]	18 hours after ingesting aldrin	DALE *et al.* (1966)
Lindane	1	—	290	6 hours after ingesting lindane	DALE *et al.* (1966)

[a] Also reported 36 p.p.b. aldrin in blood plasma.

Table V. *Organochlorine compounds in whole blood and fat of human beings with no known occupational exposure*

Country	No.	*p.p'*-DDT & *p.p'*-DDE		Dieldrin		Reference
		Whole blood (p.p.b.)	Fat (p.p.m.)	Whole blood (p.p.b.)	Fat (p.p.m.)	
UK	12	—	—	0.5-4	—	ROBINSON (1963)
UK	20	—	—	0.5-10	—	BROWN *et al.* (1964)
UK	1[a]	—	—	6.0	—	KAZANTZIS *et al.* (1964)
UK	44	5-38	1.0-11.7	0.6-5.0	0.1-0.73	ROBINSON and HUNTER (1966)
USA	18[b]	2-58	2.2-31	Traces	0.002-0.8	SCHAFER and CAMPBELL (1966)
USA	63	<8-60	—	—	—	SCHAFER and CAMPBELL (1966)
USA	10	6.9-34	—	<0.01-2.8	—	DALE[c] *et al.* (1966)
USA	20	12.5-121	—	0.6-12.9	—	DALE[d] *et al.* (1967)

[a] One "nonexposed" subject excluded from table on basis of positive E.E.G. abnormalities.

[b] One subject excluded from table on basis of author's conclusion that he had been exposed to DDT.

[c] Also reported *o,p'*-DDT, *β*-BHC, and heptachlor epoxide.

[d] Also reported levels of *o,p'*-DDT, *o,p'*-DDE, *p,p'*-DDD, *β*-BHC, *γ*-BHC, and heptachlor epoxide.

clinical signs of intoxication had blood levels in excess of this value for a considerable period of time after clinical symptoms had subsided. The range in levels observed for the occupationally exposed persons with no clinical signs of intoxication probably are a reflection of the degree of exposure and the personal habits of the worker.

V. Studies relating blood levels to total body burden in the general population

The proposed threshold concentration for dieldrin in the blood of occupationally exposed human beings stimulated interest in the measurement of levels of insecticides in the blood of the general population. Two studies have been reported in which the levels of DDT and dieldrin have been measured in the blood and fat of unexposed human beings. ROBINSON and HUNTER (1966) reported finding for 44 individuals geometric mean values of 0.013 p.p.m. of DDT (range 0.005 to 0.038) in the blood and 4.0 p.p.m. (range 1.0 to 11.7) in the fat, and dieldrin levels of 1.4 p.p.b. (range 0.6 to 5.0) in the blood and 0.22 p.p.m. (range 0.10 to 0.73) in the fat. The ratio of mean concentrations of dieldrin in whole blood and depot fat, 159, was significantly different than the ratio, 306, for DDT. The concentrations of total DDT compounds in the blood and depot fat were not significantly correlated; a similar lack of correlation was found for dieldrin. A significant correlation was found, however, when these dieldrin data on the "general population" ($n=44$) were summarized on a plot of $\log_{10}$ p.p.m. of blood versus $\log_{10}$ p.p.m. of fat along with data on occupationally exposed formulation plant operatives ($n=2$), volunteers receiving 50 µg. of dieldrin/day for three months ($n=3$), and volunteers receiving 211 µg./day for three months ($n=3$). SCHAFER and CAMPBELL (1966) reported a range of 0.002 to 0.094 p.p.m. of DDT plus DDE in the blood of 19 individuals from Montgomery County, Ohio. The fat levels ranged from 2.2 to 31.0 p.p.m. Traces of dieldrin in blood (minimum detectable limit with their method was about 0.001 p.p.b.) and a range of 0.002 to 0.8 p.p.m. of dieldrin in the fat was reported. The authors also reported a correlation value, R^2, of 0.77 between the blood and fat levels of DDT plus DDE for 18 of the 19 subjects. The one subject that was excluded from statistical analysis had the highest blood level observed in the experiment (0.094 p.p.m.) and a fat level approximately equal to the average of the group (5.4 p.p.m.). According to these authors, available evidence concerning this subject pointed to a recent exposure to DDT.

Caution must be exercised in drawing conclusions from data in these two studies. The subjects for the two surveys were not obtained by a planned random subsampling plan, the number of observations was too small, and there was a large source of unexplained variation in both studies. The 44 British subjects sampled by ROBINSON and HUNTER were hospital patients. The 19 U.S.A. subjects examined by SCHAFER and CAMPBELL were victims of accidental deaths. Emphasis in exposure studies, such as that of BROWN *et al.* (1964), was on the concentration of insecticide in the blood at the time intoxication was observed. Their data indicated a correlation between the blood levels and symptomatology of intoxication. They observed considerable individual variation, however. The variables causing this variability in blood levels at the time of intoxication might also be those that produced the wide range for unexposed human beings that was observed in the two studies described above.

To attain results that can be generalized for the whole population, future studies having a more general sampling plan are needed. The mathematical models used to observe correlations in the above mentioned studies were empirical curve fits. Additional data may suggest a more reliable mathematical form.

All of the data on levels of insecticides in the blood of the general human population were summarized, along with available corresponding data for fat, in Table V. It will be observed that none of the 188 blood assays exceeded levels of 0.15 p.p.m. of *p,p'*-DDT plus *p,p'*-DDE and 0.015 p.p.m. of dieldrin.

A generalization similar to this was used by LAIRD (1967) in evaluating the health hazard from the 1958 larvicidal experiments in the Tokelau Islands of the South Pacific. Briquettes containing dieldrin had been in the water supply of some of the households for a five-year period prior to assaying samples of human serum. Since dieldrin concentrations of 0.1 to 4.7 p.p.b. were observed in the blood serum of twenty natives using this water supply, it was assumed that a minimum health hazard was associated with this experiment.

VI. Conclusions

The general world population now carries a body burden of organochlorine compounds with insecticidal activity. The level in any given individual depends upon his history of exposure. Correlations are being sought between blood levels of these insecticidal compounds plus their metabolites and the total body burden. After known exposures to these organochlorine compounds, measureable amounts are observed in the blood of animals along with metabolites for such compounds as *p,p'*-DDT, aldrin, and heptachlor. After a single ingestion the level of each insecticide reaches a maximum in the blood and then decreases. The time required to produce this maximum concentration is dependent on the chemical and the animal species. Feeding studies with different levels of intake of dieldrin or DDT produce blood levels that reflect both the time between exposure and testing and the exposure level.

Several species-dependent blood concentrations have been suggested as indicative of known exposures. Critical values have been suggested for blood levels of one pesticide, endrin, in fish. All catfish killed from exposure to endrin had blood concentrations in excess of 0.30 p.p.m. while the critical value for gizzard shad was 0.100 p.p.m. A blood level of 15 p.p.b. has been suggested for human beings as indicative of industrial exposure to aldrin and/or dieldrin. A threshold concentration of 150 to 200 p.p.b. for dieldrin has been proposed for human beings and dogs. With a blood level higher than the threshold concentration, the man or dog is in critical condition requiring some as yet unidentified factor or factors to precipitate intoxication.

Two surveys of the general population of human beings have been conducted to ascertain possible correlations between blood levels of insecticides and the total body burden. There was a large source of unexplained variation in both surveys. The number of observations was small and subjects for these surveys were not obtained by a planned random subsampling plan. However, none of the 188 blood assays reported in these two surveys exceeded levels of 0.15 p.p.m. of *p,p'*-DDT plus *p,p'*-DDE and 15 p.p.b. of dieldrin.

Summary

Many of the organochlorine compounds useful as pesticides are not readily degraded in the environment, nor are they detoxified *in vivo*. The public health significance of these compounds in our environment is under close scrutiny. Although these compounds are very insoluble in water, two of them, DDT and dieldrin, have been observed in blood during exposure episodes at concentrations $>$ 1,000 p.p.m. for DDT and $>$ 0.5 p.p.m. for dieldrin. Several "threshold" concentrations have been proposed. Dieldrin levels in blood of 0.015 p.p.m. are indicative of industrial exposure to aldrin and/or dieldrin. Blood levels of 0.15 to 0.20 p.p.m. of dieldrin indicate that an exposed man or dog may display symptoms of severe intoxication from the insecticide. Channel catfish that died from exposure to endrin had blood levels in excess of 0.3 p.p.m., while another species of fish, gizzard shad, had a threshold endrin concentration of 0.1 p.p.m. The ingestion of sublethal concentrations of organochlorine compounds such as DDT and dieldrin produces vague symptomatology similar to that caused by a wide variety of other substances. With the development of sensitive specific assay methods for these compounds, it has become possible to measure the concentration of DDT and dieldrin in the blood of the human population that is not occupationally exposed. It has been proposed that blood levels of these materials might be used to indicate exposure levels or total body burden. Evidence to date indicates that levels of DDT in human blood in excess of 0.15 p.p.m. and dieldrin in excess of 0.015 p.p.m. are indicative of either a recent exposure over and above that normally assimilated from the environment or the mobilization of fat depots associated with a loss in total body weight. Insufficient data are available to determine if blood concentrations are reliable indications of the total body burden of the general population.

References

ARCHER, T.E., and D.G. CROSBY: Gas chromatographic measurement of toxaphene in milk, fat, blood, and alfalfa hay. Bull. Environ. Contamination Toxicol. 1, 70 (1966).

BARNES, J.M., and D.F. HEATH: Some toxic effects of dieldrin in rats. Brit. J. Ind. Med. 21, 280 (1964).

BROWN, V.K.H., C.G. HUNTER, and A. RICHARDSON: A blood test diagnostic of exposure to aldrin and dieldrin. Brit. J. Ind. Med. 21, 283 (1964).

BRUNGS, W.A., and D.I. MOUNT: Lethal endrin concentration in the blood of gizzard shad. J. Fisheries Research Board 24, 2 (1967).

CROSBY, D.G., and T.E. ARCHER: A rapid analytical method for persistent pesticides in proteinaceous samples. Bull. Environ. Contamination Toxicol. 1, 16 (1966).

DALE, W.E., M.F. COPELAND, and W.J. HAYES,JR.: Chlorinated insecticides in the body fat of people in India. Bull. World Health Org. 33, 471 (1965).

—, A. CURLEY, and C. CUETO: Hexane extractable chlorinated insecticides in human blood. Life Sciences 5, 47 (1966).

—, —, and W.J. HAYES, JR.: Determination of chlorinated insecticides in human blood. Ind. Med. and Surg. 36, 275 (1967).

—, T.B. GAINES, and W.J. HAYES, JR.: Poisoning by DDT: Relation between clinical signs and concentration in rat brain. Science 142, 1474 (1963).

DURHAM, W.F.: Pesticide exposure levels in

man and animals. Arch. Environ. Health 10, 842 (1965).

EGAN, H., R. GOULDING, J. ROBURN, and J. O'G. TATTON: Organo-chlorine pesticide residues in human fat and human milk. Brit. Med. J. 2, 66 (1965).

GOODWIN, E.S., R. GOULDEN, and J.G. REYNOLDS: Rapid identification and determination of residues of chlorinated pesticides by crops by gas-liquid chromatography. Analyst 87, 169 (1962).

HAYES, W.J., JR.: Review of the metabolism of chlorinated hydrocarbon insecticides especially in mammals. Ann. Rev. Pharmacol. 5, 27 (1965).

HEATH, D.F., and M. VANDEKAR: Toxicity and metabolism of dieldrin in rats. Brit. J. Ind. Med. 21, 269 (1964).

HOFFMAN, W.S., W.I. FISHBEIN, and M.B. ANDELMAN: Pesticide storage in human fat tissue. J. Am. Med. Assoc. 188, 819 (1964).

HUNTER, C.G.: Human exposure to organochlorine pesticides. Proc. Roy. Soc. Med. 60, 27 (1967).

JAIN, N.C., C.R. FONTAN, and P.L. KIRK: Simplified gas chromatographic analysis of pesticides from blood. J. Pharm. Pharmacol. 17, 362 (1965).

JUDAH, J.D.: Studies on the metabolism and mode of action of DDT. Brit. J. Pharmacol. 4, 120 (1949).

KADIS, V.W., and O.J. JONASSON: The detection and persistence of chlorinated insecticides in human and animal blood. Canadian J. Public Health 56, 433 (1965).

KAZANTZIS, G., A.I.G. MCLAUGHLIN, and P.E. PRIOR:

Poisoning in industrial workers by the insecticide aldrin. Brit. J. Ind. Med. 21, 46 (1964).

LABEN, R.C., T.E. ARCHER, D.G. CROSBY, and S.A. PEOPLES: Lactational output of DDT fed prepartum to dairy cattle. J. Dairy Sc. 48, 701 (1965).

LAIRD, M.: A Coral Island experiment. A new approach to mosquito control. WHO Chronicle 21, 18 (1967).

LAUG, E.P.: 2,2-Bis (p-chlorophenyl)-1,1,1-trichloroethane (DDT) in the tissues, body fluids and excreta of the rabbit following oral administration. J. Parmacol. 86, 332 (1946).

MOSS, J.A., and D.E. HATHAWAY: Transport of organic compounds in the mammal. Partition of dieldrin and telodrin between the cellular components and soluble proteins of blood. Biochem. J. 91, 384 (1964).

MOUNT, D.I., L.W. VIGOR, and M.L. SCHAFER: Endrin: Use of concentration in blood to diagnose acute toxicity to fish. Science 152, 1388 (1966).

QUINBY, G.E., W.J. HAYES, JR., J.F. ARMSTRONG and W.F. DURHAM: DDT storage in the United States population. J. Amer. Med. Assoc. 191, 109 (1965).

REYNOLDS, J.G.: The determination of traces of dieldrin in blood. Boletin de la Oficina Sanitaria Panamericana 43, 527 (1957).

RICHARDSON, L.A., J.R. LANE, W.S. GARDNER, J.T. PEELER, and J.E. CAMPBELL: Relationship of dietary intake to concentration of dieldrin and endrin in dogs. Bull. Environ. Contamination Toxicol. 2, 207 (1967).

ROBINSON, J.: The determination of dieldrin in the blood by gas liquid chromatography. Forensic Immunology, Med. Path., and Toxicol. Rept. 3rd Internat. Meeting. Abridged Proc. Excerpta Medica Intern. Congress Series No. 80, 135 (1963).

—, and C.G. HUNTER: Organochlorine insecticides: Concentrations in human blood and adipose tissue. Arch. Environ. Health 13, 558 (1966).

ROTHE, C.F., A.M. MATTSON, R.M. NUESLEIN, and W.J. HAYES, JR.: Metabolism of chlorophenothane (DDT). Arch. Ind. Health 16, 82 (1957).

SASCHENBRECKER, P.W., and D.J. ECOBICHON: Extraction and gas chromatographic analysis of chlorinated insecticides from animal tissues. J. Agr. Food Chem. 15, 168 (1967).

SCHAFER, M.L., and J.E. CAMPBELL: Distribution of pesticide residues in human body tissues from Montgomery County, Ohio. Adv. Chem. Series 60, 89 (1966).

SMITH, M.I., and E.F. STOHLMAN: The pharmacologic action of 2,2-bis (p-chlorophenyl)-1, 1,1-trichloroethane and its estimation in the tissues and body fluids. Public Health Reports 59, 984 (1944).

STIFF, H.A., and J.C. CASTILLO: The determination of 2,2-bis (p-chlorophenyl)-1,1,1-trichloroethane (DDT) in organs and body fluids after oral administration. J. Biol. Chem., 545 (1945).

WASSERMANN, M., D. WASSERMAN, L. ZELLERMAYER, and M. GON: Pesticides in people. Storage of DDT in the people of Israel. Pesticides Monitoring Journal 1, 15 (1967).

WHITING, F.M., and J.W. STULL: Rate of transfer of DDT from the blood compartment. Bull. Environ. Contamination Toxicol. 1, 187 (1966).

ESTIMATING PESTICIDE EXPOSURE IN MAN AS RELATED TO MEASURABLE INTAKE; ENVIRONMENTAL VERSUS CHEMICAL INDEX

LLOYD A. SELBY, KENNETH W. NEWELL, CARMEL WAGGENSPACK, GEORGE A. HAUSER AND GLADYS JUNKER[1]

INTRODUCTION

The toxic potential of pesticide intake upon human populations has been suggested (1–4). Although toxic effects have most frequently been described for persons with massive occupational or accidental exposure, other individuals in a general population may be exposed continuously to small quantities of pesticides from a wide variety of environmental sources.

The effects of massive pesticide exposure and intake on occupational groups may be specific (3, 5, 6) or non-specific (4, 7–9). Specific effects have been observed in formulators and spray operators with heavy organophosphate exposure and intake. It is believed the impact on the kidney resulted in a change in its function or structure (5).

The effects of a low-level pesticide exposure and intake upon a general population may, with time, involve subtle insult to the same target organs as those effected by high level exposure. This type of exposure might not cause overt, acute disease, but could cause discrete physiological changes conducive to a chronic disease. However, this remains to be proven. Long-term effects could be measured in a general population by a comparison of subgroups with different levels of estimated pesticide intake. The problem of estimating intake is thus basic to such studies.

An available method of estimating pesticide intake is a mechanism which traps the pesticide (10, 11), but this is expensive and impractical on general population studies. Other methods presently available for estimating the pesticide intake of an individual in a general population require the laboratory analysis of body tissues. Although laboratory methods exist for the analysis of pesticide residues, it is impractical to obtain large numbers of biopsy specimens from individuals in a general population or to analyze large numbers of specimens for pesticide residues without major laboratory facilities.

If epidemiologic studies are to be done on the pesticide intake of a general population, a practical, subjective method must be devised to estimate an individual's levels. This method could then define within this population, subgroups with different pesticide levels.

The Pesticide Exposure Index (PEX)[a] was used to estimate individual pesticide intake. A mechanism was evolved to obtain all data relevant to an individual's pesticide exposure. These data were obtained by direct interview from the user rather than the supplier.

The PEX values were compared to the residual pesticide levels in tissue specimens of the patient. The body tissues were collected from the individuals before they were interviewed. The analysis of specimens and the interview were carried out independently, i.e., in a "double blind" manner.

Lafourche Parish was chosen for the study because of its well developed hospital facilities, active medical society and health department. Also, it was probable that the population in Lafourche had a variation in its exposure to pesticides.

MATERIALS AND METHODS

Study criteria

Pesticide Exposure Indices were determined only for individuals from whom tissue specimens were obtained. These individuals were over 20 years of age; had resided in the study area at least one year, prior to the date the initial specimen was collected; and were patients in one of the three hospitals in Lafourche Parish, Louisiana. Adipose and blood specimens were obtained from elective surgical patients. Placental and maternal blood specimens were obtained from women who delivered spontaneously.

[a] **PEX will be used in this paper to denote the Pesticide Exposure Index, Study, Interview or Value calculated from the Index.**

Persons from whom specimens were collected, or their relatives, were interviewed to determine their pesticide usage soon after the patient was discharged from the hospital. Written consent was obtained for the use of clinical material.

Because specific confirmation for the organophosphate compounds is difficult, specimens were analyzed only for the chlorinated hydrocarbon pesticides. For the latter group, specific and sensitive methods of analyses have been described (12–14). Experimental studies in our laboratory indicated that with the methods employed and described (14), most chlorinated hydrocarbons were detected in artificially spiked lard samples, i.e., spiked at a level of 1.0 ppm or 0.1 ppm respectively.[7]

Collection of specimens and data

Five to ten grams of adipose tissue were collected by the surgeon, placed in a 1 oz. wide-mouth glass jar with a metal screw-cap lid and frozen without preservatives. These jars had been washed and rinsed with hexane. Whole placentas were obtained from obstetrical patients, wrapped in aluminum foil, and frozen without preservatives.

At least 10 cc of whole blood was collected from both elective surgical patients and the women from whom placentas had been collected. The blood specimens were obtained, using 10 cc vacutainers containing 25 mg of potassium oxalate, and refrigerated until delivered to the laboratory.

Each hospital in Lafourche Parish was visited weekly by a study nurse to obtain specimens collected, and to determine if the patient met the study criteria. If so, arrangements were made for a nurse to interview him at home. A nurse interviewed the patient, using the direct interview technique.

Information obtained by the nurse was then recorded by use of a standard record system (15). The data collected included demographic information on the patient, his major daily activities, the location and use of the unit of land on which he worked, resided, and/or enjoyed recreation, and the pesticide usage on each of these units.

Specifically the "Case Interview Data" record included data on the individual's occupation, length of residence in Lafourche Parish, and his daily activities, including where he spent most of his waking hours. A large section of the record was devoted to data relevant to the known pesticide exposure of the individual for the 12 month period immediately preceding the collection of the specimens. Detailed questions were asked concerning the mixing, application or physical presence when a pesticide was used. For each pesticide additional data was obtained including the name and percentage of each active ingredient, total amount used, amount and name of the diluent and the location where the pesticide was mixed and applied. Finally confirmatory data was obtained from the pest control companies which serviced one or more of the areas in which the individual was considered to be exposed.

Laboratory analyses of specimens

All specimens that had detectable chlorinated hydrocarbon pesticides in the initial gas chromatography screening test, were subjected to further confirmatory tests. These tests were the official methods presented in *Guide to the Analysis of Pesticides Residues*, Vol. I and II (14) or modifications. Microcoulometric-GLC, gas chromatography, or thin layer chromatography were used as confirmatory tests in which pesticide levels of 0.1 ppm or higher were recovered from the tissue. A majority of the blood and placental specimens were confirmed only on a second GLC column because of the extremely low levels (parts per billion range) of pesticides recovered.

Adipose and placental tissue were extracted by the "Mills petroleum ether extraction procedure" (16). The extract was cleaned using either a "Florisil Column" or "Acetonitrile Partitioning Method" (14).

[7] Unpublished data.

Pesticide residues in the extract were quantitated using electron capture gas liquid chromatography (EC-GLC) with a SE30, QF1 column or DC-200 column. If the residue level was 0.1 ppm or greater, the level was confirmed by microcoulometric-GLC or thin layer chromatography. If the residue level was less than 0.1 ppm but 0.1 ppb or greater, the extract was assayed using both the SE30, QF1 and DC-200 columns. The adipose results were reported in ppm (10^{-6} gm/gm), while the placental results were reported in ppb (10^{-9} gm/gm).

Blood specimens were extracted using the "Mills petroleum ether extraction procedure" (16) or the "Hexane Extraction Procedure" as reported, 1966, by Dale et al. (17). The blood extract was not cleaned. Residues were quantitated using EC-GLC with a SE30, QF1 column and confirmed by EC-GLC with a DC-200 column if 0.1 ppb or greater. These results were also reported in ppb (10^{-9} gm/gm).

Study data

During the study period (August 5, 1966–March 6, 1967), a total of 115 patients met the study criteria. Adipose and blood specimens were collected from 62 elective surgical patients, and placental and maternal blood specimens were collected from 53 women who delivered spontaneously. An additional placenta was collected from one woman who gave birth to twins.

The major land use areas of the study area and the geographic distribution of residence for the individuals in the population surveyed were defined as: Area I, Rural cane; Area II, Rural Non-cane; and Area III, Urban. For analytical purposes, each person in the study population was located geographically by residence into one of these three areas.

Fewer males were drawn into the study population, with approximately one male to 10 females in the elective surgical patients. In addition, only two (1.7 per cent) of the specimens were obtained from the non-white population. The majority of those surveyed were white, female housewives.

Using "occupation group" as defined by the U.S. Census Bureau, categories of "farmers" and "sales personnel" were not represented in the study population. While 97 (84.3 per cent) individuals stated they were unemployed (housewives and one retired male), 18 (15.7 per cent) stated they were gainfully employed at the time they entered the hospital.

In presenting the laboratory findings of the study, Total Equivalent (TE) values were calculated for DDT, BHC and heptachlor. Total Equivalent DDT was defined as the sum of the values for DDT, DDE and DDD; while the TE of BHC was defined as the sum of the values for Lindane and BHC; and the TE of heptachlor as the sum of the values for heptachlor and heptachlor epoxide. Because of molecular weight differences between the equivalent compounds and the other metabolites or epoxides, correction factors (1.115 and 1.108) were used with DDE and DDD respectively to calculate Total Equivalent DDT. A correction factor (0.959) was used with heptachlor epoxide to calculate the Total Equivalent Heptachlor (18). The descriptive results of the laboratory findings for placental and maternal blood specimens by pesticide recovered will be presented elsewhere (19).

The PEX model, however, was not based on the individual properties of the pesticide, nor on laboratory analysis of the body tissues. The PEX represented the relationship between man, his pesticide usage, and his environment (here defined in terms of land or buildings). It was based on information collected, by the questionnaire, used collectively with correction factors to determine a PEX value, for each active pesticide ingredient.

Variables considered important in the development of the PEX were those related to exposure, the individual, the pesticide and the environment. Factors considered practical to measure with a questionnaire were duration; frequency and location of exposure; the activity of the individual during exposure, e.g., mixing the pesticide; and the percentage of active ingredient(s), physical state, area of application and method of applying the pesticide. Certain correction factors were necessary because part of the data collected was considered essential in the calculation of the PEX value, but not usable in its original form. The correction factors, determined prior to the study, included the potential routes of exposure; physical state of the pesticide; location of application, i.e., indoors or outdoors; an estimate of the individual's degree of exposure based on his pesticide activity, e.g., applying the pesticide; and an estimated degradation period for the product by general class of pesticide, e.g., chlorinated hydrocarbons. The PEX was computed initially in gram-hours; evaluation of the data, however, suggested kilogram-days as a more meaningful unit of measurement.

The relationship between the PEX values and tissue-blood levels of pesticides detected in each individual was evaluated by a χ^2 test of Kendall's rank correlation coefficient (20). This included the ranking of each individual by region and type of paired tissue-blood specimens for Total Equivalent DDT. Total Equivalent BHC, Total Equivalent heptachlor, dieldrin and aldrin. The rationale for the χ^2 evaluation of Kendall's tau was that by definition $\chi^2_{(\alpha)} = 2 \Sigma \log_e Pij$, where Pij was the probability of occurrence of each calculated tau value. Degrees of freedom was considered to be twice the number of tau values not equal to one. When an individual tau value equaled one, the probability was zero and the log of the probability was equal to minus infinity. Each tau score was calculated and its probability was determined from a table which accounted for tied ranks (21). Then the observed χ^2 value was calculated.

Also, the relationship between the DDT PEX values and tissue-blood levels of the population as a whole was analyzed by Kendall's rank correlation coefficient (20).

Results

A frequency distribution of the pesticides detected in the adipose-blood specimens is presented in figure 1. As illustrated in this figure, DDE, DDT, Lindane, BHC, and heptachlor epoxide were detected most frequently. A majority of the 21 adipose specimens that had detectable levels of heptachlor also had detectable levels of heptachlor epoxide at a higher level. In the 10 blood specimens with detectable heptachlor, heptachlor epoxide either was not detected or it was detected at a lower level. However, heptachlor or heptachlor epoxide was also detected in nine of the 10 paired adipose specimens. In six of the nine adipose specimens, the detected level of heptachlor epoxide was greater than the heptachlor level. Of the eight blood specimens with detectable levels of aldrin, six also had detectable levels of dieldrin and in four of the six, the dieldrin level was greater than the aldrin level.

The distribution of the laboratory values was skewed, therefore a logarithmic transfer of the data was used to obtain the geometric means. The geometric mean, variance and range of the chlorinated hydrocarbons detected are presented in table 1 and 2 for adipose and blood specimens respectively.

The major land use areas of the study and the geographic distribution of residence for each individual in the population surveyed were depicted (figure 2). These data revealed that 54 (46.9 per cent) resided in a rural cane area, 47 (40.9 per cent) in an urban area and 13 (11.3 per cent) in a rural non-cane area. The area of residence was unclassified for one (0.9 per cent) individual.

Sufficient data was obtained to compute PEX values for 95 of the 115 individuals in the study population. Three persons

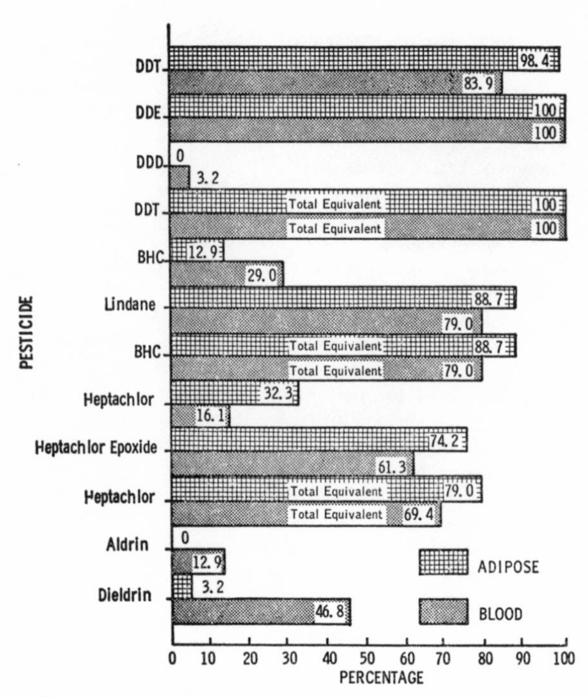

FIGURE 1. Percentage of adipose and blood specimens containing detectable levels of pesticide residues.

TABLE 1

Concentration of chlorinated hydrocarbon pesticides detected in 62 adipose specimens

Compound	No. of positive individuals	Concentration expressed as logarithm$_{10}$		Geometric means (ppm)	Range of observation (ppm)
		Mean	Variance		
DDT	61	0.1202	0.4895	1.32	0.199–16.884
DDE	62	0.5685	0.3675	3.70	0.195–27.302
DDD	0	—	—	—	—
DDT$_{T.E.}$*	62	0.7097	0.3988	5.13	0.195–41.418
BHC	8	−1.2389	0.2496	0.06	0.018–0.128
Lindane	55	−0.8140	0.3444	0.15	0.037–4.156
BHC$_{T.E.}$*	55	−0.8140	0.2852	0.15	0.037–0.815
Heptachlor	20	−0.9203	0.5989	0.12	0.021–2.264
Heptachlor Epoxide	46	−0.8476	0.3457	0.14	0.024–0.928
Heptachlor$_{T.E.}$*	49	−0.7010	0.4104	0.20	0.033–2.264
Aldrin	0	—	—	—	—
Dieldrin	2	−0.4682	0.9046	0.34	0.078–1.484

* T.E. = Total Equivalent.

TABLE 2

Concentration of chlorinated hydrocarbon pesticides detected in 62 blood specimens

Compound	No. of positive individuals	Concentration expressed as logarithm$_{10}$		Geometric means (ppb)	Range of observation (ppb)
		Mean	Variance		
DDT	52	0.3895	0.4792	2.45	0.3–28.8
DDE	62	0.4084	0.3347	2.56	0.3–22.4
DDD	2	−0.4112	0.0632	0.39	0.3–0.4
DDT$_{T.E.}$*	62	0.6749	0.4210	4.73	0.3–48.1
BHC	18	−0.5542	0.3579	0.28	0.1–0.9
Lindane	49	−0.3931	0.3646	0.40	0.1–4.1
BHC$_{T.E.}$*	49	−0.2984	0.3793	0.50	0.1–4.1
Heptachlor	10	−0.3846	0.7276	0.41	0.1–0.9
Heptachlor Epoxide	38	−0.6849	0.3004	0.21	0.1–3.5
Heptachlor$_{T.E.}$*	43	−0.5631	0.4548	0.27	0.1–6.5
Aldrin	8	−0.1730	1.0440	0.67	0.1–65.3
Dieldrin	29	−0.5539	0.4538	0.28	0.1–5.6

* T.E. = Total Equivalent.

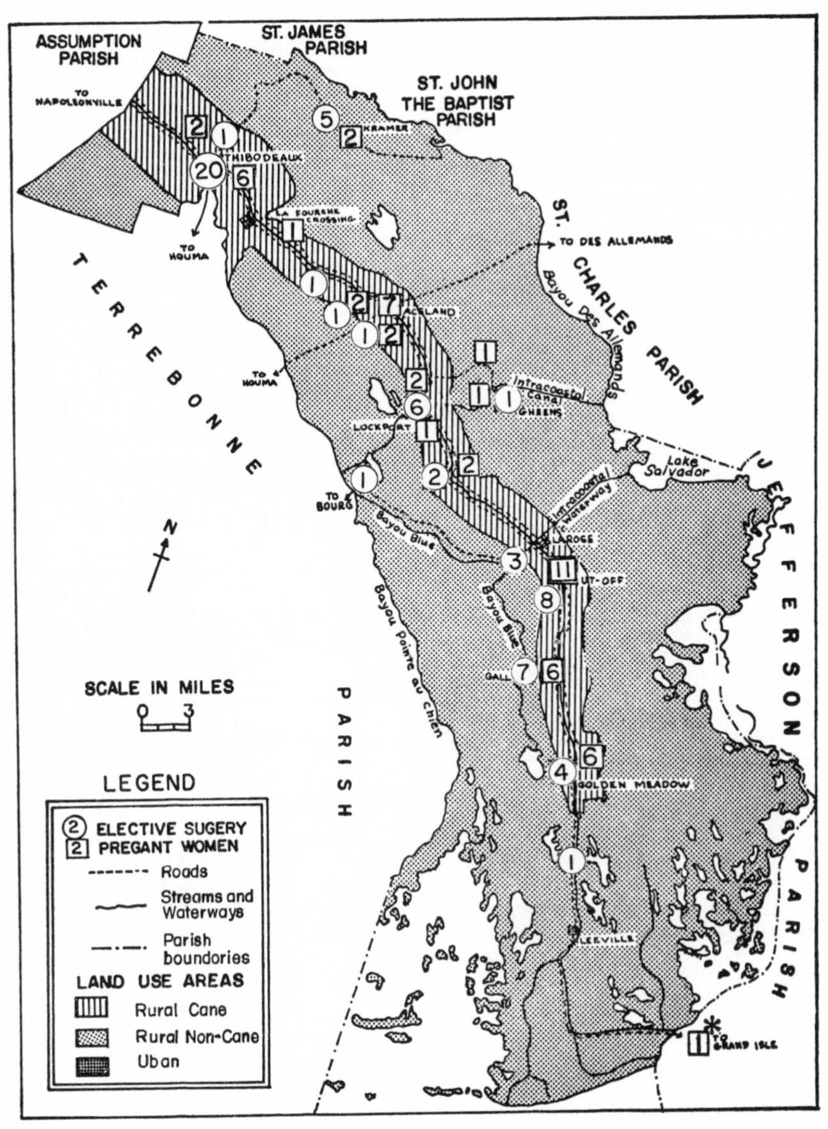

FIGURE 2. Geographic distribution by area of residence of the 115 individuals surveyed in the population of Lafourche Parish, Louisiana.

Table 3

*Chi Square values based on rank correlation between individual adipose-blood specimens for each person for total equivalent DDT, total equivalent BHC, total equivalent heptachlor, dieldrin and aldrin**

Sex	Area of residence	No. of persons	Degrees of freedom	χ^2 Point of rejection H:$_0$† if χ^2 (α = .05)	Observed χ^2	Conclusion α = .05
Female	I. Rural cane	18	32	46.2	72.8	Reject H:$_0$
	II. Rural non-cane	9	12	21.1	29.4	Reject H:$_0$
	III. Urban	26	48	73.9	107.2	Reject H:$_0$
	All areas	53	92	114.4	209.4	Reject H:$_0$
Males	All areas	9	16	26.3	44.9	Reject H:$_0$

* Values for categories containing less than 5 individuals were not calculated. However, their values were included in calculating the total for all areas.

† There is no correlation between the rank of the detected levels of total equivalent DDT, total equivalent BHC, total equivalent heptachlor, dieldrin, and aldrin.

Table 4

*Chi Square values based on rank correlation between individual placenta-blood specimens for each person for total equivalent DDT, total equivalent BHC, total equivalent heptachlor, dieldrin and aldrin**

Sex	Area of residence	No. of persons	Degrees of freedom	χ^2 Point of rejection H:$_0$† if χ^2 (α = .05)	observed χ^2	Conclusion α = .05
Female	I. Rural cane	32	46	76.3	121.9	Reject H:$_0$
	II. Rural non-cane	4	—	—	—	—
	III. Urban	16	30	53.8	73.9	Reject H:$_0$
	Unclassified	1	—	—	—	—
	All areas	53	86	113.2	212.8	Reject H:$_0$

* Values for categories containing less than 5 individuals were not calculated. However, their values were included in calculating the total for all areas.

† There is no correlation between the rank of the detected levels of total equivalent DDT, total equivalent BHC, total equivalent heptachlor, dieldrin and aldrin.

stated they had no known pesticide exposure and PEX values were not computed for three individuals who were unaware of exposure to pesticides. Of the PEX values computed, 132 (60.6 per cent) were calculated for the chlorinated hydrocarbon compounds and 86 (39.4 per cent) for the organic phosphates.

Of the 95 persons with a known pesticide exposure, 66 stated they applied but did not mix the pesticide, and 49 stated they were present when someone else mixed or applied the pesticide. In addition, 31 (32.6 per cent) persons stated they utilized the services of a Pest Control Company. Also, 65 (68.4 per cent) of the 95 stated that aerial spraying of the fields adjacent to their residence had occurred but they could not provide the essential data necessary to compute a PEX value.

The rank correlation between the tissue-blood levels for adipose-blood and placental-maternal blood tissue is presented in tables 3 and 4 respectively. In addition, the rank correlation between the PEX values and tissue-blood levels of individuals surveyed by area of residence, sex and type of specimen is presented in table 5. There was no association α = .05 between an individual's PEX and tissue-blood values while there was a relationship observed between the tissue-blood values for each individual.

In the comparison between the PEX and tissue values, 45 of the 132 PEXs computed for the chlorinated hydrocarbons were for

TABLE 5

*Chi Square values based on rank correlation between individual PEX values—specimen residual values for each person for total equivalent DDT, total equivalent BHC, total equivalent heptachlor, dieldrin, and aldrin**

Specimen type	Sex	Area of residence	No. of persons	Degrees of freedom	χ^2 Point of rejection $H:_0$† if χ^2 ($\alpha = .05$)	Observed χ^2	Conclusion $\alpha = .05$
Adipose	Female	I. Rural cane	18	36	58.9	26.7	Accept $H:_0$
		II. Rural non-cane	9	18	28.9	12.6	Accept $H:_0$
		III. Urban	23	46	65.2	32.5	Accept $H:_0$
		All areas	50	100	109.2	71.8	Accept $H:_0$
	Male	All areas	9	18	28.9	12.7	Accept $H:_0$
Blood	Female	I. Rural cane	42	78	110.3	54.3	Accept $H:_0$
		II. Rural non-cane	13	26	38.9	27.2	Accept $H:_0$
		III. Urban	33	64	83.7	37.5	Accept $H:_0$
		All areas	88	168	199.3	119.0	Accept $H:_0$
	Male	All areas	9	18	28.9	12.7	Accept $H:_0$
Placenta	Female	I. Rural cane	24	46	65.3	18.7	Accept $H:_0$
		II. Rural non-cane	4	—	—	—	—
		III. Urban	9	18	28.9	10.4	Accept $H:_0$
		All areas	37	70	92.9	40.0	Accept $H:_0$

* Values for categories containing less than 5 individuals were not calculated. However, their values were included in calculating the total for all areas.

† There is no correlation between the PEX rank and detected levels of total equivalent DDT, total equivalent BHC, total equivalent heptachlor, dieldrin and aldrin.

compounds recovered by the laboratory methods used in this study. Seventeen (37.8 per cent) of the 45 PEXs were computed for DDT whereas 16 (35.6 per cent) were computed for dieldrin. In examining the paired specimens from persons for whom the homologous index was computed, nine had a DDT PEX and DDT and/or DDE residues in their adipose-blood specimens, and eight had DDT and/or DDE residues in their placenta-blood specimens while only two of the 16 with a PEX to dieldrin had dieldrin residues in their tissue-blood specimens.

Analysis of the PEX and tissue-blood levels, of the population as a whole, included a comparison of Total Equivalent DDT values in adipose and blood specimens. In order to include those individuals with no PEX or tissue-blood levels for DDT. Kendall's rank correlation coefficient tau was used. At a level of less than $\alpha = .05$, the hypothesis of no correlation between adipose and blood levels was rejected. In like manner, a rank correlation was done between the DDT, PEX values and the DDT adipose residual values.

The level of Total Equivalent DDT for all persons from whom an adipose specimen was obtained was used to test the hypothesis of no correlation between the PEX value and the adipose level of Total Equivalent DDT. Of the 62 persons used for this test, nine had a PEX other than zero. The range of the Total Equivalent DDT PEX values for those with a positive PEX was 10.50–2251.69 kilogram-days. Their median PEX value was 22.34 kilogram-days and mean value was 305.06 ± 748.6 kilogram-days. Each PEX was ranked with respect to its magnitude and compared to all PEX values. The adipose levels of Total Equivalent DDT were ranked and a test was conducted based on the ranks of the PEX and adipose levels of each of the 62 persons taking into account the tied ranks. The hypothesis of no correlation between the PEX value and the adipose level of Total Equivalent DDT at less than $\alpha = .05$ was accepted (20).

Discussion

Aldrin has been detected in human tissues in the United States (22). It is to our knowledge, however, the first reported detection of heptachlor in human tissue (9, 23), and the first reported detection of aldrin in blood specimens, collected from individuals in a general population without occupational or accidental exposure to aldrin (9). Although none of the individuals in this study gave a history of occupational or accidental exposure to either aldrin or heptachlor, 7 (6.0 per cent) individuals stated they applied or had been present in an area where heptachlor was used. Of these 7 individuals, with a known exposure to heptachlor, heptachlor epoxide was detected in the blood of 5 individuals and in the adipose of 2 individuals. In addition, heptachlor was detected in one of these individuals' blood and one individual's adipose; however, heptachlor epoxide was also detected in both their blood and adipose specimen.

The detection, in this study, of either aldrin or heptachlor could be questioned because statements in the literature suggest that it would be unlikely for one to detect either of these compounds in mammalian tissue due to a rapid conversion to their respective epoxides (4, 23). Nevertheless, the results obtained in this study support and agree with this basic premise. In a majority of cases, where either aldrin or heptachlor was detected, their respective epoxides were also detected and in most instances, at higher levels.

In addition, we feel that the laboratory methods used strengthen these findings. Peak areas and relative retention times of the detected pesticides in the samples were compared with those of known pesticide standards for each specimen. The concentration of the pesticide in the standard was adjusted to equal approximately that concentration detected in the sample. In addition, the detection of pesticides was confirmed by duplication on a second chromatographic column, or by another chromatographic method. Finally, to eliminate errors, calculations were done independently by two chemists.

These findings, however, were considered to be secondary, for the major objective of this study was to develop a subjective method for estimating an individual's pesticide exposure intake (PEX), and to compare this computed value with pesticide residues in body tissue and fluid specimens.

If a correlation existed between these values, then the PEX might be considered comparable to the tissue residual levels for estimating the relative level of pesticide intake. This analysis was therefore a comparison between two systems of measuring pesticide intake. The PEX is an environmental index, and the pesticide residual levels are a chemical index. The collection of tissue-blood specimens prior to obtaining the pesticide usage data predetermined the experimental design and population of the study.

A reason for developing a method of estimating individual pesticide intake was to allow for the subdivision of individuals in a general population into groups with different levels of pesticide intake for further study. Specific epidemiological studies on comparable subgroups of a population would be more meaningful in defining causal relationships than in defining associations.

In the development of the PEX index, a method to account for the variation in the individual and his daily activities as related to his environment was never completely resolved.

As the PEX evolved, it was apparent that certain information could not be obtained by direct interview with a questionnaire. A pretest of the record system resulted in a reevaluation of some of the components of the PEX. For example, if a person was present during or after a pesticide was applied, the duration of exposure was not determined, instead, it was based on a predetermined, estimated number of hours of exposure per week for the person depending on his major daily activity, i.e., home, work or school.

In the tabulation and analysis of the PEX values, a closer evaluation was made of all records and a general pattern of each component of the index. It was concluded that, in the development of an environmental index more accurate data must be obtained about exposure from the standpoint of daily activities.

The record system designed for this study was successful in obtaining sufficient information to compute PEX values. Sixty per cent of the computed PEXs were for the chlorinated hydrocarbons; this was probably determined by the pesticide usage pattern of the population studied. However, this fact was important because only the chlorinated hydrocarbon compounds were recovered by the laboratory methods used. Also the collection of the specimens prior to the PEX interview increased the amount of data available for cross tabulation and analysis.

Dorn (24) points out that if individuals in a study are from a hospital population, the distributions may represent characteristics of the population as well as selected effects of the cause of hospitalization. The comparative analysis of the residual levels, in this study did not include the reason for hospitalization. The population chosen was selected for observation on the basis that they furnished the study with tissue and blood specimens, but this selection was not, and is not, related to pesticide usage, exposure or intake.

An initial examination of the data suggested that the values for both the environmental index (PEX) and chemical index (tissue-blood values) were not normally distributed. For this reason, some types of analysis, including linear correlation, were not satisfactory methods for comparing the two indices.

Evaluation of the data suggested that the variation of a particular pesticide in blood did not depend on the level that is observed in adipose tissue, at least not on a linear scale. What was observed in this study was a multivariant biological system with values not normally distributed. For these reasons, the two index systems were analyzed by rank correlation (20, 21). By using this method of analysis, individuals were accounted for who did not have a PEX or whose laboratory values for a particular product were extremely low or undetected.

The results of χ^2 values based on rank correlation (table 5) demonstrated, at less than $\alpha = .05$, no relationship between the PEX values and the tissue-blood specimens. In addition, there were inherent weaknesses in the index: PEX values could not be computed for certain groups of pesticides, e.g., pyrethrum; with commonly used pesticides, variables essential in the complete picture of an individual's pesticide intake could not be accounted for by questionnaire. As a result, the PEX is not suitable for estimating pesticide intake either for an individual or for the population as a whole.

Campbell et al. (25) maintain that 90 per cent of the DDT intake in man comes from his food and drink while the remaining 10 per cent comes from other environmental sources. Although this may be true, other pesticides than DDT must be considered. That was the reason for attempting to develop an Index such as the PEX to include other pesticides and their estimated intake. However, 68.5 per cent of the patients gave environmental sources of pesticide exposure with insufficient data to compute a PEX. Thus, a number of sources of environmental exposure and pesticide intake could not be taken into account.

Eight of ten cases with a tissue-blood level for Total Equivalent DDT had a PEX of zero. Even if the predictive power of the PEX had been increased by revision, to improve the relationship between the questionnaire and the laboratory findings, the use of a questionnaire interview in a similar study population would be inappropriate, for the distribution of the detected pesticide levels in the tissues of the PEX zero group was similar to those of the PEX positive group.

Blood was comparable to adipose or placental tissue as an indicator of the relative level of pesticide intake for these pesticides. Thus a chemical index was demonstrated, by χ^2 values based on rank correlation, to be a more logical choice than the environmental index (PEX) for estimating pesticide intake with the chlorinated hydrocarbons.

While these findings are important, they are secondary to the main objective which was to evolve a subjective method for dividing a general population into subgroups with various defined levels of estimated pesticide exposure and intake. Our conclusion is that if the measurement of exposure and intake is made by adipose, placenta and blood levels, PEX values (an environmental index) as measured in this study are ineffectual. The toxic potential of pesticides upon persons in a general population cannot be examined by correlation of which pesticide the individual states he has a known exposure.

References

1. Hoogendam, L., Versteeg, J. P. J., and de Vlieger, M. Electroencephalogram in insecticides toxicity. Arch. Environ. Health, 1962, *4:* 86–94.
2. Durham, W. F. Assessment of environment exposure to pesticides, Proceedings from the short course on occupational health aspects of pesticides. Edited by Link and Whitaker, University of Oklahoma Press, Norman, Oklahoma, 1964.
3. Hoogendam, L., Versteeg, J. P. J., and de Vlieger, M. Nine years' toxicity control in insecticide plants. Arch. Environ. Health, 1965, *10:* 441–448.
4. Chichester, C. O. Research in pesticides. Academic Press, New York, 1965.
5. Davies, J. E. Personal communications, May, 1965.
6. Fisher, R. Toxic renal failure induced by the pesticide aldrin. Literature Bull., 1966, *1:* 12.
7. Hayes, W. J., Jr. Un estudio sobre la intoxication producida por Dieldrin. Bol. Ofic. Sanit. Panamer., 1957, *43:* 534.
8. Hayes, W. J., Jr. Review of the metabolism of chlorinated hydrocarbon insecticides especially in mammals. Ann. Rev. Pharm., 1965, *5:* 27–52.
9. Scientific Aspects of Pest Control. National Academy of Sciences, National Research Council, Washington, D.C., 1966.
10. Durham, W. F. and Wolfe, H. R. Measurement of the exposure of workers to pesticides. Bull. W. H. O., 1962, *26:* 75–91.
11. Durham, W. F. Pesticides exposure levels in man and animals. Arch. Environ. Health, 1965, *10:* 842–846.
12. Bobbitt, J. M. Thin layer chromatography. Reinhold Publishing Corporation, New York, 1963.
13. Zweig, Gunter. Analytical methods for pesticide plant regulators and food additives. Vol. I, Principles, methods and general application. Academic Press, New York, 1963.
14. Burchfield, H. P., Johnson, D. E., and Storrs, E. E. Guide to the analysis of pesticides residues. Vol. I and II. U.S. Department Health, Education and Welfare, Public Health Service, Office of Pesticides, Washington, D.C. 1965.
15. Waggenspack, C. and LeBlanc, D. R. Unpublished data, 1966.
16. Mills, P. A. Petroleum ether extraction of tissue, Training course for chemists, Perrine, Florida, Winter 1965.
17. Dale, W. E. Curley, A. and Cueto, C., Jr. Hexane extractable chlorinated insecticides in human blood, Life Sci., 1966, *5:* 47–54.
18. Negherbon, W. O. Handbook of Toxicology. Volume III Insecticides, W. B. Saunders Co., Philadelphia, 1959.
19. Selby, L. A., Newell, K. W., Hauser, G. H., and Junker, G. A comparison of chlorinated hydrocarbon pesticides in maternal blood and placental tissues. Accepted for publication, Environmental Research.
20. Kendall, M. G. Rank Correlation Methods. Charles Griffin and Company Limited, London, 1955.
21. Sillitto, G. P. The distribution of Kendall's coefficient of rank correlation in rankings containing ties. Biometrika, 1947, *34:* 36–41.
22. Radomski, J. L. and Fiserova-Bergerova, V. The determination of pesticides in tissues

with the electron capture detector without prior cleanup. Industr. Med. Surg., 1965, *34:* 934–939.
23. Durham, W. F. Body burden of pesticides in man. N.Y. Acad. Sci. In press.
24. Dorn, H. F. Some applications of biometry in the collection and evaluation of medical data. J. Chronic Diseases, 1955, *1:* 638–664.
25. Campbell, J. E., Richardson, L. A. and Schaefer, M. A. Insecticide residues in the human diet. Arch. Environ. Health, 1965, *10:* 831–836.

ORGANIC PHOSPHORUS INSECTICIDE POISONING IN CHILDREN

JOHN SILVERIO, M.D., F.A.A.P., F.A.S.H.A.

Accidental poisoning has increased alarmingly and is now the second most important cause of morbidity and mortality in American children (1). Each year, from 500,000 to 2,000,000 ingest toxic substances, and about 500 of those under 5 years of age die (2). Nearly all must be given emergency treatment, and those who ingest corrosive materials such as caustics or irritants usually require hospitalization.

Insecticides are responsible for 15,000 to 60,000 of these yearly accidents (Table I). Those containing the organic phosphorus esters are perhaps of greatest interest to the pediatrician. These extremely toxic compounds are absorbed through the skin and mucous membranes, and incapacitate or kill without having to be swallowed. Too often the victims are school-age children, especially youngsters who innocently handle "empty" containers and teen-agers who work part-time during vacations, helping, for example, to load spraying planes. In both instances the poisoning occurs simply by contact.

Exposure to these materials is seasonal (3) and is much more frequent in rural and agricultural than in urban areas. To illustrate, 70 such poisonings were documented in a four-year period (1960 to 1963) among adolescents helping with crop dusting at the Valley Baptist Hospital in Harlingen, Texas (4), while only 3 were documented in a similar four-year period (1964 to 1967) among 10,077 children treated at two large metropolitan hospitals (Philadelphia General and the University of Pennsylvania) (5).

TABLE I

INGESTION ACCIDENTS—1966: THE TEN MOST FREQUENT TYPES*

Substance	Per cent of total
1. Aspirin	24.9
2. Soaps, detergents, cleansers	4.0
3. Vitamins and iron	3.8
4. Bleach	3.5
5. Insecticides (excluding mothballs)	3.0
6. Plants	2.6
7. Analgesics and antipyretics	2.3
8. Disinfectants and deodorizers	2.2
9. Hormones	2.1
10. Polishes and waxes	2.1
(all others)	49.5

*Reference (1).

The symptoms of poisoning are due exclusively to the inhibition of cholinesterase, the enzyme that normally prevents acetylcholine from accumulating at postganglionic nerve-endings and overstimulating the parasympathetic nervous system. The esters inactivate cholinesterase by attaching their free phosphoryl groups to it. The symptoms of parasympathetic overstimulation appear promptly within 20 to 30 minutes of exposure and usually last at least 24 to 48 hours. They consist principally of increased secretions (sweat, saliva, tears, bronchial fluid), gastrointestinal disturbances (nausea, vomiting, cramps, diarrhea), ocular disturbances (miosis, blurred vision—one of the earliest signs), and respiratory difficulties (bronchiolar spasm, pulmonary edema). Other symptoms often encountered include muscle disturbances (weakness, fasciculation, tremors, twitching, paralysis), ataxia, vertigo, dyspnea and cyanosis. Convulsions, coma, loss of reflexes and of sphincter control are ominous signs, although they do not necessarily preclude total recovery if resuscitative measures are energetically applied.

Poisoning by phosphoric ester insecticides should be suspected whenever the child exhibits any of these symptoms within 6 hours of exposure to a toxic substance. The diagnosis is confirmed by demonstrating a depression of the plasma and red blood cell cholinesterase, or, if this procedure is not feasible, by an increase in urinary metabolites such as paranitrophenol. If these parameters are unaffected or if exposure occurred more than 12 hours before symptoms were noted, another basis should be sought for the disorder.

The therapeutic management of this type of poisoning has been detailed in standard texts (6–8). In general, treatment will depend on the severity of the symptoms. In all cases, the insecticide should be washed off the skin or mucous membranes or extracted from the stomach. If poisoning has been severe, a patent airway must be maintained and artificial respiration applied. After the cyanosis has been corrected, atropine is administered to neutralize the excess of accumulated acetylcholine. The dose recommended for children is from 1 to 2 mg. intravenously or intramuscularly every 10 to 15 minutes until the signs of atropinization appear (dry, flushed skin; pulse 140 or more per minute). Cholinesterase activators such as 2-pyridine aldoxime methiodide (2-PAM) and diacetyl monoxime (DAM) are also available. If used, they should be administered intravenously; 2-PAM should be given slowly over a period of 10 to 15 minutes, DAM even more slowly over 15 to 30 minutes. Reflexes are ordinarily restored within one-half to one hour. Morphine, meperidine, barbiturates, and other respiratory depressants must be avoided. Treatment should

TABLE II

TOXICITY AND STRUCTURAL RELATIONSHIP OF THREE REPRESENTATIVE ORGANIC PHOSPHORUS INSECTICIDES

INSECTICIDE	(ETHYL-O)(ETHYL-O)P(=S)O--p-NITROPHENYL PARATHION	(METHYL-O)(METHYL-O)P(=O)O--2,2-DICHLOROVINYL DDVP	(METHYL-O)(METHYL-O)P(=S)S--1,2-DICARBETHOXYETHYL MALATHION
DERIVED FROM	(HO)(HO)P(=S)OH THIOPHOSPHORIC ACID	(HO)(HO)P(=O)OH PHOSPHORIC ACID	(HO)(HO)P(=S)SH DITHIOPHOSPHORIC ACID
TOXICITY	HIGH	MODERATE	SLIGHT
ORAL LD_{50} In RATS (Approx. values)	10 mg./kg.	100 mg./kg.	1000 mg./kg.

be continued until serum cholinesterase values return to normal. Total recovery from severe poisoning may require several months.

The three most widely employed organic phosphorus insecticides are parathion, DDVP (dimethyldichlorovinyl phosphate), and malation. The toxicity ratio of these compounds is about 1:10:100 (Table II). Parathion, a chemical warfare agent developed by the Germans in World War II (9), is extremely toxic; one drop in the eye may be fatal (6). An oxidation product, paraoxon, is even more toxic—about 10 times as toxic as parathion itself (10); it is this substance that is often responsible for severe reactions in young children who have played with the discarded containers. DDVP is much less hazardous but is more widely distributed, being used, for example, to eradicate the phorid fly from mushroom cultures and the cigarette beetle from tobacco warehouses in Pennsylvania (5). Malathion is the least toxic as shown by the fact that a report in the literature (11) showed that no adverse effects in humans were observed after 2500 tons of it were sprayed on inhabited areas of Florida in 1956 to eradicate the Mediterranean fruit fly from citrus trees.

The probable lethal dose of any given drug can be estimated for an adult by extrapolating roughly from the comparative LD_{50}'s in animals (12). Using the LD_{50}'s in rats, the probable lethal doses of parathion, DDVP, and malathion for the human adult are, respectively, about a "pinch," a teaspoonful and a fluid ounce (Table III). For children, the corresponding doses (by weight) probably range from one-twentieth the adult dose at birth to about one-half the adult dose at 10 years of age (Young's rule). Since they are so toxic, the phosphate ester insecticides have no place in the home or school. Their use must be confined to commercial agriculture.

TABLE III

PROBABLE LETHAL DOSE OF A DRUG IN THE HUMAN ADULT, ESTIMATED FROM ITS LD_{50} IN THE RAT*

LD_{50} (mg./kg.)	Probable lethal dose
<5	A few drops
5–50	A "pinch" to a teaspoonful
50–500	A teaspoonful to two tablespoonsful
500–5000	One ounce to one pint
5000–15,000	One pint to one quart

*For parathion, DDVP, and malathion the respective LD_{50}'s are about 10, 100, and 1000 mg./kg.; thus they fall near the lower limits of the second, third, and fourth lethal ranges shown here. Children are more sensitive to these insecticides than adults; thus, lethal doses are comparatively lower.

In summary, the three organic phosphorus esters most commonly found in insecticides are malathion (least toxic), DDVP (intermediate), and parathion (extremely toxic). Each year in the U.S., thousands of children are accidentally exposed to these compounds, which are rapidly absorbed through the skin or mucous membranes after mere contact. The signs and symptoms of poisoning are those of parasympathetic nervous system overstimulation, based on depression of the enzyme cholinesterase. Treatment consists in maintaining respiration and in administering atropine and cholinesterase activators. Recovery follows the reestablishment of normal serum cholinesterase levels.

REFERENCES

1. "Survey of Products Most Frequently Named in Ingestion Accidents," Washington: *Bulletin of the National Clearinghouse for Poison Control Centers*, September-October, 1967.
2. Verhulst, H. L., and Crotty, J. J. "Childhood Poisoning Accidents," *Journal of the American Medical Association* 203: 1049–50, March 18, 1968.

3. Washington: *Bulletin of the National Clearinghouse for Poison Control Centers*, July-August, 1967.
4. Gallaher, G. L. " 'Low Volume' Insect Control and Parathion Poisoning," *Texas Medicine*, 63: 39–41, October, 1967.
5. Silverio, J. Personal data.
6. Nelson, W. E. *Textbook of Pediatrics*, 8th Edition, Philadelphia: Saunders & Co., 1964, p. 1574.
7. Shirkey, H. C. *Pediatric Therapy*, St. Louis: C. V. Mosby Co., 1964.
8. Gellis, S. S., and Kagan, B. M. *Current Pediatric Therapy*, Philadelphia: Saunders & Co., 1964.
9. Metcalf, R. L. *Organic Insecticides. Their Chemistry and Mode of Action.* New York: Interscience, Inc., 1955.
10. Milby, H. T., Ottoboni, F., and Mitchell, H. W. "Parathion Residue Poisoning Among Orchard Workers," *Journal of the American Medical Assocation* 189: 351–6, August 3, 1964.
11. Shepherd, D. R. "Eradication of Mediterranean Fruit Fly in Florida," *FAO Plant Protection Bulletin*, 5: 101–3, April, 1957.
12. Hayes, W. J., Jr. *Clinical Handbook on Economic Poisons*, Atlanta: U.S. Communicable Disease Control Center, Toxicology Section, 1963.

Insecticides
Household Use and Respiratory Impairment

BETSY P. WEINER, M.D., and ROBERT M. WORTH, M.D., Ph.D.

TO LOOK FOR possible harmful effects associated with frequent home use of pesticides, in 1965 we asked the Hawaii State Department of Health to include a pesticide questionnaire in their ongoing health interview survey. Demographic, health, and pesticide data from 1567 households visited by this survey cover a random sample of about 1 per cent of the population of the island of Oahu (City and County of Honolulu). The evidence from these 1965 data shows a correlation between frequent insecticide use in the home and increased prevalence of asthma, chronic bronchitis, and sinus trouble. Apart from respiratory disorders, we could find no suggestive statistical evidence linking any other chronic condition[1] and insecticide use.

METHOD

Insecticide use and asthma are both common in Hawaii. One-fourth of our population uses insecticides daily (5.5%) or weekly (19.0%). Three-fourths report less frequent use, monthly to yearly (68.7%) or not at all (6.8%).

Selecting the two extremes, daily use and no use, as being at greatest risk and least risk to possible adverse effects from insecticides, we concentrated on these two groups. The daily-use group consists of 376 persons, and there are 328 non-users. The former have a slightly larger proportion of children under 15 years of age. All ethnic groups in Hawaii are equally represented in both groups. Income and educational levels, occupation, and place of residence are similar in the two groups.

Asthma, chronic bronchitis, or both, are reported for 99 of 376 persons in the daily use group as compared with only 61 of 328 persons reporting no use ($p<.05$). Repeated attacks of sinus trouble are also reported significantly more frequently in the daily use group ($p<.01$).

To test the hypothesis that frequent use of insecticides is associated with increased frequency of respiratory difficulties, a blind follow-up study two years later of the same population gives us a more complete assessment for comparison. All people still living at the same address are included. Of these, 350 persons, or 93% of those who had not moved, were reached in their homes in the summer of 1967. An additional 30 people born or moved into these homes in the interim are included. At this visit an extensive history was obtained for each individual, as well as environmental observations of vegetation, animals, and mold odor for each household. Respiratory symptoms, past and present medical conditions, smoking habits, and occupational and all household exposures were tabulated. Respiratory function tested by Vitalor spirometer provided a permanent tracing for comparison with normative standards. Auscultation of the lungs was also done. To avoid bias, questions on insecticide use came at the end of each visit.

From the School of Public Health and Pacific Biomedical Research Center, University of Hawaii.

	NUMBER OF PERSONS	NUMBER OF PERSONS BY CLASS			
VENTILATORY FUNCTION					
AMA class		*Class 1*	*Class 2*	*Class 3*	*Class 4*
Respiratory impairment		0%	20-30%	40-50%	60-90%
$FEV_{1.0}$ & FVC-%*		85+%	70-84%	55-69%	$< 55\%$
INSECTICIDE USE					
A. Consistent heavy use	50	16	19	9	6
B. No use to heavy use	23	5	10	6	2
C. Daily use to light use	20	6	7	6	1
A+B+C. Any heavy use in past 2 years	93†	27	36	21	9
D. Consistent light use	49†	25	13	10	1
	142				
INSECTICIDE USE IN 81 OF THE ABOVE ADULTS WHO DO NOT SMOKE CIGARETTES					
A. Consistent heavy use	29	10	9	6	4
B. No use to heavy use	10	2	2	4	2
C. Daily use to light use	10	2	5	3	0
A+B+C. Any heavy use in past 2 years	49‡	14	16	13	6
D. Consistent light use	32‡	18	8	6	0
	81				

* % of predicted values for normal population by age, sex, and height.
† Statistical comparison of 93 persons in the 4 classes of respiratory impairment in households with heavy insecticide use (A+B+C) with 49 persons in households with consistent light insecticide use D. $\chi^2 = 8.486$, 3df, $p < .05$.
‡ Statistical comparison of same two groups without cigarette smokers. $\chi^2 = 14.872$, 3df, $p < .01$.

DEFINITIONS

Although habits and needs do vary in time, most people had a consistent pattern of insecticide use. To classify use, once a week or oftener is defined as heavy use. Less than once a week is defined as light use. By this definition, 30 of 44 original daily use households were found two years later to be still applying insecticides heavily. The other 14 households, using insecticides less frequently, either had fewer insects or became more tolerant of those they had. Similarly, 32 of 46 original no-use households were consistent in that they were still not applying insecticides frequently, but were applying as often as the large majority of our population, monthly to yearly. This consistent light use group is thus representative of the total original sample. The other 14 original no-use households were found two years later to be applying insecticides frequently, about half having been reluctant to admit the need or the practice when first questioned, though using heavily all along. The other half had only recently begun frequent use.

Respiratory impairment is classified according to the AMA Committee's rating scale.[2] It is based on two tests of respiratory function performed on the spirometer; flow rate or forced expiratory volume in the first second ($FEV_{1.0}$) and forced vital capacity (FVC). Normative standards give the expected performance based on age, sex, and height. Class 1 includes those with least impairment and 85 to 100 per cent of expected function. Class 2 includes those with 20-30 per cent impairment and 70-84 per cent of expected performance. Class 3 is defined as 40-50 per cent impairment with 55-69 per cent of expected function. Class 4 includes those with most impairment (60-90 per cent) and less than 55 per cent of expected respiratory function.

OBSERVATIONS

All individuals 20 years of age or older and without acute respiratory illnesses are compared in Table 1, which gives the rating by respiratory impairment for persons in each category of insecticide use. The consistently light-use group is found to perform significantly better than those in households using insecticides heavily in the past two years ($p < .05$). When persons in the same groups are compared, excluding those who smoke cigarettes, a greater difference is found ($p < .01$).

There was no definite correlation of impairment with any specific agent. Multiple and changing use of different types of insecticides in spray form was the common pattern observed. No other environmental variable was found to correlate with the observed differences in respiratory impairment.

For those persons eligible for household visits in the summer of 1967, but relocated or not seen, records available from the original survey (1965) were reviewed to see if any obvious biases were present. Cases of asthma and chronic sinusitis reported in 1965 but not visited in 1967 were found to be distributed between heavy-use and no-use

INSECTICIDE USE	Total	NUMBER OF PERSONS WITH CURRENT SYMPTOMS*			
		Asthma	*Chronic sinusitis*	*Chronic bronchitis*	*Perennial nasal allergy*
A. Consistent heavy use	156	16	6	5	18
B. No use to heavy use	50	3	3	0	5
C. Daily use to light use	62	2	1	4	4
A+B+C. Any heavy use in past 2 years	268	21	10	9	27
D. Consistent light use	112	1	1	0	9
	380				

* Statistical comparison of persons with and without respiratory conditions in the two groups, 268 heavy insecticide users and 112 consistent light users. $\chi^2 = 15.191$, 4df, $p < .01$.

households in proportions similar to those we report below. We conclude therefore that no important out-migration biases are present.

Respiratory conditions defined by presence of symptoms during the previous year are listed in Table 2 by insecticide use. Asthma is defined as having more than one spell of wheezing during this time. Chronic sinusitis refers to continuing symptoms of sinus pain and discharge. Chronic bronchitis is defined as chronic productive cough throughout the year. Perennial nasal allergy is the designation for sneezing or nose-itching plus nasal discharge most days of the year. Persons are placed in only one category. Of seven adult asthmatics, five have associated bronchitis. Half of all asthmatics have perennial nasal allergy as well. Asthma and chronic sinusitis are found preponderantly in the consistently heavy-use group. Perennial nasal allergy is twice as common in this group. Persons living in households using insecticides frequently have significantly more respiratory conditions than those persons living in households where insecticides are used infrequently ($p < .01$). Physical findings and history of insecticide use are consistent with the evidence associating pesticide use and respiratory conditions.

Both reduced ventilatory function and a higher prevalence of asthma are found to be associated with frequent application of spray insecticides. No single compound or combination used has been implicated. Most commonly used is a combination of pyrethrum, piperonyl butoxide, and petroleum distillates. The next most common type of combination used contains DDVP, an organic phosphate, and dieldrin, a chlorinated hydrocarbon, each in 0.5 per cent concentration, plus petroleum distillates.[3] Pyrethrum as mosquito punk or coil is often used in addition to other products. Pyrethrum has long been used and is available in about 2,000 varieties of sprays.[4] It is thought to have allergic effects.[5,6,7] Organic phosphate compounds are reported to be capable of producing nasal discharge and occasional wheezing.[8,9]

MUST WE SPRAY?

While flies and mosquitoes are a seasonal problem on the mainland, they are present the year round in a tropical climate. Cockroaches are a year-round problem in both areas, but spraying individual cockroaches is a wasteful, ineffective, and irrational way of suppressing them. Poisoned bait in tablet form is far more effective. No space sprays are recommended by the U.S. Public Health Service for regular indoor use. Sprays are convenient to use, and may satisfy certain hostile impulses, but their use rarely reduces insect infestation. Baits and repellents are far more effective, and safer.

Our evidence shows association between frequent use of household insecticides and respiratory impairment. To test for causal effects of specific components of the sprays, inhalant challenge to volunteers is necessary.

REFERENCES AND NOTES

1. No suggestive findings were observed for the following chronic conditions: tuberculosis, rheumatic fever, hardening of the arteries, high blood pressure, stroke, varicose veins, hemorrhoids, benign tumor, chronic gall bladder or liver trouble, kidney stones or chronic kidney trouble, mental illness, arthritis, diabetes, thyroid trouble, other allergy, epilepsy, chronic nervous trouble, cancer, chronic skin trouble, hernia, prostate trouble, palsy, paralysis, or congenital condition.
2. Committee on Rating of Mental and Physical Impairment: The respiratory system, JAMA 194:919-932 (Nov. 22) 1965.
3. DDVP is 2,2-dichlorovinyl dimethyl phosphate. Dieldrin is not less than 85% of 1, 2, 3, 4, 10, 10-hexa-chloro-6, 7-epoxy-1, 4, 4a, 5, 6, 7, 8, 8a-octa-hydro-1, 4-endo-exo-5, 8-dimethanonaphthalene.
4. Gleason, M. N., Gosselin, R. E., and Hodge, H. C.: Clinical Toxicology of Commercial Products, Baltimore, Williams & Wilkins Co., 1963, pp. 131-132.
5. Garratt, J. R., and Bigger, J. W.: Asthma due to insect powder, Brit. Med. J. 2:764 (Oct. 27) 1923.
6. Feinberg, G. M.: Pyrethrum sensitization: its importance and relation to pollen allergy, JAMA 102:1557-1558 (May 12) 1934.
7. Sheldon, J. M., Lovell, R. G., and Mathews, K. P.: A Manual of Clinical Allergy, ed. 2, Philadelphia: W. B. Saunders Co., 1967, p. 451.
8. Koelle, G. B.: Anticholinesterase Agents, in Goodman, L. S., and Gilman, A.: The Pharmacologic Basis of Therapeutics, ed. 3, New York: Macmillan Co., 1965, Chapt. 22, pp. 447-454.
9. Grob, D.: Manifestations and treatment of poisoning due to nerve gas and other organic phosphate anticholinesterase compounds, Arch. Intern. Med. 98:221-239 (Aug.) 1956.

10. Supported by USPHS grant PH 86-65-79. We thank Eileen Shimatsu Cabente and the Hawaii State Department of Health for their help. ■

Pesticide-Induced Illness

Public Health Aspects of Diagnosis and Treatment

IRMA WEST, M.D.

MAN HAS MANIPULATED his environment on so large a scale that he has inadvertently invented and produced a multitude of the most complicated new problems ever to confront the health professions. Unfortunately, we have been slow to realize that plans for health and safety should be built into technologic advances in the planning stages. By the time technical tools are in operation and their use results in undesirable and unexpected effects upon people and their environment, the best opportunity to minimize these effects efficiently and humanely is largely lost. So it is with many of our new environmental health problems whether they are air pollution or other environmental contamination with modern chemicals, including pesticides.

Pesticides are materials which mitigate or kill unwanted animal or plant life. About 650 have been invented in the last 25 years. These new chemicals, plus a few older ones, are formulated into over 57,000 trade name products registered for sale in the United States. Never before have hundreds of new chemicals possessing such varying degrees and kinds of potential for good and harm been introduced into the environment in so short a time.

It is important that problems in environmental medicine be viewed in context. With the possible exception of drugs, pesticides have been the first great experiment in the mass use of chemical technology. About one hundred million pounds of pesticides are now applied annually in California. Ten years ago about half that amount was applied and 20 years ago use of the new synthetic pesticides was just beginning. By far the greatest portion of pesticides is used in agriculture. Only 1 per cent is applied for control of disease vectors. About 59 per cent of the pesticides used are insecticides, 15 per cent fungicides, 15 per cent defoliants and herbicides, 10 per cent fumigants and 1 per cent rodenticides.

Pesticides have brought great benefits—and, with them, disturbing adverse side effects which are summarized in Table 1.

A question which naturally arises is, do the benefits outweigh the adverse side effects? The answer depends, of course, on what values one assigns to the items listed. The food technologist, the agriculturist and the chemical manufacturer will point to the sizable benefits as the more important—the food surpluses, the economic importance of the commodities where pesticides play a significant role in production or preservation. On the other hand, the biologist, the conservationist and the wildlife expert will look with alarm at the chain of events arising from the worldwide contamination of the environment with the persistent chlorinated hydrocarbon pesticides, including the insidious build-up of these chemicals in the food chains. Medicine and public health have interests in both sides of the picture. Worldwide, pesticides have been one of the methods used in successfully combating malaria and other vector-borne diseases. Millions of lives have been saved. In California the threat of Western and St. Louis Equine viral encephalitis is held in check by the mosquito abatement activities. On the other hand, deaths and serious illness from acute pesticide poisoning and other recognized adverse effects upon California citizens which occur regularly are a tragic and unnecessary waste of human health and life (See Tables 2 and 3). Further, the uncertainties about long-term effects upon people arising from a contaminated environment and from the storage of chlorinated hydrocarbon in human fat, are reason for considerable uneasiness in the medical profession and among public health workers. There is, of course, no objective answer to the question of the relative value of the benefits versus the adverse side effects from pesticides. Information on which to make such a judgment is far from complete and may never be available. However, as time goes on the relative importance of these benefits and side effects may become more obvious.

A question of greater significance from the standpoint of public health and environmental medicine is, can we have the benefits of pesticides without the undesirable side effects? It is technically feasible and well within the realm of possibility to use pesticides in a manner which will reduce undesirable side effects to almost zero. However, such a program would call for revolutionary changes in our standards for research and field testing, and in our control over developing technology. We can obtain as much protection against the adverse side effects as we are willing to insist upon and pay for.

The prevention of untoward effects upon the health of the population arising out of our technology is emerging as a most important and difficult public health function. Our society has never really faced the issue of what would be necessary to prevent the undesirable side effects arising from the use of pesticides and still enjoy their benefits. First, before a pesticide was put into general use it would be necessary to know, through research and field tests, what all of the potential undesirable effects are. Second, these chemicals must really be controlled so that they will not be used in a manner producing adverse effects. Third, because methods for predicting adverse effects cannot be expected to be perfect, they must be continually evaluated and a monitoring system for human health must be established with built-in power to stop and revise uses of pesticides when they become suspect of producing undesirable effects.

This kind of system may seem insurmountably difficult, but that is because our administrative vision has never been big enough for our environmental health problems. A good control program is technically feasible. It has been routine, for

TABLE 1.—*Benefits and Undesirable Side Effects Arising from the Use of Pesticides*

Benefits	*Undesirable Side Effects*
Enhance production of food and fiber.	Human poisoning and other diseases from pesticides.
Help preserve stored food and other commodities.	Contamination of the environment with destruction of beneficial plant and animal life such as bees and wildlife; concentration of chlorinated hydrocarbons in food chains.
Help control vector-borne disease.	Pesticide residues on food.
Help control nuisance pests.	Storage of chlorinated hydrocarbon pesticides in human and animal fat.
Protect economically and aesthetically valuable resources (forests, parks, trees, lumber, flowers, gardens, etc.).	Development of resistance to chemicals by pests.

example, in the development of our space program. However, when it comes to down-to-earth matters involving the general population, too little too late is more often the case. Air pollution is another example of a situation in which our administrative imagination and machinery has never been big enough to catch up and come to grips with the problem.

Using hindsight in analyzing the genesis of undesirable and often unexpected side effects from pesticides, it is apparent that considerable research and control went into some aspects of pesticide usage and very little into others. A great deal of research and attention was paid to establishing tolerances for pesticides on crops in which pesticides were applied. Applications of pesticides were carefully prescribed and the crops monitored. Crops with more than the legal tolerances were condemned. However, pesticide poisoning among workers applying these materials and contamination of the environment, for example, received very little if any research attention or effective control. It is also apparent that the skills and technical knowledge employed in the development of pesticides were in fields related to the intended uses of pesticides. Technical skills related to the adverse side effects were often not included.

It should be stated at this point that there are very distinguished scientists who are not optimistic about our ability to control unwanted side effects arising from our technological tools. For example, Dr. Rene Dubos (15 April 1965 *Journal of Occupational Medicine*) is quoted as follows:

"Present programs for controlling potential threats to health from new substances and technologic innovations are doomed to failure because we lack the scientific knowledge to provide a sound basis for control.

"Current testing techniques have been developed almost exclusively for the study of acute, direct toxic effects.

"In contrast, most untoward effects of the technological environment are delayed and indirect. . . . Yet little is being done in schools of medicine and public health or in research institutes or government laboratories to develop the kind of knowledge that is needed for evaluating the long-range effects on man of modern ways of life.

"The dangers associated with ionizing radiation, or with cigarette smoking should have sensitized the public as well as scientists to the importance of delayed effects. But, surprisingly, this knowledge has not increased awareness of the fact that most other technological innovations also have delayed effects.

"The slow evolution of chronic bronchitis from air pollutants, the late ocular lesions following use of chloroquine, the accumulation of the tetracyclines in the fetus, and of course all the carcinogenic effects, are but a few of the countless objectionable results of new substances or technologies which appeared at first essentially safe."

The same scientist spoke to the same point in a statement appearing in *Bioscience*, 14:11, January 1964:

"There is no need to belabor the obvious truth that, while modern science has been highly productive of isolated fragments of knowledge, it has been far less successful in dealing with the complexity of natural phenomena, especially those involving life. In order to deal with problems of organized complexity, it is therefore essential to investigate situations in which several interrelated systems function in an integrated manner. Multifactorial investigations will naturally demand entirely new conceptual and experimental methods, very different from those involving only one variable, which have been the stock in trade of experimental science during the past 300 years and to which there is an increasing tendency to limit biological research."

The time has come to state that the medical profession has been placed in a difficult position with respect to the recognition and treatment of untoward effects upon human health from pesticides and other modern chemicals. Since the field is one of growing potential for liability, it is somewhat surprising that so little protest has been heard from physicians. The present mechanisms for bringing information and education to physicians are not geared to meet today's rapid introduction of hundreds of new chemicals of potentially dangerous effect. It is entirely unrealistic to expect every physician in general practice to keep up with what is known and unknown about the toxicologic properties of modern chemicals. Since technology produced this urgent problem, it is only fair to expect it should be used to devise imaginative new procedures to bring to the physician, when he needs it, effective help and up-to-date information to assist in the diagnosis and treatment of poisoning and other conditions re-

TABLE 2.—*Accidental Deaths Attributed to Poisoning from Pesticides and Other Agricultural Chemicals California 1951-1965**

Year	Total	Children: Total	Children: Arsenic	Children: Organic Phosphates	Children: Other	Workers: Total	Workers: Organic Phosphates	Workers: Methyl Bromide	Workers: Other	Others: Total	Others: Arsenic	Others: Other
1951-65 Total	128	76	44	13	19	29	12	7	10	23	0	13
*1965	5	2	1	1	0	1	0	1	0	2	0	2
1964	2	1	1	0	0	1	1	0	0	0	0	0
1963	6	3	1	1	1	1	1	0	0	2	0	2
1962	5	4	1	2	1	1	1	0	0	0	0	0
1961	6	3	2	0	1	3	2	0	1	0	0	0
1960	4	4	4	0	0	0	0	0	0	0	0	0
1959	18	10	4	2	4	5	1	2	2	3	1	2
1958	13	6	2	0	4	3	1	1	1	4	3	1
1957	12	8	5	1	2	2	1	0	1	2	1	1
1956	18	11	9	1	1	4	0	2	2	3	2	1
1955	6	3	1	0	2	1	0	0	1	2	2	0
1954	12	9	3	4	2	2	1	0	1	1	1	0
1953	10	4	3	1	0	4	3	0	1	2	0	2
1952	6	5	4	0	1	1	0	1	0	0	0	0
1951	5	3	3	0	0	0	0	0	0	2	0	2

(Number of Deaths)

Source: State of California, Department of Agriculture, Annual Reports, Bureau of Chemistry, 1951-1960; State of California, Department of Public Health, Death Records and Occupational Disease Attributed to Agricultural Chemicals, 1951-1963 by Bureau of Occupational Health.

*All 1965 data preliminary, a small number of death certificates yet to be processed.

Note: Suicides now outnumber accidental deaths from pesticides (13 in 1964, 19 in 1965).

sulting from exposure to pesticides. Also urgently needed is the development of many more clinical chemical laboratory tests to help the physician confirm a diagnosis. For many chemicals, such tests either do not exist or are not available locally. Circumstantial and clinical evidence alone are often quite inadequate to arrive at a sound diagnosis. As a result there are probably more unsound diagnoses and more missed diagnoses in chemical poisoning than in most other areas of medicine.

One of the most immediate problems in public health is the lack of information about what effects pesticides and other modern chemicals are having in the population. Only a part of this information is available through death certificates, the Doctor's First Report of Work Injury (See Tables 2 and 3) and an occasional research project.

These problems of the practicing physician and public health suggest the possibility of a joint solution. A state or national center to provide

TABLE 3.—*Reports of Occupational Disease Attributed to Pesticides and Other Agricultural Chemicals for All Conditions and Systemic Poisoning, California 1953-1964*

Year	Total	Systemic Poisoning: Total	Attributed to Organic Phosphates (Phosphate Esters)	Attributed to Chlorinated Hydrocarbons	Attributed to Other Agricultural Chemicals
Total	9,894	3,296	2,412	141	743
1964	1,328	230	135	24	71
1963	1,013	345	267	14	64
1962	827	219	140	21	58
1961	911	268	194	14	60
1960	975	368	283	7	78
1959	1,093	499	407	10	82
1958	910	328	227	14	87
1957	749	252	189	12	51
1956	789	281	197	11	73
1955	531	183	126	4*	53
1954	391	122	101	1*	20
1953	377	201	146	9*	46

(Reports of Occupational Disease)

*DDT only.

Source: State of California, Department of Industrial Relations, *Doctor's First Report of Work Injury.*
Statistics compiled by Bureau of Occupational Health, Department of Public Health.

medical and toxicological consultation to physicians with diagnostic and treatment problems could, at the same time, record, mechanize and analyze case data for study and dissemination. Any number of benefits to physicians, their patients and the public could arise from such a center. For example, unexpected effects upon health could be picked up and documented much earlier and appropriate preventive measures taken; the physician would have help in obtaining and interpreting laboratory tests; and liability problems would be minimized. This proposal is for an advanced stage in the development of the Poison Information Centers now in use. It is time we recognized the value of augmenting a valuable service of this kind to meet the needs of the day.

Lindane and Hematologic Reactions

Irma West, MD

THE AMERICAN Medical Association's Council on Drugs maintains a registry on blood dyscrasias in which are listed 18 reports of major blood dyscrasias where exposure to lindane or benzene hexachloride (BHC) has been implicated directly or circumstantially.[1] (Hexachlorobenzene or hexachlorocyclohexane is a mixture of its various isomers of which lindane, the gamma isomer, is one.) Cases have been reported from other countries: 20 from Mexico implicating lindane or chlorophenothane (DDT),[2] or both, 2 from Czechoslovakia,[3] 5 from Greece,[4] 1 from Sweden implicating DDT and lindane,[5] 2 from France,[6,7] and 2 from Australia.[8] Individual case reports also appear in the American medical literature.[9,11]

Serious damage to the bone marrow attributed to chemicals has occurred before both following the use of specific drugs, and among workmen using certain industrial chemicals. The possibility of its occurrence in the wake of the tremendous increase in the number and use of pesticides should not come as a surprise. Lists of medicines and other chemicals presenting blood dyscrasia hazards have been accumulating as epidemiological and circumstantial evidence are reported. The antibiotic, chloramphenicol, is the best known of the drugs,[12-14] and the solvent, benzol, the most frequent offender among industrial chemicals.[15] There is no laboratory or other safe test available at this writing which will confirm or deny in the individual case a cause and effect relationship between exposure to a particular chemical and bone marrow failure. Only when a large number of people are involved can epidemiological studies provide evidence of an association between a chemical and the disease. It appears that for many of the chemicals reported to adversely affect the bone marrow, only a small proportion of the persons exposed become victims of irreversible bone marrow failure. The amount of exposure is not necessarily related to the extent of the damage.[16] At this writing, this phenomenon is not predictable through animal or other tests.

Benzene hexachloride was first synthesized in 1825 by Michael Faraday,[17] but it was not introduced as a pesticide until early in the 1940's. It consists of a mixture of isomers of which the gamma isomer, lindane, is the most active as a pesticide. A comprehensive summary of what was known about the pharmacology and toxicology of benzene hexachloride and its principal isomers was published in 1951.[17] Damage to the bone marrow was not mentioned. It was noted that very little information was available on inhalation toxicity. Ingestion and cutaneous toxicity were considered to be about that of DDT, and the acute toxic effects, as with DDT, were from central nervous system stimulation. Liver damage was present in fatally poisoned animals. The BHC isomers have been found stored in the fat of the general population of the United States,[20-22] England,[23] France,[24] and India.[25]

United States production of BHC has varied from about 25 million pounds in 1961 to 7 million pounds in 1963. The demand has since been relatively stationary in this country but rising rapidly in other countries.[19]

California Cases

Since 1954, lindane exposure in California has been implicated directly or circumstantially in cases of serious blood dyscrasias, primarily pancytopenia (aplastic anemia). The following cases which have come to the attention of the California State Health Department are selected because a substantial

From the Bureau of Occupational Health, California Department of Public Health, Berkeley.

Reprint requests to 2151 Berkeley Way, Berkeley, Calif 94704 (Dr. West).

exposure to lindane preceded the onset of the illness and there was no exposure to any other agent known to cause bone marrow hypoplasia. Additional cases have been reported but the history of exposure was incomplete or not clear-cut. Large quantities of a number of different pesticides are used in California but no similar series of cases of blood dyscrasias implicating other pesticides have been reported.

It should be emphasized that outside of the American Medical Association's registry[1] and the Food and Drug Administration's program on adverse reactions to drugs there is no comprehensive program of human health surveillance specifically designed to detect, investigate, and take action when unexpected health problems arise from the use of all of the chemicals in the environment. It is mostly by chance that such effects are suspected and reported.

CASE 1.—A 39-year-old physician's wife first became ill in 1952 and a diagnosis of aplastic anemia was made in 1954. In 1951, a pest control operator began treating her home every three months with lindane. The family consisting of the father, mother, and son would move out for about two hours while lindane was vaporized throughout the house. On most occasions they noted a white "film" on the furniture and other surfaces for some time afterward. In 1954, a lindane treatment was repeated because an insect was found inside. The odor and film on the furniture was so noticeable the housewife cleaned the house thoroughly, including walls and woodwork, rugs, and upholstery. Soon afterward she became very weak and the diagnosis of aplastic anemia was made by her physicians. Later that year she experienced a sudden febrile episode after eating in a restaurant where a lindane vaporizer was operating. A precipitous drop in circulating red blood cells was noted following this episode. She was maintained on treatment with transfusions until her death in 1960.[9]

CASE 2.—A 34-year-old white man died in July 1958. The death certificate lists cerebral hemorrhage with aplastic anemia as the cause of death. His living quarters had been heavily fumigated with lindane three months before onset of his illness in November 1957 when a diagnosis of aplastic anemia had been made. The hematologist attending the case reported it to the Health Department.

CASE 3.—A 4-year-old child became seriously ill in 1955 and a tentative diagnosis of atypical leukemia was made. Four other members of the family had developed complaints during this same period. "Abnormal" blood counts and mild anemia were reported in the mother, father, and the 10-month-old and 6-year-old siblings. The 6-year-old developed allergic-type respiratory complaints. A lindane vaporizer had been operating continuously in the kitchen of their home for 1½ years. (Lindane vaporizers are thermal generators or small electrical heating devices attached to the wall and plugged into a wall socket. Pesticide pellets are placed in the generator. As the pellets are heated they emit vapor into the room air.) When the diagnosis of a serious blood dyscrasia was made, the vaporizer was removed. The 4-year-old recovered and the complaints of the rest of the family abated.

CASE 4.—An 8-year-old girl developed scattered ecchymosis over her lower extremities in August 1960 at which time her physician made a diagnosis of aplastic anemia. For about two hours daily for two years before onset of her illness she had played in the kitchen of a neighbor's home where a lindane vaporizer was continuously operating. On three occasions acute exacerbations followed visiting restaurants where lindane vaporizers were subsequently found to have been operating. She died in November 1961. The death certificate lists cerebral hemorrhage, thrombocytopenia, and hypoplastic bone marrow as the cause of death.[10]

CASE 5.—In February 1963, a diagnosis of aplastic anemia was made in a 53-year-old man. For three months before onset of his illness a lindane vaporizer had been continuously operating in his den where he spent several hours a day. He died November 1964. The death certificate lists pulmonary hemorrhage from aplastic anemia as the cause of death.[10]

CASE 6.—In August 1963, a diagnosis of aplastic anemia was made in a 58-year-old man. About nine days before onset of his illness a pest-control operator treated with lindane the yard and interior of the home of his daughter where he was staying. The patient had been engulfed in a "cloud" of the spray inadvertently expelled from the spray apparatus. He died September 1963. The death certificate lists subdural and intratracheal pulmonary hemorrhage, and thrombocytopenia due to hypoplasia of the bone marrow as the cause of death.

Possible Mechanisms of Action

The mechanism by which chemicals cause damage to the bone marrow is unknown. Three theories have been advanced most often. The first assumes a direct toxic effect upon the "stem" cells of the bone marrow which occurs to some extent to everyone receiving a significant exposure. In most persons the damage is transient and recovery occurs, but for a small number it progresses and is fatal. The second theory supposes that an allergic response to the chemical occurs in a small proportion of persons after several exposures. In the two California lindane cases, and in a report from Mexico,[2] sudden exacerbations were reported following inadvertent reexposure to the same pesticide, an observation which is most consistent with the allergic theory but does not necessarily rule out the first theory. A third theory supposes that a small number of those exposed are deficient in the usual detoxification or excretion mechanisms in the metabolism of the chemical in question. This deficiency may be due to inborne errors, poor function of the liver or kidney, or the presence of other interfering chemicals, such as drugs, in the body at the same time. Thus the offending chemical changes or accumulates in such a manner as to damage the primitive undifferentiated "stem" cells in the bone marrow from which a good portion of the circulating blood cells arise. It is worth noting that primitive cells are considered particularly vulnerable to toxic agents. Another factor of interest is the ability of lindane and certain chemicals to accumulate in body fat. There is considerable fatty tissue in the bone marrow. Whatever the mechanism, both the etiology of bone marrow failure and the role of specific chemical agents should be the subject of intensive research. We are in the unfortunate position of introducing a host of new chemicals into the environment with no method of predicting whether or not, or under what conditions, they might produce a fatal disease. Furthermore, we do not possess a method of proving or disproving that the chemical was responsible in the individual case.

In addition to blood dyscrasias other questions concerning safety of lindane have arisen. Two California children have died and many others seriously poisoned after swallowing the lindane pellets made for vaporizers. Allergic responses among persons working in areas where vaporizers are operating have been reported in the literature[26] and allergic responses among persons whose homes were treated by pest control operators with lindane have been reported to the Health Department in California.

Regulations and Recommendations on the Safety of Lindane Vaporizers

The first statement on the subject of thermal generators continuously emitting lindane or DDT appeared in Washington, DC, in September 1951 from the Interdepartmental Committee on Pest Control[27] which was composed of representatives of the Departments of Agriculture, Interior, Defense, and the Federal Security Agency. This statement listed the specific strict operational controls as a condition of use, and recommended that these devices not be used in homes, sleeping quarters, or where food could become contaminated. They recommended that the vaporizers be used only in commercial or industrial premises.

In Chicago in May 1952, the Committee on Pesticides of the Council on Pharmacy of the American Medical Association issued its first statement: "The Health Hazards of Electric Vaporizing Devices for Insecticides."[28] Its supplementary report appeared on July 25, 1953,[29] and another report "Abuse of Insecticide Fumigating Devices" appeared in 1954.[26] These reports emphasized the concern of the medical profession for the safety of persons using these devices in the home and reported upon cases of anemia among those who had been exposed to lindane from these devices. The extravagant promotional schemes for selling the wall generators were decried. A minority report[30] from this Committee on Pesticides went even further in stating that the marketing of lindane or DDT vaporizers violated principles of good public health and medicine, since there was no evidence that continued human exposure was without harm. Use of these devices in any occupied area was unwise and, furthermore, there was no demonstrated need for this type of insect control to justify its use in the face of its potential danger to human beings.

In November 1952, the California State Board of Health passed a resolution with respect to lindane vaporizers.[31] It expressed concern about the lack of technical information on the toxicity of lindane and recommended that these vaporizers be used under carefully controlled conditions or not at all. They urged that vaporizers not be used where people sleep, work, or where food is packaged. The lack of technical data on which to base sound regulations was noted. This recommendation is still in effect.

In August 1963, the California Department of Agriculture no longer registered labels for lindane pellets for vaporizers where use in the home was directed. Several months later Congress passed Senator Ribicoff's bill (S 1605) which repealed the federal law which had allowed registration of pesticides under protest. Lindane for vaporizers for home use had been registered under protest by the Federal Department of Agriculture. However, lindane vaporizers may still be sold for use in occupied areas such as schools, hospitals, and work places provided the label does not recommend household use. Lindane may still be vaporized in living quarters by pest control operators.

Comment

When the vapor of a chemically stable pesticide like lindane is dispensed into a dwelling, it can recirculate and condense as the air currents and the household heating and ventilating equipment dictate. Whether it is dispensed continuously by a small wall vaporizer or at regular intervals and in larger amounts by a pest control operator, the circulating lindane can produce a continuous exposure. It could be greater in duration and amount than that of workers exposed to lindane in industry.[28]

Continuous pollution of the indoor air in occupied areas by a chemical for pest control purposes can be questioned from several standpoints. First, other methods of controlling pests are available where human exposure is not continuous and is of a much lower order of magnitude. Second, when the pests are merely a nuisance rather than a disease vector the benefits are not commensurate with risks. Third, the standard animal tests consisting of ingestion of chemicals in the diet which have been used as an indicator of the toxicity of lindane, for example, are not adequate to predict long-term human response to chemicals by inhalation, which is often a much more efficient route of absorption because the inhaled material can enter the bloodstream directly. Fourth, the population subject to continuous inhalation of pesticides from dispensing devices in the home includes its most vulnerable segment: the infants, the sick, and the elderly. Before continuous exposure to a given pesticide in the home is permitted, some method for determining beforehand its effects on the infant and unborn child must be developed. As yet unexplored are the health effects of continuous inhalation of a pesticide upon persons with various chronic diseases. Also requiring study are effects of continuous or repeated inhalation of a pesticide upon those taking drugs, medicines, or alcohol. Some method of predicting through animal experiments or other tests which chemicals under what circumstances may damage the human bone marrow is another example of techniques which should be developed if we are to reduce our present limitations in predicting the safety of chemicals.

Summary

From a number of independent sources throughout the world circumstantial evidence has been accumulating which places lindane, a chlorinated hydrocarbon pesticide, suspect in the cause of serious blood dyscrasias, primarily aplastic anemia. Cases have been reported in other countries as well as in the United States. Six cases from California are discussed where the exposure to lindane came either from thermal generators designed to continuously dispense lindane vapor into the indoor air to control flying insects, or from lindane vaporized periodically into the household by pest control operators.

During the past 15 years, widely publicized statements from medical and public health agencies have recommended against the use of lindane vaporizers particularly in the home. However, these devices were registered under the provisions of the Federal Insecticide, Rodenticide, and Fungicide Act and millions were sold over the past 15 years. This Act allowed pesticides to be registered "under protest" even when the registering agency considered a product to be unsafe or ineffective in some respect but the allegation was unproved in court. This provision was repealed in 1963. However, lindane vaporizers are still registered. Only the label must recommend against use in the home. They may be used in schools, hospitals, or places of work. There is no bar to vaporizing lindane into living quarters by pest control operators.

This experience with lindane is an example of the limitations in our technical ability to predict and administer chemical safety. At this writing there is no method of predicting through animal or other tests which chemicals under what circumstances may damage the bone marrow. There is no method yet developed for proving a cause and effect relationship between a chemical exposure and bone marrow failure in the individual case which may occur. Research is urgently needed to correct these deficiencies as well as to determine if there is a cause and effect relationship between exposure to lindane and adverse effects on the bone marrow. Regardless of whether or not a cause and effect relationship exists, the burden of proof, particularly involving a fatal disease, should not rest upon the public. Practical and flexible administrative provisions which can take heed of widespread recommendations from medical and public health agencies should be available so that further exposure of the public to a chemical can be curtailed promptly until research can satisfactorily answer important health questions which have arisen.

Generic and Trade Names of Drug

Chloramphenicol— *Chloromycetin.*

References

1. Best, W.R.: Drug-Associated Blood Dyscrasias, *JAMA* **185**:286, 1963.
2. Sanchez-Medal, L.; Castanedo, J.P.; and Garcia-Rojas, F.: Insecticides and Aplastic Anemia, *New Eng J Med* **269**:1365, 1963.
3. Jedlicka, V.L., et al: Paramyeloblastic Leukaemia Appearing Simultaneously in Two Blood Cousins After Simultaneous Contact With Gammexane (Hexachlorcyclohexane), *Acta Med Scand* **161**:447, 1958.
4. Danopoulos, E.; Melissinos, K.; and Katsas, G.: Serious Poisoning by Hexachlorocyclohexane, *Arch Indust Hyg* **8**:582, 1953.
5. Friberg, L., and Martensson, J.: Case of Panmyelophthisis After Exposure to Chlorophenothane and Benzene Hexachloride, *Arch Indust Hyg* **8**:166, 1953.
6. Marchand, M.; Dubrulle, P.; and Goudemand, M.: Agranulocytose Chez un Sujet Soumis a des Vapeurs D'Hexachlorocyclohexane, *Arch Mal Prof* **17**:256, 1956.
7. Albahary, C.; Dubrisay, J.; and Guerin: Pancytopenie Rebelle au Lindane (Isomere y de L'Hexachlorocyclohexane), *Arch Mal Prof* **18**:687, 1957.
8. Woodliff, H.J.; Connor, P.M.; and Scopa, J.: Aplastic Anemia Associated With Insecticides, *Med J Aust* **1**:628, 1966.
9. West, I., and Milby, T.H.: Public Health Problems Arising From the Use of Pesticides, *Residue Rev* **11**:140, 1965.
10. Loge, J.P.: Aplastic Anemia Following Exposure to Benzene Hexachloride (Lindane), *JAMA* **193**:110, 1965.
11. Mendeloff, A.I., and Smith, D.E. (eds): Exposure to Insecticides, Bone Marrow Failure, Gastrointestinal Bleeding, and Uncontrollable Infections, *Amer J Med* **19**:274, 1965.
12. Scott, J.L., et al: A Controlled Double-Blind Study of the Hematologic Toxicity of Chloramphenicol, *New Eng J Med* **272**:1137, 1965.
13. Smick, K.M., et al: Fatal Aplastic Anemia: An Epidemiological Study of Its Relationship to the Drug Chloramphenicol, *J Chronic Dis* **17**:899, 1964.
14. Sharp, A.A.: Chloramphenicol-Induced Blood Dyscrasias: Analysis of 40 Cases, *Brit Med J* **1**:735, 1963.
15. Vigliani, E.C., and Saita, G.: Benzene and Leukemia, *New Eng J Med* **271**:872, 1964.
16. Osgood, E.E.: Hypoplastic Anemias and Related Syndromes Caused by Drug Idiosyncrasy (Report to the Council on Pharmacy and Chemistry), *JAMA* **152**:816, 1953.
17. American Medical Association Committee on Pesticides: Toxic Effects of Technical Benzene Hexachloride and Its Principal Isomers (Report to the Council on Pharmacy and Chemistry), *JAMA* **147**:571, 1951.
18. Agricultural Stabilization and Conservation Service, US Department of Agriculture: *The Pesticide Situation For 1963-1964,* 1964, p 3.
19. Agricultural Stabilization and Conservation Service, U.S. Department of Agriculture: *The Pesticide Review, 1966,* 1966, p 29.
20. Dale, W.E., and Quinby, G.E.: Chlorinated Insecticides in the Body Fat of People in the United States, *Science* **142**:593, 1963.
21. Hoffman, W.S.; Fishbein, W.I.; and Andelman, M.B.: The Pesticide Content of Human Fat Tissue, *Arch Environ Health* **9**:387, 1964.
22. Hayes, W.J.; Dale, W.E.; and Burse, V.W.: Chlorinated Hydrocarbon Pesticides in the Fat of People in New Orleans, *Life Sci* **4**:1611, 1965.
23. Egan, H., et al: Organo-Chlorine Pesticide Residues in Human Fat and Human Milk, *Brit Med J* **2**:66, 1965.
24. Hayes, W.J., and Dale, W.E.: Storage of Insecticides in French People, *Nature* **199**:1189, 1963.
25. Dale, W.E.; Copeland, M.F.; and Hayes, W.J.: Chlorinated Insecticides in the Body Fat of People in India, *Bull WHO* **33**:471, 1965.
26. American Medical Association Committee on Pesticides: Abuse of Insecticide Fumigating Devices (Report to the Council on Pharmacy and Chemistry), *JAMA* **156**:607, 1954.
27. Interdepartmental Committee on Pest Control, Department of Agriculture, Interior, Defense, and Federal Security Agency, Washington, D. C.: A Statement on the Health Hazards of Thermal Generators as Used for the Control of Flying Insects, *J Econ Entom* **44**:1027, 1951. (Revised: *J Econ Entom* **46**:181, 1953.)
28. American Medical Association Committee on Pesticides: Health Hazards of Electric Vaporizing Devices For Insecticides (Report to the Council on Pharmacy and Chemistry), *JAMA* **149**:367, 1952.
29. American Medical Association Committee on Pesticides: Health Problems of Vaporizing and Fumigating Devices for Insecticides: A Supplementary Report (Report to the Council on Pharmacy and Chemistry), *JAMA* **152**:1232, 1953.
30. The Vaporizer Hazard, *Consumer Reports* **18**:320 (July) 1953.
31. California State Board of Public Health: Resolution by the State Board of Public Health With Respect to the Use of Lindane Vaporizers, Berkeley, Calif, November 1952.

Diagnostic and Therapeutic Problems of Parathion Poisonings

DAVID W. WYCKOFF, M.D., JOHN E. DAVIES, M.D., M.P.H., ANA BARQUET, PH.D., and JOSEPH H. DAVIS, M.D.

POISONINGS DUE TO organophosphate pesticides are not infrequently fatal. For example, in Dade County, Fla., 50 deaths caused by these agents have occurred during a 7½-year period. Of these 48% have been suicides, 42% have been accidents, and the balance have been homicides or were of undetermined cause (1).

Parathion, which has been and still is that pesticide most commonly responsible for these often avoidable incidents (2), produces symptoms that often simulate other more common diseases. Accurate and early diagnosis requires the judicious use and interpretation of both clinical and laboratory procedures. These procedures are illustrated in this paper by the case of a 14-year-old girl who had ingested approximately 450 mg of parathion.

CASE REPORT

At about 5 PM on the afternoon of April 24, 1967, Patient G. L. complained to her mother of dizziness, vomiting, and "weakness." She progressively worsened, complained that she "couldn't see," and developed severe weakness of the lower extremities, causing her to fall several times. At approximately 7 PM her mother took her to the hospital emergency room. While in the lobby she developed what were considered to be epileptoid convulsions, accompanied by urinary and fecal incontinence. She was treated with 260 mg of phenobarbital intravenously, over a period of 15 min, for status epilepticus and subsequently became apneic. Ventilation was maintained by endotracheal intubation and the use of a Bird respirator. At this time no history of exposure to toxins was obtained, nor was there any other suggestion of possible cause. Her mother stated that she was a girl of normal intelligence, with no known mental problems or depressive tendencies.

On admission the pulse was regular at 104/min; the blood pressure, 110/80 mm Hg; and the temperature, 100 F. She was comatose and unresponsive to all stimuli. Her pupils were constricted and unreactive to light. Urinary and fecal incontinence, profuse perspiration, and excessive salivation were

---

From the Community Studies on Pesticides–Dade County, Miami, Fla.

This study was supported by the Pesticides Program, National Communicable Disease Center, Atlanta, Ga., through the Florida State Board of Health, under contract PH-86-65-26, U. S. Public Health Service, Washington, D. C.

Requests for reprints should be addressed to John E. Davies, M.D., Director, Community Studies on Pesticides–Dade County, 1390 N. W. 14th Ave., Room 401, Miami, Fla. 33125.

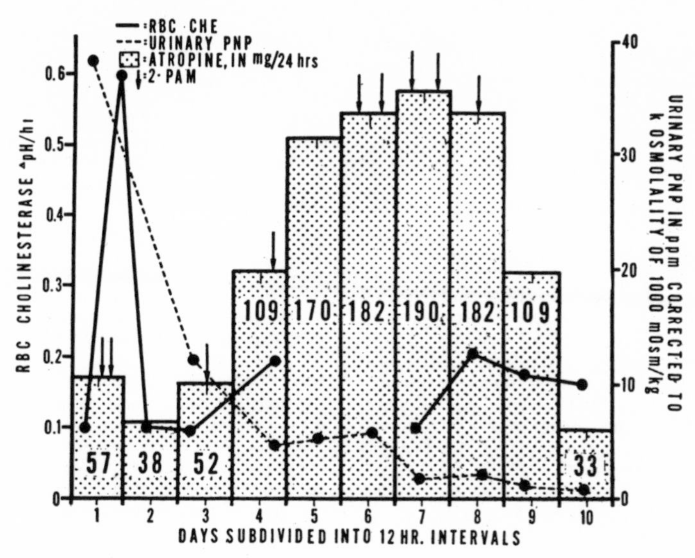

FIGURE 1. Clinical and therapeutic response to atropine and 2-pyridine aldoxime methylchloride (*2-PAM*) 14 hr after ingestion of 450 mg parathion. PNP = paranitrophenol.

noted. She had flaccid paralysis of all extremities. Deep tendon reflexes were absent.

Laboratory examination showed a white blood cell count of 16,000/mm³, with 72% segmented neutrophils, 19% band neutrophils, and 9% lymphocytes. The blood glucose concentration was 360 mg/100 ml before the intravenous administration of 10% glucose in water. The arterial P_{CO_2} was 30, with a pH of 7.56. Other laboratory values, including spinal fluid analysis, were within normal limits.

These signs and symptoms persisted throughout the night, and poisoning was increasingly suspected. Further questioning of the mother at approximately 10 AM the following morning brought out the presence of "roach powder" in the home. This powder was submitted to the office of the medical examiner, where it was immediately characterized by odor and appearance to be consistent with parathion. Oxalated blood and gastric content obtained 14 hr after admission were likewise submitted. The gastric content had no unusual odor or appearance. A bromthymol blue cholinesterase screening test (3) on the oxalated blood revealed no inhibition of cholinesterase activity. Rapid ultraviolet spectrophotometric screening of gastric content (4) demonstrated no absorbance curve for parathion or any other organophosphate.

The blood cholinesterase activity result was discounted because oxalated rather than heparinized blood was submitted. The negative ultraviolet spectrophotometric test was likewise discounted because 14 hr had elapsed since admission to the hospital. This could allow evacuation of the stomach to leave organophosphate pesticide levels below the levels of sensitivity of the test, which is 0.05 mg/100 ml or 500 parts per million (ppm). The same oxalated blood was found to have an erythrocyte cholinesterase activity of 0.52 ΔpH/hr (change in pH units per hour) using the Michel method (5). This level, consistent with the lower limits of normal, was likewise discounted as it was performed upon oxalated blood.

Heparinized blood was obtained at approximately noon on the same day and was also analyzed by the bromthymol blue and the Michel methods. The bromthymol blue test showed marked cholinesterase inhibition, and the Michel analysis revealed an erythrocyte cholinesterase activity of 0.10 ΔpH/hr, both results indicating significant cholinesterase depression. Analysis of the same heparinized blood, using the method of Fleisher and Pope (6), showed total inhibition of cholinesterase activity. The urinary paranitrophenol (PNP) concentration was 42 ppm as measured by the Elliott method (7). This and all subsequent PNP values were corrected to a uniform osmolality of 1,000 mOsm/kg. Figure 1 illustrates the sequential changes of PNP and erythrocyte cholinesterase values, together with the therapeutic regimen observed throughout the course of the illness.

Analysis by gas chromatography of the initial gastric wash revealed a parathion level of 68 parts per billion (ppb). The initial heparinized blood sample contained parathion at a concentration of 115 ppb. Similar analysis of heparinized and oxalated bloods failed to show detectable concentrations of paraoxon. A PNP level of 1.75 ppm, as determined by a modified Elliott method (8), was detected in the oxalated blood.

Subsequent analysis of the powder brought from the patient's home identified it as parathion at a concentration of approximately 15%.

A home visit was made. It was learned that the unlabeled "roach powder" had been obtained 2 years previously from a street peddler. It had been periodically scattered, in sizable amounts, in areas that the mother considered to be inaccessible—that is, behind the stove, sofa bed, radiophonograph console, etc. The mother was given instructions as to how to thoroughly clean up the remains of the material. Blood and urine samples from the mother and her remaining five children, ranging in age from 3 to 16 years and all asymptomatic, were obtained. Erythrocyte cholinesterase tests showed normal activity. Urinary PNP determinations were negative, except for the finding of a PNP concentration of 1.13 ppm in the urine of the mother. This was considered to be due to her handling of the substance during a prior house cleaning.

Therapy with atropine sulfate was begun 14 hr after the onset of symptoms. The first dose was at 9 AM on April 25, and by 11:30 AM atropine sulfate was being given intravenously at the rate of 2.0 mg every 15 min. At this time the patient was also thoroughly washed, hair included. After 3 hr of atropine therapy, some alleviation of the muscarinic effects of the intoxication had occurred, but marked improvement failed to take place until 1 PM the same day, when 1 g of 2-pyridine aldoxime methylchloride (2-PAM) was administered intravenously. This was followed by prompt restoration of voluntary respiration and somatic motor function. At this time a heparinized blood erythrocyte cholinesterase level was found to be 0.58 ΔpH/hr, this being a post-PAM response. This biochemical response to 2-PAM therapy was not again observed during the course of the illness. With further atropine and 2-PAM therapy, as well as continued gastric lavage, the patient's improvement was maintained and her sensorium largely cleared. Cathartics were not used because of continued massive diarrhea.

The following day she said that she had ingested a heaping teaspoonful, approximately 450 mg, of the parathion powder at about 2 PM on the day of her admission, because of her despair over what she believed to be her pregnant condition—having missed a menstrual period the previous month.

The clinical condition of the patient deteriorated on the afternoon of April 26, when her temperature reached 106 F. This was thought to be due to the atropine therapy, which was first decreased and then discontinued. Over the next 18 hr the symptoms of parathion toxicity recurred, and the following day atropine and 2-PAM therapy was restarted. The next 5 days of her illness were stormy, owing to the onset of pneumonia and a *Staphylococcus aureus* septicemia. Respiratory embarrassment as well as hypersecretion necessitated a tracheostomy. Atropine dosage was maintained at 2 mg intravenously every 15 min, a regimen sufficient to prevent the miosis and hypersecretion that were the principal clinical indicators of toxicity. 2-PAM was given when respiration appeared to be compromised. Some increase in respiratory excursion was thought to occur after the injections, but the dramatic clinical response after the initial dose of 2-PAM did not recur. Persistent toxicity was confirmed by the continued presence of PNP in the urine, this being consistent with continued metabolism of parathion. Erythrocyte cholinesterase levels remained severely depressed.

Urinalyses done on the second and fourth days of the intoxication revealed a moderate albuminuria (100 mg/100 ml). Gross aminoaciduria, as detected by paper chromatography (9), was present on the seventh day, and an increase of the alpha-amino nitrogen/creatinine ratio from 0.32 and 0.42 was also observed between the fourth and seventh days.

During the final 2 days of her illness the patient appeared near recovery from the effects of the parathion. This marked clinical improvement was reflected in the decreased and less frequent dosage of atropine needed to maintain pupillary dilation as well as the virtual elimination of urinary PNP. At 2 AM in the morning of the twelfth day of her illness she suffered a catastrophic hemorrhage from the tracheostomy and died. An autopsy disclosed the cause of death to be an erosion of the innominate artery by the cannula, a recognized complication of tracheotomy (10). The pupils were markedly constricted, probably because the patient died just before a scheduled dose of atropine. No evidence of pregnancy was found.

Discussion

A number of epidemiological and diagnostic lessons can be learned from this case. With regard to the former, parathion is still the leading cause of pesticide poisoning in Dade County. In spite of vigorous local legislation its availability is still in part the result of the activity of the itinerant peddler of illegal "roach powder." This is especially common in the nonwhite population. In such areas, therefore, apparent convulsions in the young negro should call for serious consideration of parathion as the logical agent.

SYMPTOMS AND SIGNS

The diseases that parathion poisoning simulates include pneumonitis, gastroenteritis, encephalitis, brain injury, and hypertensive encephalopathy. As was the case here, alternative central nervous system disease was the first diagnostic choice in three of our previous poisonings. In the absence of a history of exposure, signs stemming from parasympathetic hyperactivity are the clearest clinical indicators of cause. Miosis, as was present in this case, is always especially emphasized. Mydriasis, however, has been observed in approximately 13% of local cases of parathion intoxication (11).

Leukocytosis and hyperglycemia have been reported previously and were also seen in this case (12). These findings, understandably, can contribute to errors of diagnosis.

CHEMICAL LABORATORY TESTS

Although the initial diagnosis of organophosphate poisoning is usually clinical, laboratory tests provide additional confirmation. Pitfalls in the interpretation of these are lessened if the right specimen is collected at the right time.

CHOLINESTERASE DETERMINATIONS

Gastric washings and urine specimens should be the first available materials. Cholinesterase measurements should be obtained from heparinized pre-PAM (pyridine aldoxime methylchloride) blood. The marked discrepancy between the initial erythrocyte cholinesterase level determined from heparinized blood and that from oxalated blood points up the increased accuracy of the former and stresses the importance of using heparinized blood for this analysis. The initial post-PAM increase in cholinesterase activity to the low normal level of 0.58 ΔpH/hr shows the necessity of obtaining blood before oxime therapy, if this parameter is to be correctly evaluated in the establishment of a diagnosis.

Even with these precautions further diagnostic and management problems occur. The demonstration of cholinesterase inhibition is of great value in the diagnosis of organophosphate poisoning. Accuracy in blood cholinesterase laboratory results is essential, and pitfalls in laboratory methodology have been well-documented in the review by Witter (13). The prolonged duration of erythrocyte cholinesterase inhibition, however, precludes its use as a measure of either persistent toxicity or of therapeutic response. The lack of correlation with clinical findings after the initiation of therapy was again demonstrated in this case. A rise in cholinesterase levels was noted after the first dose of 2-PAM, but throughout the remainder of the illness they remained in the range of 0.1 to 0.2 ΔpH/hr despite the marked clinical improvement of the patient during the last 2 days of her illness. Similarly, no correlation between cholinesterase level and presence of toxin was noted. On the final day of her illness, when the urinary paranitrophenol (PNP) was only 0.50 ppm, the erythrocyte cholinesterase activity was still severely depressed, the value being 0.15 ΔpH/hr.

URINARY PARANITROPHENOL

In contrast, the urinary PNP concentration was shown to be an excellent indicator of toxicity, as has been recorded before (11). While useful as a supplement to cholinesterase inhibition in the diagnosis of a certain class of organophosphate poisonings (PNP is found as a metabolite in cases of parathion, methyl parathion, ethyl *p*-nitrophenyl thionobenzenephosphate, dicapthon, and Chlorthion® intoxication only), its measurement is particularly valuable as a quantitative index of intoxication throughout the course of illness. In this case, the elevated values found throughout most of the illness correlated well with the persistence of clinical signs of toxicity, such as miosis and hypersecretion, and the finding of only 0.50 ppm of PNP on the final day of her illness was accompanied by marked clinical improvement of the patient. The usefulness of this metabolite as an indicator is apparent.

IDENTIFICATION OF THE PESTICIDE

Valuable as it is in management, the lack of specificity of PNP necessitates the development of other diagnostic procedures for the accurate identification of the specific chemical causing the intoxication. This is especially true in instances where there may be medicolegal implications. In this case preliminary use was made of the pesticide itself as a diagnostic tool.

Gas chromatography, using the Radomski non-clean-up method (14), was the technique used in analysis for parathion. Analyses were performed on the following material: heparinized and oxalated blood samples drawn within 24 hr of ingestion of poison, heparinized postmortem blood, gastric washing collected on the first day of illness, and the "roach powder" brought from the patient's home. Gas chromatographs equipped with electron capture de-

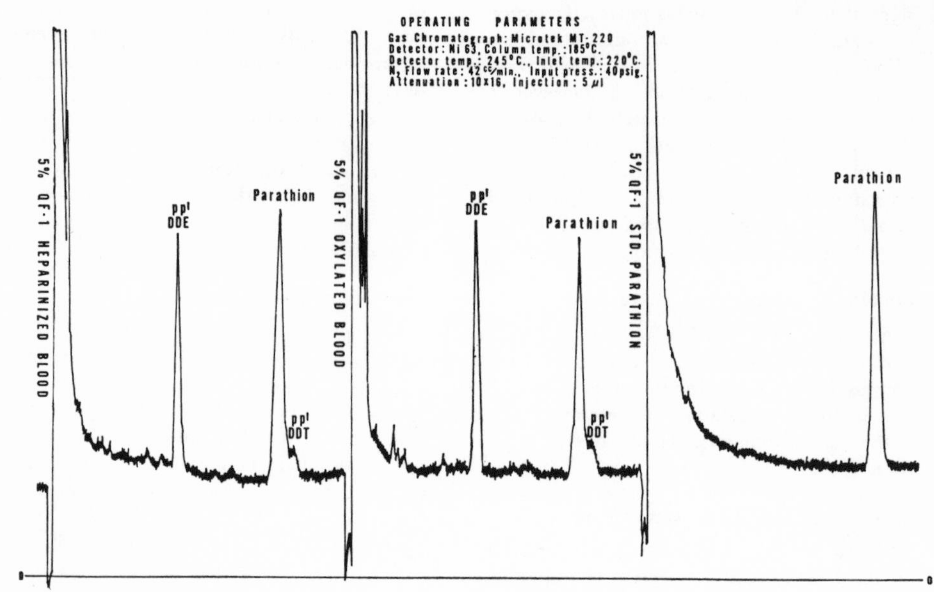

FIGURE 2. Demonstration of pesticides in blood samples from Patient G. L. on first day of illness.

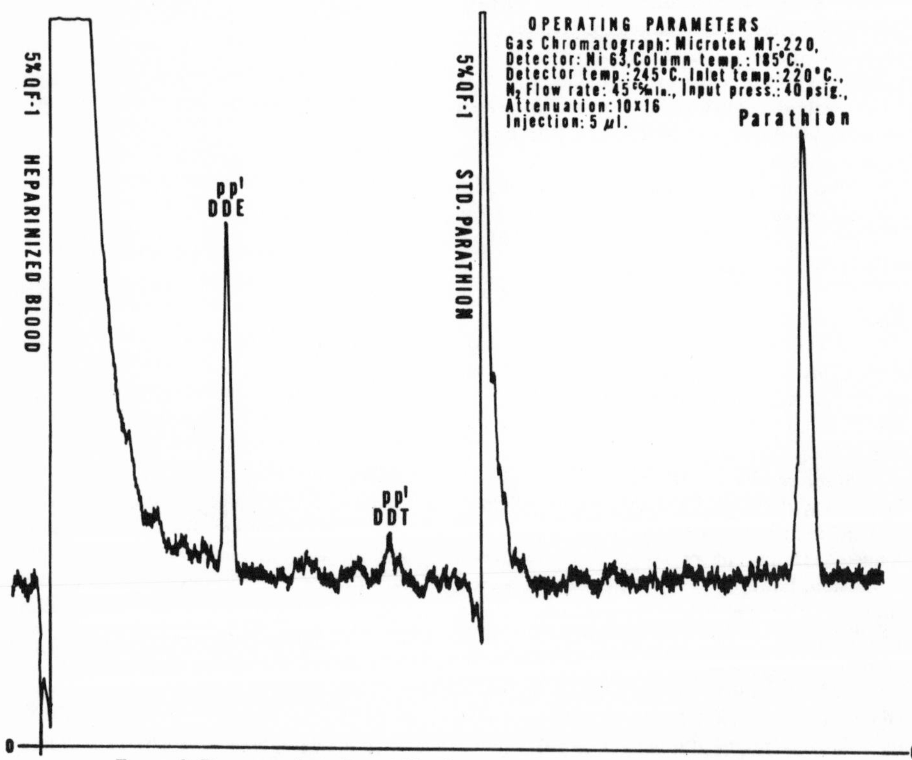

FIGURE 3. Demonstration of pesticides in postmortem blood from Patient G. L.

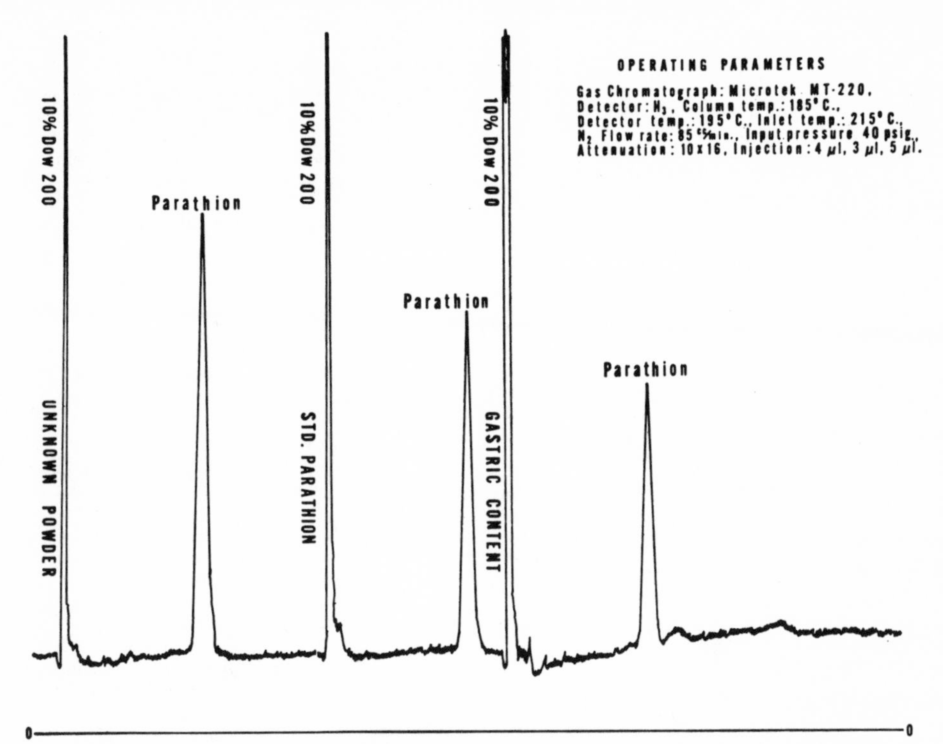

FIGURE 4. Demonstration of parathion content in unknown "roach powder" and in gastric washings taken from Patient G. L. on first day of illness.

tectors were used, the detectors being either ^{63}Ni or ^{3}H. In order to insure accuracy in the identifications, two analyses, each using a different column, were done on each sample. The columns were 10% DOW 200 on Chromport XXX, and 5% QF-1 on Chromsorb W. Other operating parameters are listed on Figures 2, 3, and 4.

Figure 2 reproduces the chromatogram of the antemortem blood samples, and both samples demonstrated peaks identical with that of the parathion standard. This represents the first reported detection of parathion in human blood. The other two peaks coincided with those of p,p′-DDE* and p,p′-DDT* standards. Contrast this with Figure 3, the postmortem blood chromatogram. The DDE and DDT peaks are still present, but the parathion peak has disappeared. This correlates well with the patient's condition on the day before death, when she was clinically virtually free of toxicity and the urinary PNP level was negligible. Figure 4 demonstrates the presence of parathion in the initial gastric content and the "roach powder." Both samples display peaks identical with that of the parathion standard. Confirmation of parathion content in the above blood samples was established by thin-layer chromatography, using the method described by Walker and Beroza (15).

Quantitation of material under analysis usually by calculation of peak height and comparison with standard is strikingly sensitive. The lower limit of sensitivity to parathion using gas chromatography is approximately 10 ppb as compared, for instance, with 100 ppb using thin-layer chromatography. This sensitivity and its accuracy in identification suggest a more prominent role in the future for gas chromatography, and an increasing use of analysis for specific pesticide in the diagnosis of organophosphates intoxication can be anticipated.

* DDE = derivative of dichloro-diphenyl-dichloroethane stored in fat; DDT = chlorophenothane.

THERAPEUTIC PROBLEMS

In addition to the diagnostic problems discussed above, four therapeutic problems were well-illustrated in this case.

1. Marked clinical improvement may follow introduction of drug therapy, but when massive ingestion of toxin has occurred the continuing absorption of the material will cause rapid recurrence of clinical intoxication should this therapy be prematurely reduced. Thus, the need for constant surveillance of both clinical and laboratory indicators of toxicity is emphasized.

2. The prolonged course of illness was thought to be due to continued absorption of poison from the gastrointestinal tract and demonstrated the need for action directed at evacuation of the bowel. PNP excretion was inordinately long and was probably due both to continued absorption and possibly to the presence of some impairment of renal excretion. Renal damage was evident by the occurrence of moderate albuminuria, significant aminoaciduria, and elevation of the alpha-amino nitrogen to creatinine ratio. The latter sequels of pesticide poisoning have been noted in other cases (16).

3. The necessity of continuous high atropine dosage to counteract toxicity was again demonstrated (2, 17, 18). The total dose of atropine given was 1,122 mg. Pupillary mydriasis and absence of secretion were found to be the most useful clinical indicators of sufficient dosage. It should be remembered that undertreatment with atropine represents a far greater threat to the patient than overtreatment.

4. Whether the total dosage of 7.5 g of 2-PAM administered in this case was adequate is uncertain. More frequent 1-g doses have been given in previous cases with good results, and the possibility that this might have resulted in less reliance on the Bird respirator and a less prolonged course of illness cannot be excluded. The dramatic response noted after the first injection was not seen in subsequent 2-PAM infusions, but very definite improvement in respiratory excursion was observed despite absent erythrocyte cholinesterase elevation. In view of the relative lack of side effects or toxicity associated with this drug (19), vigorous therapy should be pursued in those cases where absorption of the pesticide is suspected.

Acknowledgment

The authors wish to acknowledge with gratitude the assistance of Dr. John Price in affording us the opportunity for pesticide analyses in this case. Appreciation is also due to Mr. Robert L. Roeth for the illustrations and to Mrs. Gigi Nachman and other members of the Community Studies on Pesticides laboratory for the thin-layer chromatography and other analyses.

References

1. Office of the Medical Examiner, Dade County, Statistical Report, 1962. Subsequent Data from the Office of the Medical Examiner, unpublished.
2. Davies, J. E., Davis, J. H., Frazier, D. E., Mann, J. B., Reich, G. A., Tocci, P. M.: Disturbances of metabolism in organophosphate poisoning. *Industr. Med. Surg.* 36: 58, 1967.
3. Limperos, G., Ranta, K. E.: A rapid screening test for the determination of the approximate cholinesterase activity of human blood. *Science* 117: 453, 1953.
4. Fisk, A. J., Czerwinski, G. R., Kenhart, J. H.: A rapid screening technique for the identification of organic phosphate insecticides in gastric lavage fluid or dermal residues. *J. Forensic Sci.* 10: 473, 1965.
5. Michel, H. O.: Electrometric method for determination of red blood cell and plasma cholinesterase activity. *J. Lab. Clin. Med.* 34: 1564, 1949.
6. Fleisher, J. G., Pope, E. J.: Colorimetric method for determination of red blood cell cholinesterase activity in whole blood. *Arch. Industr. Hyg.* 9: 323, 1954.
7. Elliott, J. W., Walker, K. C., Penick, A. E., Durham, W. F.: A sensitive procedure for urinary p-nitrophenol determination as a measure of exposure to parathion. *J. Agr. Food Chem.* 8: 111, 1960.
8. Barquet, A., Davies, J. E., Davis, J. H.: Death due to parathion? In preparation.
9. Efron, M. L., Young, D., Moser, H. W., MacCready, R. A.: A simple chromatographic screening test for the detection of disorders of amino acid metabolism. A technic using whole blood or urine collected on filter paper. *New Eng. J. Med.* 270: 1378, 1964.
10. Willerson, J. T., Fred, H. L.: Delayed fatal hemorrhage after tracheotomy. *Arch. Intern. Med. (Chicago)* 116: 138, 1965.
11. Davis, J. H., Davies, J. E., Fisk, A. J.: Occurrence, diagnosis and treatment of organophosphate pesticide poisoning on man. Presented at the New York Academy of Sciences Symposium on Biological Effects of Pesticide in Mammalian Systems, May 1967, New York, N. Y.

12. Arterberry, J. D., Durham, W. F., Elliott, J. W., Wolfe, H. R.: Exposure to parathion. Measurement by blood cholinesterase level and urinary p-nitrophenol excretion. *Arch. Environ. Health (Chicago)* 3: 476, 1961.

13. Witter, R. F.: Measurement of blood cholinesterase. A critical account of methods of estimating cholinesterase with reference to their usefulness and limitations under different conditions. *Arch. Environ. Health (Chicago)* 6: 537, 1963.

14. Radomski, J. L., Fiserova-Bergerova, V.: The determination of pesticides in tissues with the electron capture detector without prior clean-up. *Industr. Med. Surg.* 34: 934, 1965.

15. Walker, K. C., Beroza, M.: Thin layer chromatography for insecticide analysis. *J. Ass. Offic. Agr. Chem.* 46: 250, 1963.

16. Davies, J. E., Mann, J. B., Tocci, P. M.: Renal tubular dysfunction and amino acid disturbances under conditions of pesticide exposure. Presented at the New York Academy of Sciences Symposium on Biological Effects of Pesticides on Mammalian Systems, May 1967, New York, N. Y.

17. Hayes, W. J., Jr.: Parathion poisoning and its treatment. *JAMA* 192: 49, 1965.

18. Holmes, J. H.: Clinical experiences with pesticides, in *Proceedings of the Short Course on the Occupational Health Aspects of Pesticides,* University of Oklahoma, Norman, Okla., 1964, pp. 131–143.

19. *Safety in Handling of Thimet Systemic Insecticide.* Brochure prepared by American Cyanamid Company, Wayne, N. J., 1963.

Treatment of Organophosphorus and Chlorinated Hydrocarbon Insecticide Intoxications

MITCHELL R. ZAVON, M.D.

THE NUMBER OF DEATHS resulting from accidental or deliberate misuse of insecticides does not appear to have increased in recent years in the United States, but poisoning from insecticides continues to occur. Awareness of the possibility of insecticide intoxication has occasionally exceeded the probability of intoxication from this group of chemicals and has resulted in mistreatment because of a failure to proceed beyond "insecticide" in the differential diagnosis. In this article we will discuss the newer, organic insecticides, and will not refer to the older, inorganic materials, such as the arsenicals. The physician should, nevertheless, be aware that among the pesticides, the arsenicals and mercurials still cause as many deaths each year as all of the organic materials combined.

Insecticides are used both in rural and urban areas. They may be found in the home, office, farm, and factory. An unlikely setting, however, cannot be used to exclude the possibility that an illness is the result of absorption of an insecticide.

The following general criteria must be met in order to make the diagnosis of insecticide intoxication: (1) history of exposure or a reasonable possibility that such exposure could have occurred, (2) onset of the illness within 12 hours of the last exposure, and (3) a clinical picture consistent with that described for the particular chemical involved.

When the above criteria cannot be met it is safer to look for other causes of the illness. Laboratory procedures may aid in the diagnosis but are of little value with many insecticides. Treatment must often be initiated before laboratory results are available; thus they may be of more value in confirming a diagnosis than in making it in the first place. It is best, if possible, before initiating treatment, to obtain samples of blood, urine, and gastric contents (if poisoning is due to ingestion). Analysis for cholinesterase activity or for specific insecticides or their metabolites may be possible if samples of body fluid are retained.

Specific treatment will be discussed for specific classes of insecticides, and for general principles the reader is referred to the article by Arena, p. 599. A word of caution regarding treatment is in order. Unless the dose is overwhelming, proper treatment will result in recovery even in severe cases. Failure of treatment is too frequently a result of failure to treat energetically enough, failure to keep the patient under continuous observation once treatment has been instituted, and failure to maintain a sufficient supply of oxygen due to preoccupation with the primary effects of intoxication. The only authenticated residua following recovery from insecticide intoxication are those resulting from anoxemia during the acute illness.

If illness is sufficient to warrant any treatment, it is sufficient to warrant hospitalization for at least 24 hours with careful observation during hospitalization.

Preventive drug therapy for those working with insecticides is never indicated and may only serve to confuse the clinical syndrome if intoxication does occur.

ORGANOPHOSPHORUS INSECTICIDES

Diagnosis

Organophosphorus insecticides exert their effect primarily by inhibiting the enzyme cholinesterase. This results in the accumulation of acetylcholine which causes symptoms of excessive stimulation of the parasympathetic nervous system, central nervous system effects, and effects on the myoneural junction. The signs and symptoms have been described as muscarinic and nicotinic but are better listed and then questioned, because all may or may not be present. Thus, the symptoms and signs listed below may not all be present in a mild case:

Anorexia	Excessive sweating
Nausea	Excessive salivation and tearing
Mental confusion	Headache
Vomiting	Pupillary constriction
Abdominal cramps	Excessive respiratory tract secretion
Diarrhea	Muscular twitching
"Weakness"	Convulsions

The most common symptom is apt to be abdominal cramps, headache, dyspnea, or a feeling of weakness, described by the patient as "all gone." The blurred vision caused by pupillary constriction is also common but may be absent and should not be relied on as a *sine qua non* for diagnosis of organophosphorus intoxication.

The pharmacologic action of the organophosphorus compounds is fundamentally the same as described above but the toxicity of the different compounds varies tremendously. Whereas a few drops of parathion can cause death, malathion is much less toxic.

The most commonly used organophosphorus insecticides are:

Chlorthion	Dimethoate	Parathion
Co-Ral	Di Syston	Phorate
DDVP	Fenthion	Phosdrin
Delnav	Guthion	Phosphamidon
Diazinon	Malathion	Ronnel
Dibrom	Methyl Parathion	TEPP

The mere handling of one of these compounds does not automatically validate a diagnosis of intoxication. Many members of this class of compound are absorbed through the skin. Some such as tetraethylpyrophosphate (TEPP) are water-soluble and direct-acting and, therefore, may cause symptoms very quickly, whereas a compound such as parathion must first be converted to a physiologically active form by the body before it causes symptoms. In part for this reason, symptoms of parathion intoxication among sprayers not infrequently become evident 6–8 or more hours after exposure has commenced.

In making the diagnosis of intoxication due to an organophosphorus cholinesterase inhibitor, we must consider the compound involved and the route of exposure. Though most of these compounds may be absorbed percutaneously, onset of symptoms by this route is likely to be much slower than a dose by mouth or by inhalation. Occupational exposure is most often percutaneous, whereas ingestion is most likely in suicides and children.

The diagnosis of poisoning by this class of compound is usually made by the history of exposure, presence of a typical symptom complex, and confirmation by cholinesterase determination. In the acute case, delay for confirmation by cholinesterase determination is not practical, but blood should be drawn in a heparinized tube, if available, before treatment is instituted, and the red cells and plasma separated by centrifuging and then frozen if they must be kept for awhile before analysis. A pretreatment sample of blood can be of immense help in those cases where, in retrospect, the diagnosis is questioned or the patient does not respond to treatment. Parathion, chlorthion, and EPN are excreted in the urine as paranitrophenol. A urine sample for estimation of paranitrophenol or other metabolites may also be valuable.

The differential diagnosis should include cerebral hemorrhage, heat exhaustion, heat stroke, and gastroenteritis.

The cholinesterase activity of the erythrocytes should be less than 25 per cent of the normal for the individual, or the population if a pre-exposure value is not available. If the cholinesterase activity is much above 25 per cent of normal in the erythrocytes, a diagnosis of anticholinesterase poisoning must be suspect.

Treatment

Organophosphorus intoxication caused by cutaneous contamination calls for immediate decontamination. Hospitals do not usually shower a patient on admission, but in these cases, unless the patient is in severe difficulty, it is best to remove all clothing and shower, being certain to wash the hair thoroughly and to clean the fingernails. If this procedure is not followed initially because of the patient's condition, it should be done as soon as practical. In some cases, recovery has occurred only when thorough showering removed the insecticide which contaminated the hair and fingernails. Clothing, including boots or shoes, should be removed and destroyed if heavily contaminated.

When intoxication is caused by ingestion, emesis is more effective than lavage. However, when the insecticide is diluted in a petroleum-based solvent, emesis can result in aspiration and subsequent chemical pneumonitis. Whether by emesis or lavage, the insecticide should be removed from the stomach as soon as possible. In a few cases where very small amounts of a very dilute solution of one of the less toxic materials have been ingested, lavage may not be required. However, the discomfort of lavage may be therapeutic in helping to prevent further ingestion by the young patient.

Following termination of exposure by washing with copious amounts of water and soap, emesis, or lavage, the following treatment is instituted:

1. Maintain a clear airway and give artificial respiration if necessary.
2. Administer atropine sulfate. For severe cases: 4–8 mg. (1/16 to 1/8 grain) intravenously every 10 minutes until full atro-

pinization occurs. (Atropinization is indicated by dilated pupils, dry, flushed skin, and a rising pulse rate.) [Atropine should not be used until cyanosis has been overcome since it can produce ventricular fibrillation in the presence of anoxia.—Ed.]

3. Administer 2-PAM (Protopam chloride, pyridine-2-aldoxime, chloride) only after atropine and if needed. The dose of 2-PAM is 1 gm., given intravenously, slowly, and repeated in 1 hour if needed.

The doses shown are for adults. For children, they should be reduced in proportion to body weight, but a dose of 10–12 mg./kg. can be recommended.

In severe cases, antidotal treatment is instituted before decontamination. Decontamination, lavage, or emesis should follow as soon as possible.

The recommended dose of atropine may appear large but the patient poisoned by a cholinesterase inhibitor is well able to tolerate it, and failure to use the highest dose tolerated is far more serious than the effects of overdosage. Once signs of atropinization occur, the dose should be adjusted to just maintain full atropinization for at least 24 hours. This blocks the excess acetylcholine until such time as the cholinesterase inhibitor is detoxified and excreted. After 24 hours, atropine is discontinued but only under constant, careful observation so that at the first sign of recurrence of symptoms the therapy can be reinstituted.

In the presence of severe anticholinesterase poisoning, more than 50 mg. of atropine has been given in a 24-hour period without an adverse effect. The usual problem is a failure to give enough atropine rather than giving too much.

As mentioned, atropine sulfate blocks the action of acetylcholine, particularly the muscarinic action, but does not restore the activity of the cholinesterase. Poisoning with many, but not all, of the cholinesterase inhibitors can be reversed by the use of PAM. The drug is effective against many of the most commonly used organophosphorus insecticides, such as parathion. It competes with the cholinesterase for the alkylphosphate and if given within 24 hours, before irreversible inhibition has occurred, can be extremely valuable. The dose is 1 gm. for adults, which may be repeated to a total of 4 gm. in 24 hours. It is most effective the sooner it is given and is of doubtful value if more than 24 hours have elapsed following intoxication.

Never give morphine, theophylline or aminophylline. It is doubtful that tranquilizers are ever indicated and there is some indication that promazine and chlorpromazine may be contraindicated.

Drug therapy to counteract the effects of this class of compounds can only succeed if decontamination has been thorough and the patient has been maintained in a viable condition. Pulmonary edema can occur and should be treated vigorously with positive-pressure oxygen and such other measures as are appropriate. An oxygen tent prevents good observation of the patient and should not be used.

In some cases of intoxication the effect of the solvent may prove to be as important as the insecticidal ingredient in the formulation. Such cases can be very complicated therapeutic problems and may require first-rate hospital facilities if the patient is to survive.

Any patient with a diagnosis of insecticide poisoning should be hospitalized and kept under observation for at least 24 hours, and if ill should be observed for 48 hours. If atropine has been administered, medical observation for 24 hours is obligatory.

Following intoxication, further exposure to anticholinesterase agents should be avoided until laboratory measurement of erythrocyte cholinesterase indicates reasonably complete recovery.

If the treatment described does not produce improvement, the diagnosis should be questioned and another etiology for the clinical picture sought. Only in overwhelming intoxications will the usual treatment fail to produce effective results if the patient is seen early in the acute phase.

CHLORINATED HYDROCARBON INSECTICIDES

Diagnosis

The toxicity of chlorinated hydrocarbon insecticides varies widely. When absorbed systemically in significant quantity, any member of this group can produce general malaise, headache, loss of appetite, and nausea. Vomiting, muscle fibrillation, and gross tremor may follow in rapid succession if the dose is sufficiently large. Convulsions occur only with severe intoxication. Symptoms are produced by action on the central nervous system but the exact mechanism, unlike the organophosphorus insecticides, is unknown. These compounds and some of their degradation products are stored in the body fat. There is no evidence that the stored materials are active.

The most commonly used members of this class are listed below:

Aldrin	Lindane*
Chlordane	Methoxychlor
DDT*	Perthane
Dieldrin*	DDD
Endrin	Thiodan
Heptachlor (epoxide*)	Toxaphene
Kelthane	

* Found in "normal" body fat.

Diagnosis of intoxication must be based on the criteria listed earlier. In addition to the symptom complex described, the insecticide or its derivatives may be found in stomach contents, blood, urine, or tissues, particularly fat. Finding the insecticide in the fat may not be diagnostic, as the entire population has residues of some of these compounds as a "normal" constituent of the body fat. Using gas liquid chromatography for analysis, minute traces of these materials can be detected.

The symptoms described above can be caused by other classes of pesticides, as well as many other exogenous factors. Unless symptoms occur within 12 hours after acute exposure, and preferably within 2 hours after exposure, another cause of illness should be suspected.

Treatment

Removal of the source of insecticide is primary in treatment unless prevented by the patient's critical condition. The patient should be washed with copious amounts of water and soap if the insecticide is on the skin. Emesis with syrup of ipecac is valuable and gastric lavage is indicated if a petroleum-based formulation has been ingested. Lavage or emesis should be followed by a saline cathartic, such as epsom salts, to promote excretion. Oily laxatives or food should be avoided as they promote absorption.

In the case of *severe intoxication*, pentobarbital sodium (Nembutal sodium) is given intravenously in a dosage of 0.2–0.5 gm., repeated as needed to control convulsions. Children should get an initial dose of 25–100 mg. depending on body weight and the severity of the intoxication. The pentobarbital may be repeated as needed, but a switch to phenobarbital sodium is usually advisable once the convulsions are controlled in order to provide more prolonged control.

In *mild intoxication*, phenobarbital sodium may be given by mouth for control of symptoms of headache, malaise, nausea and muscle fibrillation or it may be injected intramuscularly. One adult required 350 mg. (6 grains) of phenobarbital for control of symptoms caused by a chlorinated hydrocarbon insecticide. This dose did not result in a soporific effect and the patient was back to work in 48 hours.

Maintenance of a clear airway and sufficient oxygen are imperative. Calcium gluconate may be of help as an adjunct to barbiturate therapy but it has not been used often. It can be given either intravenously, 10 ml. of the 10% solution, or orally.

Anticonvulsive therapy may have to be continued for as long as a week in severe intoxications, but usually the patient will have passed the critical stage in 24 hours.

References

1. Durham, W. F., and Hayes, W. J., Jr. Organic phosphorus poisoning and its therapy. *Arch. environm. Hlth.* 5:1, 1962.
2. Golz, H. H., and Shaffer, C. B. *Toxicological Information on Cyanamid Insecticides.* American Cyanamid Co., Wayne, New Jersey.
3. Hayes, W. J., Jr. Review of the metabolism of chlorinated hydrocarbon insecticides, especially in mammals. *Ann. Rev. Pharmacol.* 5:27, 1965.
4. Hayes, W. J., Jr. *Clinical Handbook on Economic Poisons.* U. S. Government Printing Office, 1963.
5. Holmstedt, B. Pharmacology of organophosphorus cholinesterase inhibitors. *Pharmacol. Rev.* 11:3, 1959.
6. Lehman, A. J. *Summaries of Pesticide Toxicity.* Association of Food and Drug Officials of the United States, Topeka, Kansas.
7. McGee, L. C. Toxicology of some agricultural chemicals: A review of accidental poisoning of human beings by chlorinated insecticides. *Trans. Amer. clin. climat. Ass.* 68:69, 1956.
8. Ministry of Health, United Kingdom. Poisoning by organophosphorus compounds used in agriculture and horticulture. *Lancet* 2:201, 1960.
9. Tabershaw, I. R., and Cooper, W. C. Sequelae of acute organic phosphate poisoning. *J. occup. Med.* 8:1, 1966.

Effects of Noise on Man

Extinction factors in startle (acousticomotor) seizures

Harold E. Booker, M.D., Francis M. Forster, M.D., and Hallgrim Kløve, Ph.D.

THE EFFECT of sensory stimuli in evoking or inhibiting epileptic attacks is well known, and such effects probably are more frequent than suspected. Penfield[1] points out that the most common example of reflex precipitation of seizures is found when a sudden, unexpected sound startles the patient. Gowers[2] noted that seizures induced by sound are usually minor rather than major motor seizures and that a startling sound might produce a momentary discharge in normal motor centers. It is not surprising, therefore, that a pathological discharge in response to startling sound occurs in motor centers when a morbid excitability of the tissues exists. Most authors agree that it is the startle aspect, rather than acoustic aspects, of the stimulus that is effective. Such attacks frequently are described, therefore, as startle epilepsy. Studies in experimental audiogenic seizures[3,4] have been revealing but have not produced as yet a complete understanding of acoustically induced seizures. We have studied two patients with myoclonic or startle responses to sound. They were chosen for further study from a larger group of patients with startle seizures because neither patient showed any tendency toward spontaneous habituation of the response with prolonged or repeated stimulation. We found that controlled stimulus patterns could produce an extinction effect manifested by a decrease sensitivity to subsequent stimuli.

CASE REPORTS

Case 1. A 10-year-old male suffered from fr quent myoclonic seizures in response to sudd sounds. The attacks involved the upper extren ties and, to a lesser extent, the head and nec Mild jaundice on the fourth day of life was a companied by a generalized seizure. Three genera ized seizures occurred during the next three week but there were no recurrences after the first fo weeks of life. At age 16 months, an evaluation f delayed motor development suggested cerebr palsy, and a pneumoencephalogram at that tir showed cortical atrophy. At 8 years, focal clor attacks of the left arm and left leg occurred. Th could be avoided occasionally by the patient hol ing the left wrist firmly with his right hand. about this same time, generalized myoclonic startle responses to sound began to occur. At t time of evaluation, the patient walked with crutch and showed a marked spastic paraparesis. An ele troencephalogram revealed a nonspecific dysrhyt mia, and his anticonvulsant medication of phenylhydantoin (Dilantin®) and phenobarbi was maintained unchanged throughout the peri of observation.

Case 2. A 13-year-old male suffered from pe sistent myoclonic startle responses to sound i volving primarily the upper extremities but al the facial musculature, with extension of the ne and spine. Any sudden noise would evoke a br startle seizure. He was severely retarded secc dary to anoxia at birth. The myoclonic attac were first noted at age 9 months. A pneumoe cephalogram at 10 months showed diffuse corti atrophy. At present he is severely retarded wi flexion deformities of all extremities. His electr encephalogram shows a marked, nonspecific dy rhythmia. His anticonvulsant medication of t methadione (Tridione®) and phenobarbital w gradually discontinued during the period of o servation.

xtinction studies

Various sound stimuli were tape recorded nd delivered by independently controlled arphones. The output intensity at the seprate earphones was measured with a noise vel meter. Movement of the patient's exemities was recorded by an oscillometer ttached to one wrist. An electroencephaloraphic dysrhythmia consistently accompanied ne startle response in the first patient and was sed to monitor his responses. In the second patient, the electroencephalographic response was inconstant. When neuromuscular blockade was produced in this patient by intravenous succinylcholine chloride (Anectine®), no electroencephalographic dysrhythmia could be induced by supramaximal stimuli. Consequently, the startle response was monitored by an oscillometer attached to the patient's wrist. Records of electroencephalograms, oscillometer, and stimulus artifacts were taken on a 20-channel electroencephalograph. Laboratory

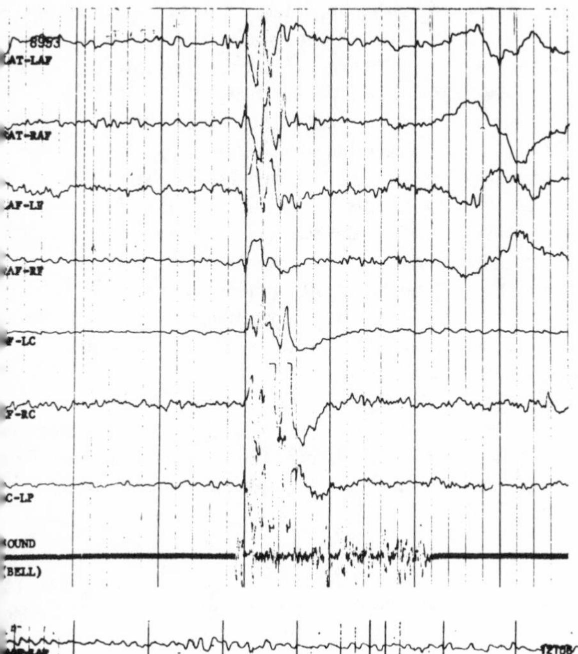

Fig. 1. Bottom tracing represents the sound (a bell) administered from the tapes; upper channels are scalp recordings. [A] The binaural administration of the sound stimulus evoked a myoclonic seizure. [B] The monaural stimulus (in this case to the left ear) evoked no clinical seizure and the brain wave pattern is undisturbed. [C] Binaural stimulus presented after 30 monaural extinction trials produced no seizure and no disturbance of the electroencephalogram.

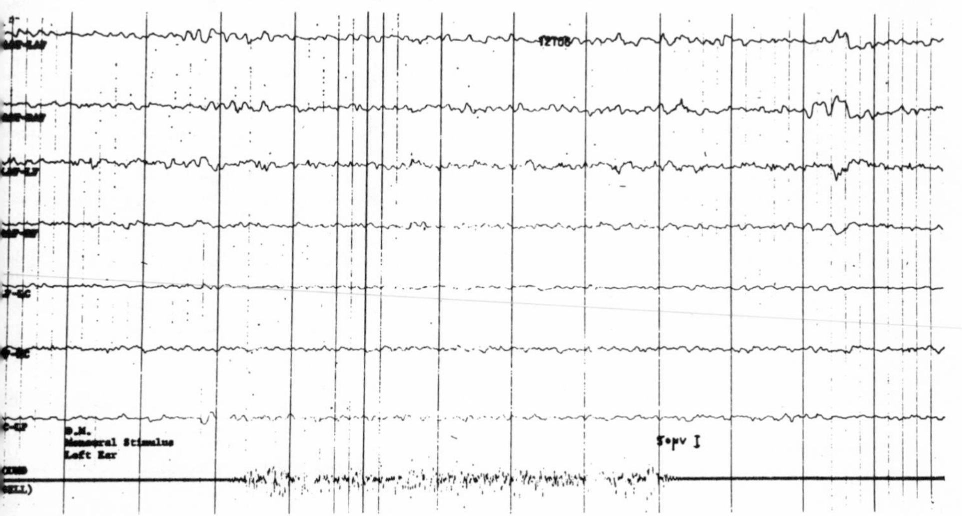

conditions and background noise were kept as constant as possible. Observations were made at various times of the day.

The patient's threshold to startling sounds was evaluated by first determining the intensity at which a sound, delivered binaurally with equal volume in each ear, would consistently evoke a startle response. The same sound then would be delivered to one earphone only. Extra padding was added to the earphones to provide a greater degree of acoustic isolation of the ears. Even with this precaution, a monaural stimulus at loud intensities could be picked up by the other ear. For evaluating monaural sensitivity, we worked at volume levels well below these higher intensities. Under these conditions, both patients were found to be monaurally insensitive. This effect cannot be attributed solely to an intensity difference, because the intensities were adjusted so that the output of the earphone delivering the monaural stimulus was approximately equal to the combined intensities of the two earphones in the binaural stimulation.

We discovered that not only were monaural stimuli ineffective in producing a response, but repeated monaural stimuli could induce a decreased sensitivity (an increased threshold) to subsequent binaural stimuli. Extinction trials were carried out after first determining the binaural threshold. One earphone was silenc and repeated trials, usually about 15, we given to one ear. The response to binau stimulation was then redetermined. This effe was most pronounced in the first patient wh 15 trials in one ear were followed by 15 tri in the other ear (Fig. 1).

While monaural stimuli were ineffective producing a startle response in the seco patient, repeated monaural trials did not co sistently produce a subsequent decrease sensitivity to binaural stimuli. We found, ho ever, that repeated subthreshold binaural tri in this patient would produce a subseque extinction effect. After determining binau threshold, the intensity of the stimulus in o earphone was maintained while the intens of the other was reduced until no startle sponse could be evoked. The stimuli then we repeatedly delivered with the different inte sities in the two ears, and the intensity of t reduced stimulus gradually was increase After 20 to 30 such extinction trials, thresho and even suprathreshold stimuli could be livered binaurally without a return of t startle responses (Fig. 2).

The general features of the extinction eff were similar, whether produced by monau or subthreshold binaural stimulation. Th was a degree of specificity for the effect. Th

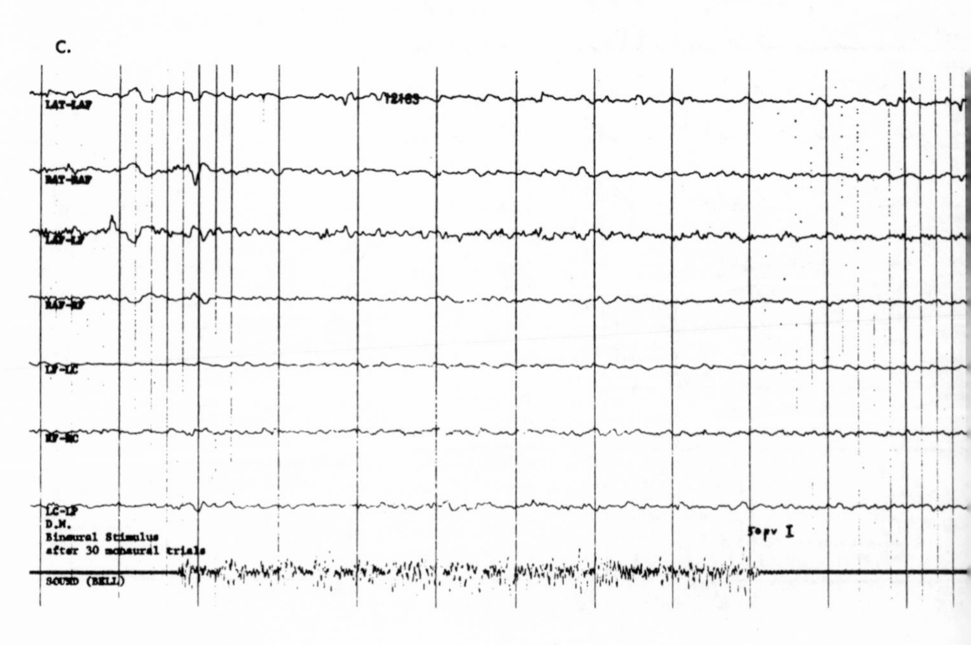

he first patient, if the extinction trials were en using a gunshot as the extinction stim-s, a decreased threshold to subsequent gun-t stimuli was observed, but there would be significant change in the threshold to ponses evoked by a bell.

The extinction effect persisted while the pa-t was in the laboratory. However, on reeval-ing the patient the next morning, a ression had occurred and the original shold levels often were obtained. An ex-imental observation tends to explain this ression effect. On some occasions, after nction trials, binaural stimuli were given increasing intensities until the new, higher eshold was obtained. Once positive re-nses were obtained at this higher threshold, itive responses tended to recur at the previous or preextinction threshold levels. Because no effective control could be obtained over the stimuli the patient received outside the laboratory, and since many of these were observed to be effective in producing a startle response, we concluded that the positive responses triggered outside the laboratory were responsible for the regression of the extinction phenomena seen the next day following each session in the laboratory. However, the general threshold level did tend to increase gradually over the period of evaluation.

CONDITIONING STUDIES

The second patient did not respond to stroboscopic stimulation. In order to further evaluate the degree to which this kind of response could be modified by sensory experi-

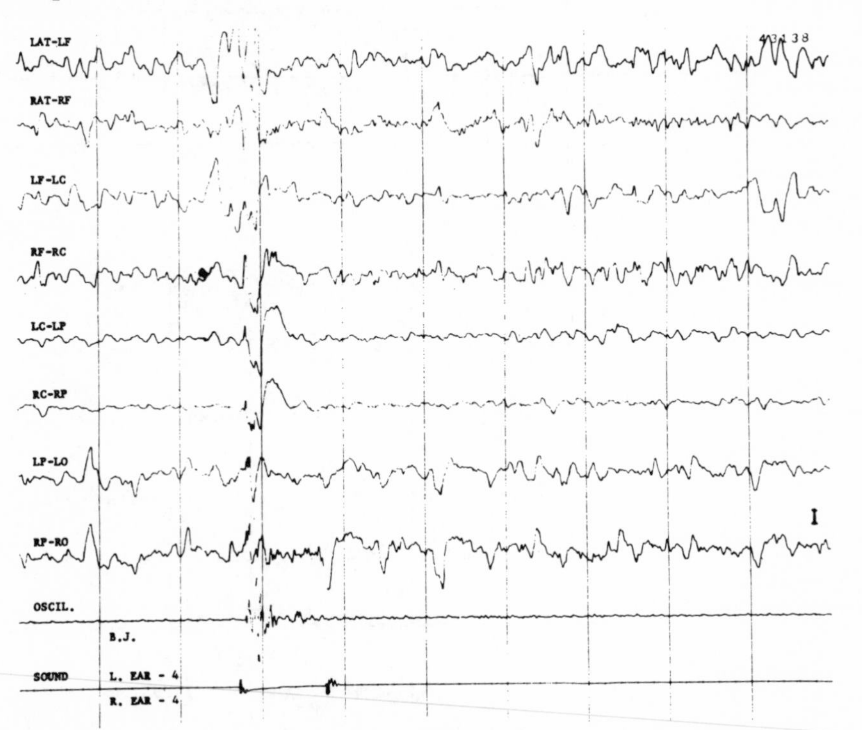

Fig. 2. The extinction effect from presentation of the sound stimuli using different sound volumes in the two ears. The bottom channel shows the presentation of the sound stimulus into both ears. The channel marked oscillometer presents the activity derived from an oscillometer on the patient's arm. [A] The myoclonic seizure elicited by the administration of sound equally into both ears at a threshold level. [B] The sound binaurally but at unequal volumes; one at the threshold, the other below the level necessary to elicit a myoclonic seizure. [C] Absence of the myoclonic seizure at volumes above the previous binaural threshold after a series of extinction trials administered with unequal volumes in both ears.

B.

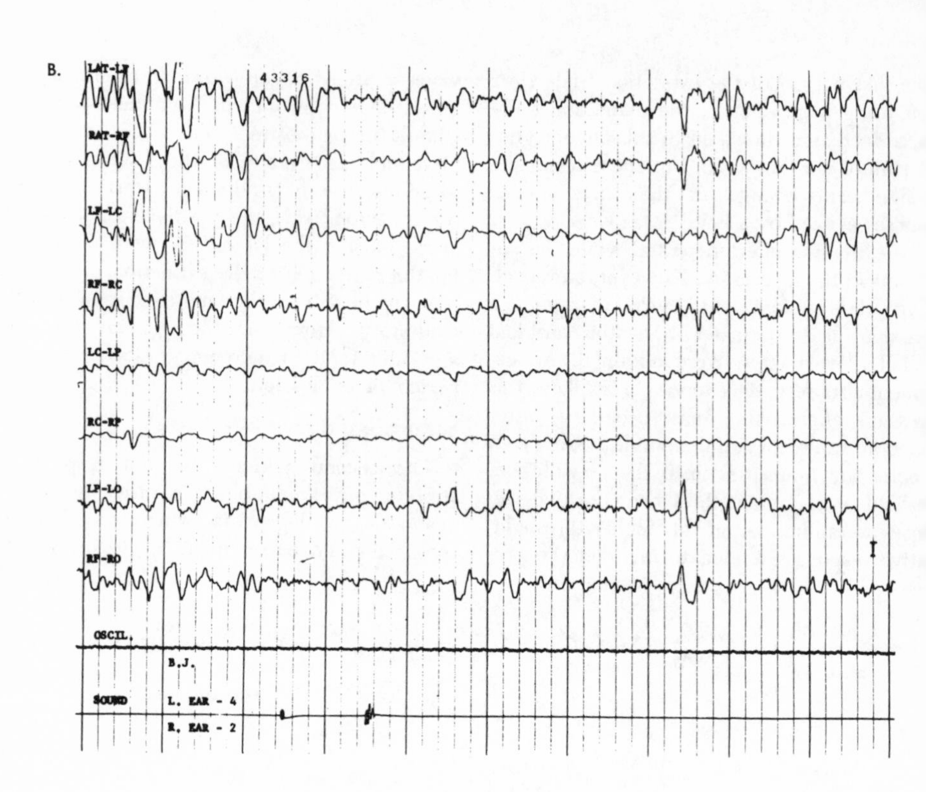
LAT-LF
43316
RAT-RF
LF-LC
RF-RC
LC-LP
RC-RP
LP-LO
RP-RO
OSCIL.
B.J.
SOUND
L. EAR - 4
R. EAR - 2

C.

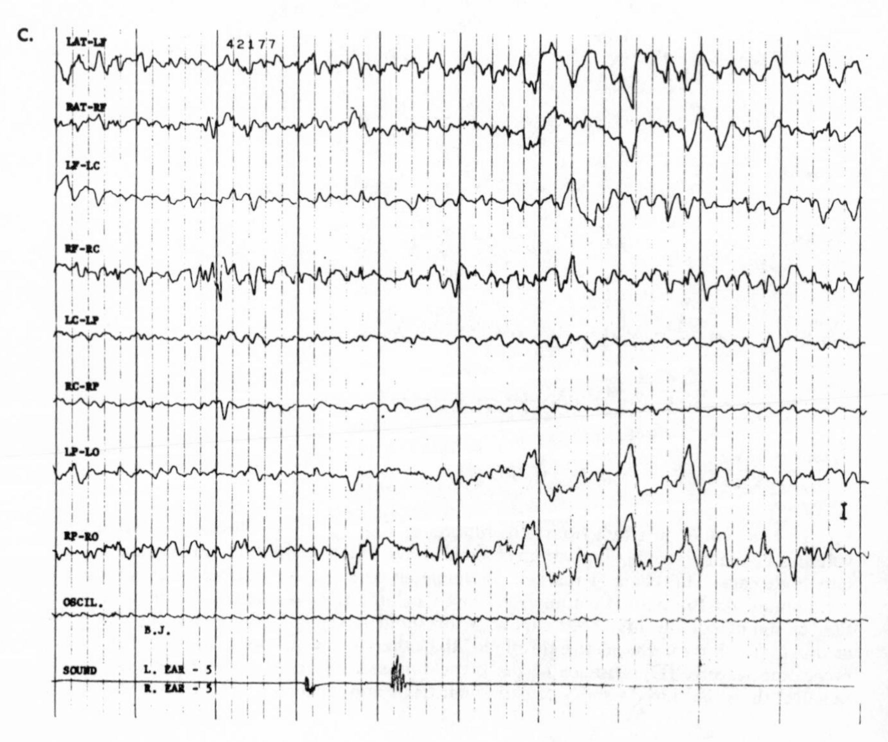
LAT-LF
42177
RAT-RF
LF-LC
RF-RC
LC-LP
RC-RP
LP-LO
RP-RO
OSCIL.
B.J.
SOUND
L. EAR - 5
R. EAR - 5

e, a classic Pavlovian conditioning proce- e was carried out in this patient. With a v to transferring the startle response from nd to light, stroboscopic stimulation at 20 es per second was used for the conditioned ulus (CS). An electric bell which con- ntly evoked the startle response served as unconditioned stimulus (US). This was vered in free fields. Control of the stimulus tionships was obtained by the use of preset e generators and relays. The CS was pre- ed for six seconds. After four seconds, the came on and persisted for two seconds. the CS and US went off together. Trials e presented at three- to five-minute inter- , and approximately 20 trials were given ach session.

fter approximately 50 presentations of the US combination, a response began to occur he onset of the CS before the US came on. By the sixty-fifth trial, the response was more marked and present over 90% of the time. This conditioned response to light involved the same movement pattern as the unconditioned response to the bell, but it had a longer latency, was less in amplitude, and was somewhat slower. On many trials the initial response was maintained in a tonic or postural fashion throughout the CS. The unconditioned response to the bell would then be superimposed on this more tonically maintained posture (Fig. 3).

After the conditioned response to the CS was well established, a differential stimulus (DS) was introduced. This was a 5-cycle-per-second stroboscopic stimulus, presented for six seconds but never paired with the US. The conditioned response occurred with the initial presentations of the DS. However, following some 20 trials during which the DS was alter-

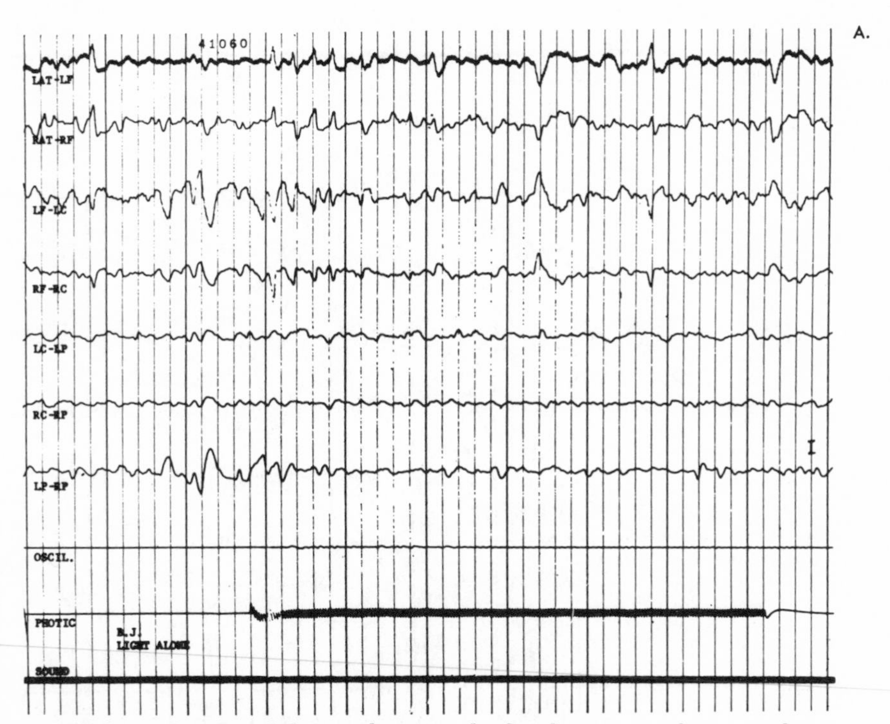

Fig. 3. The conditioning of the myoclonic sound-induced seizure to photic stimulation. The presentation of sounds are indicated in the bottom channel by the cessation of 60-cycle artifact. [A] No response to stroboscopic stimulation. [B] No response to photic stimulation during the four seconds preceding the administration of the sound stimulus. The delivery of the sound stimulus produced a myoclonic seizure. [C] Photic stimulation evoked the myoclonic seizure prior to the administration of sound stimulation after the sixtieth conditioning trial.

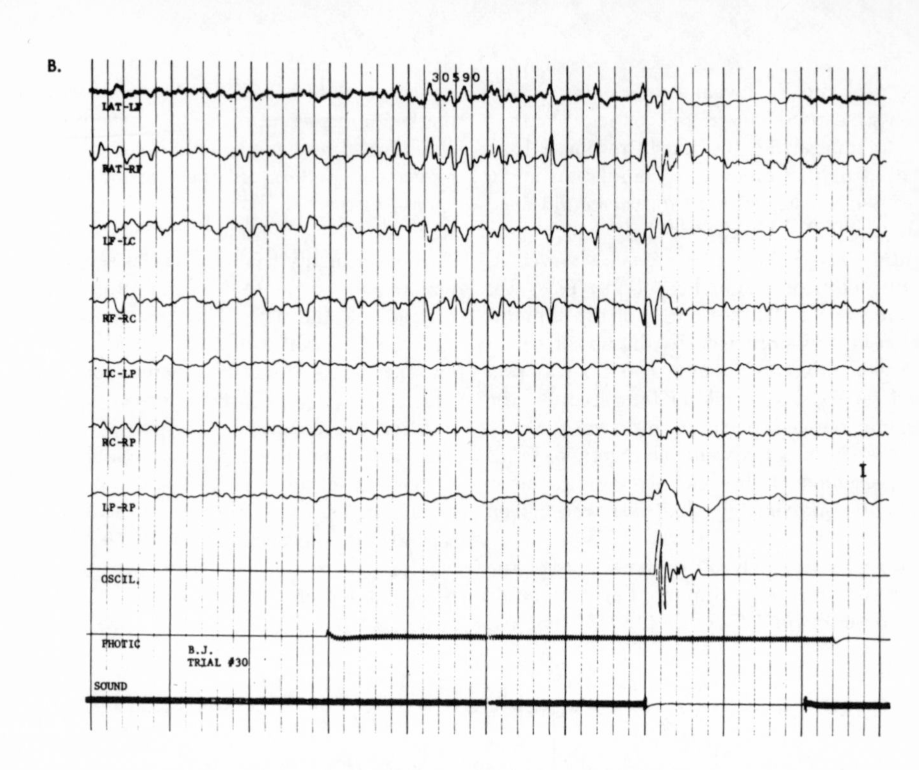
B.
30590
LAT-LF
RAT-RF
LF-LC
RF-RC
LC-LP
RC-RP
LP-RP
OSCIL.
PHOTIC
B.J.
TRIAL #30
SOUND

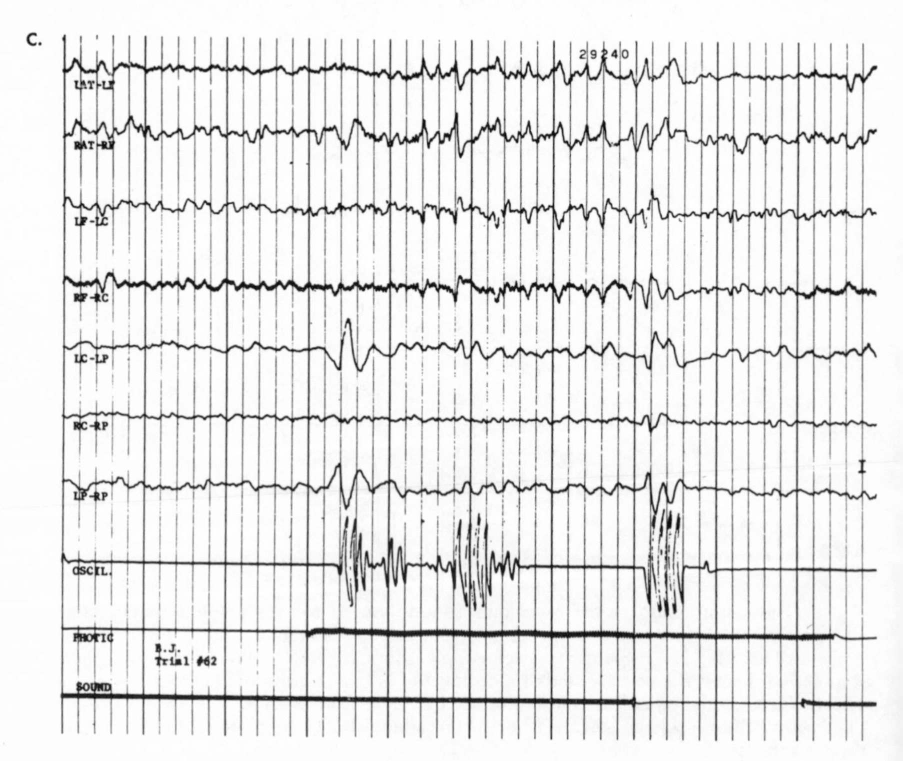
C.
29240
LAT-LF
RAT-RF
LF-LC
RF-RC
LC-LP
RC-RP
LP-RP
OSCIL.
PHOTIC
B.J.
Trial #62
SOUND

ted with the CS-US combination in a ran-
mized fashion, the patient continued to re-
ond to the CS but gradually lost all response
the DS. Thus, the conditioned response initi-
y showed stimulus generalization, but a satis-
tory discrimination could be established.

Extinction of the response was established
repeated presentation of the CS alone. In
early extinction trials, the response oc-
rred to the CS, but after approximately 20
als without the US, the response became in-
nstant. After 40 extinction trials, all response
the CS had disappeared. When the response
s thus extinguished, the patient rested for
weeks. Upon return to the laboratory, no
ponse to the stroboscope was present. The
ocedure was then repeated, this time using a
-cycle-per-second and a 10-cycle-per-second
oboscopic stimulus for the CS and DS, re-
ectively. Results were similar to those ob-
ned on the first procedure.

CUSSION

Systematic attempts to alleviate sensory-
oked seizure disorders by control of sensory
periences recently have been reported.[5,6]
evious reports from this laboratory[7-9] have
own that photoconvulsive responses could be
duced by a variety of extinction techniques.
e extinction effects observed on startle sei-
res previously described show a great degree
similarity with the extinction effects ob-
rved in our studies of photosensitive respon-
s.

The use of monaural extinction compares
th monocular extinction; neither is always
fective in producing extinction, and both are
fficult to control. The subthreshold binaural
chnique compares with the use of binocular
tinction by using differential light intensity.[8]
ere, background illumination was increased
til the stroboscope was barely perceptible
d would not evoke a response. With repeated
als, the background illumination could be
adually eliminated without the recurrence of
otosensitivity. In the binaural extinction
chnique, the intensity of the stimulus in one
r was markedly reduced and repeated trials
ere given. The intensity of the stimulus was
adually increased until finally the original
reshold and even suprathreshold levels could
be obtained without the recurrence of positive startle responses.

The general method could be described as one of approximation, and this appears to be the more consistently effective method of producing the extinction phenomena. A similar method of approximation was found to be effective in extinguishing seizures in a case of musicogenic epilepsy.[10] Other methods probably would be equally effective if they shared the general principle of repeated presentation of a noneffective stimulus which gradually approximates the effective stimulus. Moreover, the procedure can be reversed and the patient made more sensitive. In monocular extinction, if the occluded eye was incompletely covered, there was a tendency to develop monocular sensitivity. In one instance, after we had satisfied ourselves that the patient's photosensitivity could be extinguished, he was deliberately made sensitive at a frequency and intensity to which he previously had not been sensitive. This was subsequently removed by extinction trials.[8]

The extinction effect tends to regress if positive responses are triggered by random stimuli outside the laboratory. This effect also could be demonstrated in the laboratory. In two patients with photosensitivity, the regression effect has been controlled by devising methods of continuing the extinction procedures in the patient's usual environment. In one case, the patient's husband, an engineer, adjusted the television to flop over at 20 cycles per second. The patient occludes one eye and twice daily receives monocular extinction trials. In the other case, the extinction effect of the differential light intensity method was transferred to auditory clicks. A pair of glasses was devised to deliver a similar click to one ear with any sudden change in ambient light intensity. In the case of musicogenic epilepsy, alleviation of seizures was obtained only after extinction trials were carried out daily at home.[10] Neither of the patients in this report was suitable for such prolonged extinction procedures. The first patient, in whom monaural extinction was effective, could not be induced to wear an earplug outside the laboratory. The second patient was markedly retarded, and such prolonged extinction techniques were not possible.

Elucidation of the physiological mechanisms by which the extinction effect is produced requires further studies. Our data do suggest, however, a great degree of similarity of the extinction effect to conditioned responses. Both are dependent upon controlled stimulus patterns, and both are temporary unless continuously reinforced. A partial transfer of the startle response from sound to light was obtained in the second patient using a classic Pavlovian technique. The extinction effect itself could be transferred in one patient from light to sound.[9] These similarities suggest that both the extinction effects and the conditioned responses depend upon or are mediated by similar or related mechanisms. It is even possible that some effective stimuli in sensory-evoked seizures could be conditioned stimuli.

SUMMARY

Controlled stimulus patterns in startle (acousticomotor) seizures led to a rise in seizure threshold. This extinction effect is correlated with similar studies in photogenic epilepsy and in musicogenic epilepsy. Sensitivity to a previously neutral stimulus (inte mittent light) was produced by a Pavlovi conditioned procedure. These observatio enhance the concept that special sensor evoked seizures may exist on a higher nervo system basis.

REFERENCES

1. PENFIELD, W., and JASPER, H.: Epilepsy and the Fur tional Anatomy of the Brain. Boston: Little, Brown Co., 1954, p. 896.
2. GOWERS, W. G.: Epilepsy and Other Chronic Conv sive Diseases. 2nd Edition, London, 1901.
3. SERVIT, Z.: Epilepsie, Grundlager Einer Evoluntionar Pathologic. Berlin: Akademie-Verlog, 1963, p. 278.
4. STERO, J.: Experimental reflex epilepsy (audioge epilepsy). Epilepsia 3:252-273, 1962.
5. STRAUSS, H.: Jacksonian seizures of reflex origin. Arc Neurol. Psychiat. (Chic.) 44:140-152, 1940.
6. EFRON, R.: The effect of olfactory stimuli in arresti uncinate fits. Brain 79:267-281, 1956.
7. FORSTER, F. M., and CAMPOS, G. B.: Conditioning fa tors in stroboscopic-induced seizures. Epilepsia 156-165, 1964.
8. FORSTER, F. M., PTACEK, L. J., PETERSON, W. G., CHU R. W. M., BENGZON, A. R. A., and CAMPOS, G. B.: Str boscopic-Induced Seizure Discharges. Arch. Neur (Chic.) 71:603-608, 1964.
9. FORSTER, F. M., PTACEK, L. J., and PETERSON, W. C Auditory clicks in extinction of stroboscopic induc seizures. Epilepsia. (In press)
10. FORSTER, F. M., KLOVE, H. M., PETERSON, W. G., a BENGZON, A. R. A.: Modification of musicogenic ep lepsy by extinction techniques. Trans. Amer. neur Ass. (In print)

SPACE MEDICINE

EDITED BY: JOHN J. BUNTING, M.D.
F.A.C.P., F.C.C.P., F.A.C.A.

This is of particular interest to ENT specialists. Increasing noise levels to which flyers and ground workers are exposed, and the increasing threat of monetary liability for hearing impairments, have forced the military services and many airlines and private industries to undertake hearing conservation programs. The major objective of any such programs is to prevent hearing loss, through identification of noisy areas and provision of protective devices for workers exposed. A secondary objective is the avoidance of liability for hearing loss not related to the job, as detected by preemployment audiometric examinations.

Witwer and Cole have described the hearing conservation program of the U.S. Marine Corps at the Cherry Point Air Station. Hearing programs are mandatory, under Navy policies, when sound pressure levels reach 95 decibels in any or all of the common octave bands, between 300 and 4800 cycles per second. As in other studies, jet engines were found most offensive noise sources, the J-57 engine being said to cause 95 db sound pressures out to 2500 feet. For the occupants, helicopters were the noisiest aircraft.

Besides provision of protective devices for workers in noisy areas and for passengers in noisy aircraft, the Marine program included the posting of signs in noisy areas, showing the sound level in decibels likely to be encountered during engine run up. This was found to be an effective means of reminding all personnel to wear their protective devices.

The *vestibular Coriolis reaction* is a matter of real concern in aviation and space medicine. The reaction occurs when the rider on a rotating structure tilts his head with respect to the axis of rotation; it consists of uncommon confusion, nystagmus, and nausea and vomiting if severe. Otherwise unexplained aircraft accidents have been attributed to this reaction, a common hypothesis being that the pilot may have, during a turn, tilted his head to change radio frequencies. The problem gains added importance as one contemplates rotating space stations, and ponders the reaction of active occupants.

The ultimate mechanics of the reaction are not fully understood, though much has been written on the subject. Guedry and Montague have devised an apparatus by which rotation and head movement may be accurately controlled and measured, and by which the subject may with some degree of accuracy indicate the magnitude and character of the response. In this apparatus, operated in a darkened room, the subject watches a light fixed before his face, and after tilting his head uses a simulated aircraft control stick to indicate the apparent movement of the light. The stick is moved as if to cause an aircraft to pursue the light. These workers noted that both apparent velocity and apparent distance of displacement of the light were proportional to speed of rotation, as well as degree of tilt of the head, all factors being subject to fairly accurate measurement or estimation. The authors measured, as other have noted, the declining severity of responses on repetition of the acts. These reports, with the common observation of the *"Coriolis immunity"* of figure skaters, lead one to expect that occupants of rotating space stations may be able to overcome any Coriolis reactions which may occur.

Otosclerosis is of special concern to flyers, since it commonly begins to impair hearing in the late teens or early twenties, and reaches a maximum effect in 10 to 20 years. It is often undetected in pilot candidates, but reaches critical levels of impairment when pilots are in periods of active flying. Kos describes the surgical procedures for treatment of otosclerosis. The fenestration operation commonly failed to produce full restoration of hearing, and moreover left a mastoid cavity which was disqualifying for flyers. The stapes mobilization operation was commonly followed by regression. A more satisfactory procedure is described, in which the otosclerotic stapes is replaced by a venous plug, connected to the incus by steel or tantalum wire. Results have been excellent, and there have been no hazards for flyers. Kos acknowledges that an ear in which the stapes has thus been replaced may be more vulnerable to injury from high intensity noise. For pilots not commonly exposed to serious noise hazards, this possible drawback is likely to be of little concern.

SOME GENERAL IMPRESSIONS AND SOME THOUGHTS ON THE PROBLEM OF INDUSTRIAL NOISE.

By R. C. Dickson, M.B.E., M.B., Ch. B.

May I, at the outset, say how fortunate I was to be included as a member of the 1963 Travelling Fellows, as this last tour has now been geared into what must be the ultimate for our purpose of seeing the British practice of traumatic surgery.

The tour commenced with a week in London organized by Mr. Karl Nissen and consisted of visits to the London Hospital, where we had the pleasure of meeting Sir Reginald Watson-Jones, to Stanmore Royal National Orthopaedic Hospital, to the Mount Vernon Hospital at Northfields, and to the Rehabilitation Centre of the Vauxhall motor works at Luton. In addition, one morning was spent with Mr. Batchelor at his rooms, where an informal discussion on fractures of the os calcis and bones of the wrist took place. I felt that the main disadvantage of London as a centre was the time-consuming difficulties of travelling from one centre to another.

During our stay in London we attended a symposium at the Royal Society of Medicine on "Survival after Injury" at which most of the speakers advocated greater use of tracheostomy and first-aid—the standard of which is poor by comparison with our industry.

I personally was able to make contact with E.N.T. surgeons of the R.A.F. with a view to further discussions on hearing conservation, which I shall enlarge on in the second part of this paper.

The hired car was delivered to us in London and proved to be quite useless—parking and driving problems are such that it is far more convenient, cheaper and faster to use public transport, and future Fellows would be well advised to take delivery of the car elsewhere if London is still included.

The following week was spent in Oxford, where Mr. J. Scott had arranged a very interesting and informative series of meetings at the United Oxford Hospitals.

One day was spent at Stoke Mandeville, where we were most impressed with Dr. Guttman's theories and results. Oxford was chiefly notable for its anaesthetic department, whose members were most interested in our problems and who were developing apparatus for the administration of anaesthesia under non-ideal conditions. Dr. Barnes will be dealing with this more fully, so I do not propose to enlarge on the subject.

The following fortnight was spent at the Birmingham Accident Hospital, during the second week of which a course of accident surgery was held, which was most instructive. Impressions of the Birmingham Accident Hospital are that their treatment of burns was probably better than that obtaining anywhere else. However, for other forms of trauma their practice is probably not as good as most other centres—it is obvious that no one surgeon can be expected to operate on head, chest, abdominal and orthopaedic injuries with equal facility. The Birmingham Accident Hospital has served to bring to the notice of the surgeon that the old-fashioned thoughts of an out-patients' department, which was considered the Cinderella of any general hospital, were completely outdated. However, future planning in the British health services does not envisage any further accident hospitals on this line—instead each large general hospital will be equipped with a central emergency department which will be able to diagnose the full extent of injury and treat same until immediate danger to life has passed. This department will be under the control of a surgeon especially trained in traumatic surgery and will have the full facilities of the various branches of surgery which may be necessary, viz, neuro, thoracic, orthopaedic and general. Considerable thought has been given to the planning of these departments to allow of serious cases being dealt with immediately and efficiently while the normal run of minor injuries can be treated and disposed of concurrently. I visited an iron and steel foundry at Bromsgrove and saw their noise problems.

The following fortnight was spent at Bristol, where Mr. Eyre-Brook had arranged our programme. We spent several days at Frenchay watching the plastic surgeons, neuro-surgeons and thoracic surgeons. Bristol and Winford provided the pure orthopaedic side, while a more interesting morning was spent at Ham Green, where we saw the artificial respiration and artificial kidney apparatus in action. While at Bristol I was able to visit the Bristol-Siddeley aero-engine factory and see their conservation programme.

Following the customary week's break in the middle of the course, which I personally spent in Cornwall and was fortunate to have five warm sunny days out of the seven, we met again in Newcastle-on-Tyne, where Mr. Stanger had arranged the week's programme. Once again we were able to see the various branches and techniques as at previous centres. From Newcastle we travelled to Edinburgh, where Professor James kept us very busy and where we were fortunate enough to meet Dr. Boyes of America, the celebrated hand surgeon.

Edinburgh, as can be expected of a centre of medical training with its world-famous reputation, was conspicuous for its bold approach in many aspects, especially their thoughts on head injury. This latter was quite shaking, as the neurosurgery department left us in no doubt that it was our duty to raise bone flaps of half the vault and explore the brain in any obvious cranial injury.

The next week was spent at Glasgow, who by and large have much the same outlook as the Edinburgh school.

From Scotland we returned south through the industrial midlands to Sheffield for the next fortnight. Unfortunately Mr. Holdsworth went on leave within a day of our arrival, but we were well compensated by having Mr. Sharrand as our mentor. During the second week of our stay at Sheffield a further course in accident surgery was held, which was again most interesting and informative. I spent the final week in London, where I had meetings with the director of the National Coal Board, the medical personnel of the B.O.A.C. and B.E.A. and several meetings with the E.N.T. surgeons and technicians of the R.A.F.

In retrospect my impressions of this tour are that we have been misled generally with regard to the conservative attitude of the British surgeon. We found that, with very few exceptions, the practice was to employ internal fixation in fractures. This attitude is governed largely by the acute bed shortage which still exists in British hospitals. Other impressions are of the general insistence on a high standard of anaesthesia—general anaesthetics being employed for almost all procedures. Tracheostomy is advocated in all serious head, face or chest injuries. The use of blood transfusion varies from average to what appears to be ridiculous levels—we heard of one case receiving 48 pints in 24 hours.

The British health services have settled down into two well-defined grooves. On the one hand is the general practitioner who acts as a sorting house and is largely concerned with prescribing for medical conditions: most of them are keen to do midwifery, as this branch earns extra emoluments. None of them do any surgery—even minor stitching is referred to the local hospital—primary reasons being that no claim can be made on the health service for suture materials, which then come out of the practitioner's pocket; also it is time-consuming.

The other branch is the specialist service—this is entirely a hospital practice—some of the consultants do, however, have a private consulting practice. In the hospitals there is a well-defined path to the consultant level which begins as a junior house surgeon and continues through the grades of senior houseman, junior and senior registrar until final appointment to consultant is made. As can be imagined, these positions are few and far between and there must be considerable frustration during the registrar period while the years slip by and still no vacancies occur.

At consultant level salary differences are made by "merit awards" of varying grades—these are earned bv the individual's standing in the community, by his scientific contributions and various other factors. This explains why the majority of men from registrar level upward are keen to attend congresses and publish papers.

A vast rebuilding and expansion programme is in progress—this will eventually double the existing bed state of British hospitals and of course increase the demand for hospital staff. One wonders how this will be accomplished, as there is no doubt that the existing services could not function were it not for the employment of vast numbers of Commonwealth graduates. Furthermore, what will the ultimate cost be? At the moment the health service costs the British taxpayer nearly £900,000,000 per annum.

In its present form this tour is most useful to those mine medical officers who have had considerable experience in traumatic surgery, as all the courses and tutorials are designed for the registrar level. I feel that the programme, while designed to give us a true cross-section of British practice, in point of fact becomes repetitive, this, due to the comparatively limited scope of the subject and to general agreement on methods of treatment.

The benefits to be derived from these tours are both general and personal. The general has resulted and will increasingly result in future in an improvement in our surgical practice in the industry, but while one agrees that most of the thought and methods currently in practice in Britain are ideal, nevertheless many of them would be difficult if not impossible to implement in our service.

The personal benefits which I have derived are threefold, namely: Stimulation, from the contact with eminent men who are bold and thoughtful in outlook as well as adept in execution. Welcome confirmation that our practice is considered highly satisfactory in most cases and, finally, encouragement that my personal judgement and appraisal meet with approval.

I should now like to discuss some aspects of industrial noise, which is a matter of increasing importance in Britain today.

Despite the fact that British medicine was the first to recognize noise-induced deafness as an occupational hazard when John Fosbroke in 1831 described "blacksmith's deafness" in a manner which has not been bettered even to this day and when later in 1886 an equally important contribution was made by Barr on "the effects of loud sounds on the hearing of boilermakers", British laws on the control of noise and of hearing conservation in industry are almost non-existent. The reasons for this are numerous, but can largely be explained by British conservative thought. Roughly speaking, if any factory produces an unpleasant loud noise, it can only be suppressed legally by complaints to the local authority under the clause of "creating a nuisance"! Obviously this can lead to interminable litigation while complainant and defendant argue as to the degree of "nuisance"; in the meantime the factory continues to pour out its unwanted sound. British tolerance and liberty of the subject obviously extend to the production of what is picturesquely phrased "a nuisance". As for the worker in a noisy environment, there appears to be absolute agreement in their thoughts. Like his fellows the world over, the British workman is quick to demand compensation for occupational disability, but the disablement apparently must be dramatic—the loss of a digit or a limb, the loss of earning capacity due to some sudden catastrophe. When the disability is slow and insidious it is accepted as being "part of the game". The weaver in the textile industry, the steelworker in a modern smelting plant, the riveter in a shipyard or the smith in a forging plant know full well that they will be "hard of hearing" in ten years or sooner, but so were their fathers and their grandfathers and in fact it now becomes a sort of status symbol. Besides, they quote the obvious, that a mild degree of deafness is quite an advantage when working in a noisy environment. The fact that life without loss of faculty would be more pleasant if their working environment was not noisy seems to have escaped their attention. Thus it is that Britain, with its dense population and high capacity for noise production in industry, became aware of and demanded an inquiry into the problem of noise, not because of its traumatizing effect but because of the general public's annoyance at the increasing volume. The result of this inquiry has been the Wilson Report published in 1963, which covers every facet of noise but rather tends to lay emphasis on the non-auditory effects, especially annoyance. The report draws many conclusions and makes many recommendations for the control of noise which will no doubt eventually be incorporated into British law.

Prior to this report several industries in Britain had conducted research programmes and instituted their own hearing conservation measures, which naturally varied considerably with the industry concerned. A few examples serve to illustrate these different approaches.

1. *Steel Industry in Sheffield.* A survey of all processes had not been made. It was found that the electric smelt produced a maximum of 110 dB, largely percussional and broad band, sustained over half the working day. Protection of the worker by muff was rejected as being too hot and uncomfortable; some workers agreed to use glass wool ear plugs. There was no possibility of reducing noise at source. Audiograms had been done on all workers in the smelt and showed an average loss of 35 dB in those who had fifteen years' service or more. Valuable data will accumulate as a result of pre-employment audiometry on young workers who stay in the industry for their working life. No periodic audiometry is as yet being done, as it is felt that the majority of workers resent this—however, as a result of recommendations in the Wilson Report, it is felt that periodic audiometry will become obligatory in the future.

2. *Forging Plant in Birmingham.* A full survey of the forging plant is in progress; unfortunately no narrow band analysis has been made, which reduces the value of their findings, as it is obvious that the noise is largely of low frequency percussional peaks with high frequency peaks due to clutch exhaust. Overall levels 90—95 dB; peaks at presses, 125—130 dB; at hammers 120 dB. Clutch exhaust could be muffled; no reduction at source is possible for the low frequency presses and hammers—these are, however, probably only mildly traumatizing.

Pre-employment audiometry is done on all workers—no periodic audiometry done for the same reasons as above. Hearing loss for workers with more than ten years' service is an average of 40 dB in the mid-frequences.

3. *Aero Engine Factory, Bristol.* Conservation measures are of an extremely high standard. A full survey of all processes has been made by their noise control department, which is continually checking and measuring. Narrow band analysis has been made of one third octaves. The major problem here is the noise subtended by engine tests and it has been overcome by the building of relatively sound-proof buildings where the engine under review is governed by remote control in a sound-proof room. Adjustments are made only after throttling back to low revolutions —the fitters being equipped with the Canadian pattern aural dome ear protectors, which have an attenuation of 40 dB. The low revolutions reduce overall noise to about 136 dB —noise in excess of this cannot be tolerated, as bone conduction now becomes more important than air conduction, and naturally cannot be easily muffled. Extremely high intensities are recorded from some of the latest jet engines with pre-heater; for instance the Avon at take-off revolutions produces 150 dB. Such an intensity would be immediately and permanently deafening and poses problems of muffling before this engine can become a commercial proposition. Reduction of noise at source is attained by the construction of these "test beds" mentioned above, the cost of which is now in excess of £600,000 each. In addition, the exhaust is ducted out through "de-tuners". These are large silencers which reduce gas speed by expansion and baffles. The total energy remains the same, but the frequency is reduced to less than 150 C.P.S. and all one hears is a low rumble.

All personnel have pre-employment audiograms and those exposed to noise are periodically checked. The medical officer has full power to remove any employee showing signs of undue threshold shift either temporarily or permanently.

4. *Royal Air Force.* Constant survey of noise intensities both in aircraft and at various positions on the ground. Figures for most of the operational aircraft are not available, as these might give some indication of performance.

All aircrew are audiometrically examined on entry and at yearly intervals, noise susceptibility being considered grounds for disqualification from aircrew duties. An experiment in mass audiometry for all R.A.F. personnel has been abandoned because of policy changes and overall costs—it was, however, an interesting departure from normal practice in that a series of twenty individuals could be tested simultaneously. This method could not be applied to our industry as it required the individual to mark his own audiogram against a series of tape-recorded signals which varied in number and frequency as well as intensity. This demands a high degree of concentration and co-ordination and naturally literacy is a pre-requisite.

Stringent ear protection is demanded, the ground crew being required to wear ear plugs and protectors.

5. *British European Airways.* Follow closely on the pattern of the R.A.F. Flight personnel are examined at the Air Ministry. Cabin crews are examined at B.E.A. headquarters, but, strangely enough, no follow-up is made. About 1,400 of the total 6,000 employees are considered to be noise-exposed and these are given pre-employment and periodic audiometry.

Typical intensities for the Comet IV at 300 ft. are: 7,000 r.p.m. overall 111 dB, muffled 90 dB; at 8,000 r.p.m. (take-off) overall 124 dB, muffled 98 dB.

They are considering mass audiometry with an American mass automatic type which may be suitable for our conditions.

Now, however, as a result of recommendations made by the Wilson Report it seems likely that the British Government will introduce legislation for hearing conservation based on the following criteria.

(i) Exposure to broad band noise for 8 hours per day, 5 days per week for 15 years if the noise in any of the frequency bands reaches the following levels:

Frequency in C.P.S.	Sound pressure in decibels
Less than 150	100
150– 300	90
300– 600	85
600–1200	85
1200–2400	80
2400–4800	80

(ii) Doubling of the energy, i.e., the addition of 3 dB should be compensated by halving the duration of exposure.

(iii) Individual susceptibility and permanent threshold shift can only be found by pre-placement and periodic audiometry.

In addition to the adoption of these criteria, large-scale research will be undertaken by the Government in all spheres of noise, hearing loss, conservation methods and finally legislation, the long-term results of which will undoubtedly be the reduction of much unnecessary noise and the adequate silencing of processes or protection of the worker where reduction is not possible. It seems inevitable that noise-induced hearing loss and acoustic trauma will become compensatable.

It is interesting to compare the situation in South Africa at the moment. We have a more widely dispersed population—we are not heavily industrialized and as such only the urban dweller in close contact with noise from traffic and building operations has voiced any complaint as yet. The production of damaging noise in industry is, however, just as large a problem as that overseas.

Steps to combat this problem have been initiated by the scientists and industrial medical officers. The S.A. Bureau of Standards has produced a code of practice for the rating of noise for hearing conservation, which, should this be adopted into law as seems probable, means that this country would be tackling the problem by legislation first, rather than a popular demand. In the meantime the South African Air Force has conservation methods based on the Royal Air Force and American Air Force. The South African Railways have, for a number of years, been conducting surveys in their various noisy occupations and have instituted conservation measures.

Following on recommendations made by this Association in 1962, the mining industry has embarked on research into noise. A steering panel on noise and deafness hazards was formed and a programme outlined. During the past year a senior scientific officer was recruited by the Chamber and seconded to the Council for Scientific and Industrial Research for investigations of noise problems in the mining industry. A survey of noise levels and narrow band analysis has been made of various mining operations. Investigation into the silencing of pneumatic drills is in progress and certain investigations have been made with ear defenders. A report on the results will be made in the near future. So far no audiometric measurements have been made, but research into this department cannot long be delayed.

I would like to add that it is my intention to make investigation into three facets of hearing which are essential in our industry, namely, a true Bantu threshold, a susceptibility percentage and finally a true Bantu presbycusis curve. These data are essential if we are to determine how much of any hearing loss is due to noise exposure and how much to normal aging processes.

In conclusion I should like to thank the Manager of the Rand Mutual Assurance Company, the Selection Committee who chose me, and my Chief Medical Officer who submitted my name, for the privilege of having participated in this most interesting and instructive course.

Hearing Tests in Industry

• M. M. HIPSKIND, M.D., F.A.C.S.

Noise, by definition, is unwanted sound. With the industrial revolution during and after World War II, the worker has been exposed to intense noise. Prolonged exposure to the noises encountered in many industrial environments can produce a permanent hearing loss. This hearing loss is not amenable to treatment. Normal hearing cannot be restored once a noise-induced hearing loss has been acquired. Since we do not have a "pound of cure," we are forced to use the "ounce of prevention."

Before considering a program for the conservation of hearing, let us briefly review the anatomy and physiology of the ear (Figure 1). Nature protected the ear by placing the sensorineural part (the labyrinth and neural elements) in the hardest bone of the body, the petrous bone. This bone is deep within the ramparts of the calvarium and surrounded by pneumatic cavities. The petrous bone is already calcified and fully developed in the newborn.

The labyrinth communicates with the "outside world" through two windows: the oval and the round window. The oval window contains the stapes, the smallest bone in the body. Unlike any other bone, it does not grow. It has reached its full size at birth. The stapes articulates with the incus, the incus with the malleus, and the malleus with the ear drum — thus a lever system is formed to transmit sound vibrations from the ear drum to the labyrinth. Between the upper vestibular scala and the lower tympanic scala is the cochlear duct containing the organ of Corti, the mechanism which makes the miracle of hearing possible.

Physiology of the Ear

A sound wave collected by the external ear is transmitted to the ear drum, setting into motion the ossicular chain, causing the stapes to move in piston or rocking-like fashion (Figure 2). This in-and-out motion of the stapes is made possible by the counter or phase movement of the round window, i.e., when the stapes in the oval window is moved inward, the membrane of the round window moves outward. The endolymphatic fluid thus set in motion acts on the organ of Corti (Figure 3), where physical sound waves are converted to nerve impulses. These impulses are carried over the cochlear division of the auditory nerve to the transverse temporal gyri in the temporal lobe of the brain. Around this area is the cortical territory which serves the higher auditory functions. This enables us to understand the meaning of the sounds heard — the thousands of varieties of noise of everyday life, the spoken words, the sound of music, the noises in the office, home and in industry.

The normal ear can hear a sound as low as 16 vibrations per second (the rustle of leaves) and up to 16,000 vibrations per second (shrill whistle). If the ear is exposed to an extremely loud sound, two muscles protect the drum by acting on the movement of the tympanum and the stapes. The tensor tympani, by pulling inward the handle of the malleus and with it the tympanic membrane, increase the tension of the ear drum, diminishing the amplitude of the vibration. The stapedius muscle is the smallest

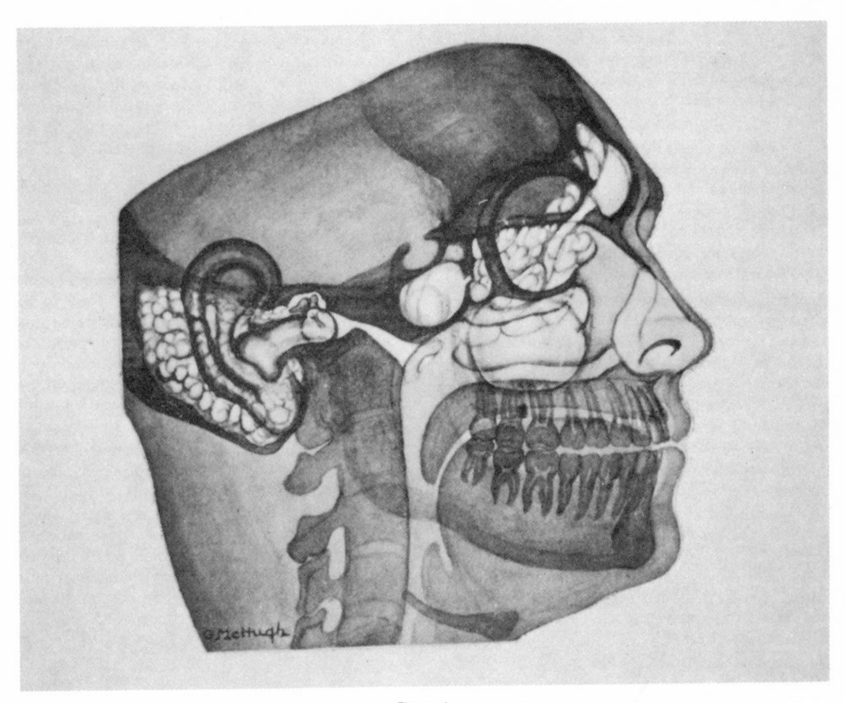

Figure I
The pneumatic cavities of the head and their relation to the auditory system. Figure shows the external and internal channels of air to the two sides of the drum membrane. (Reproduced with permission from The Human Ear, Anatomical Transparencies and Illustrations by Gladys McHugh, published under the auspices of Sonotone Corporation; distributed by T. H. McKenna, Inc., New York, N. Y.).

muscle in the body. It is attached to the stapes. Its contraction has a slowing effect or a "dampening" of low frequency vibrations.

It is significant that the ear is in a constant state of tone; day and night, awake or asleep, the ear is receiving and sending impulses. From birth until death the ear is active. We awaken from sleep when the alarm clock rings; we go through righting motions when the vestibular part of the ear signals during sleep that we are too close to the edge of the bed.

Occupational Hearing Loss

The level of noise in our industrial culture has increased in proportion to our increased technology. Dr. Samuel Rosen made a survey of the hearing of the primitive Mabaan tribesmen of the African Sudan. In this quieter civilization, his audiometric tests showed that 90-year-old men could hear nearly as well as 10-year-old boys. Hearing impairment is the most common physical handicap in the United States. One-tenth (20,000,000 people) of our entire population are classified as having defective hearing.

Occupational hearing loss has haunted the scientific, legal and industrial communities for many years. Millions of dollars are poured by our courts into the pockets of claimants for occupational hearing loss. This manna from the courts has nourished an entirely new species of worker: the individual who claims damage to his hearing as a result of his job. This situation is not uncommonly aided and abetted by suggestion of a fellow worker with strong union motivations. He implants the idea that the worker has good grounds for a suit and that he should "see a lawyer."

It is not my purpose to deny the just disposition of a just claim involving hearing loss incurred at work. In my opening remarks, I said: "Prolonged exposure to the noises encountered in many industrial environments can produce a permanent hearing loss." Occupational hearing loss probably is affecting millions of people. In 1948 the state of New York made *partial* hearing loss due to noise exposure compensable. Similar compensation laws soon extended into the states of Wisconsin and Missouri.

It is the responsibility of the physician to seek out the answer, whether the hearing loss is related to industrial noise or is related to nonoccupational causes. Only in this manner can industry protect itself against the slings and arrows of outrageous people and in particular against outrageous settlements.

During 1951, approximately 50 claims were filed before the Wisconsin Industrial Commission. A survey developed in 1963 by Mr. Ralph E. Gintz* and Dr. Meyer S. Fox** presents the comparative provisions for occupational hearing loss in Workmen's Compensation cases (Table I). A review of this helpful reference will show that all of the 50 states, except South Dakota, have statutes or rules that make either noise-induced and/or traumatic hearing loss compensable.

*Director of Workmen's Compensation, Industrial Commission, State of Wisconsin.

**Marquette University; Chief of Ear, Nose and Throat Department, Mt. Sinai Hospital, Milwaukee, Wis.; and a member of Subcommittee on Noise of the Committee on Conservation of Hearing, American Academy of Ophthalmology and Otolaryngology.

The physician in charge of an industrial program concerned with the problem of the protection of hearing will be rewarded if he directs his attention to three major parts: (1) the nature of the employee; (2) the nature of the plant environment; and (3) the nature of the clinical facilities.

The Nature of the Employee

What is the essential character of the individual, his disposition, his temperament, his intelligence? The physician must evaluate the part the employee's emotions play in absenteeism, alcohol problems and accident proneness. What is the attitude of the employee toward management and his supervisors? In the preemployment selection of applicants, a questionnaire containing sampling information of this nature can be meaningful.

Plant Environment

The plant physician, the safety engineers and the industrial hygienist can obtain valuable information by an objective walk through the plant. They will experience a degree of throat discomfort and fatigue if they must raise the voice to talk above the noise level. Can they hear the words used in an office environment or must they reduce the vocabulary to shorter phrases of conversation to be understood? Can they hear

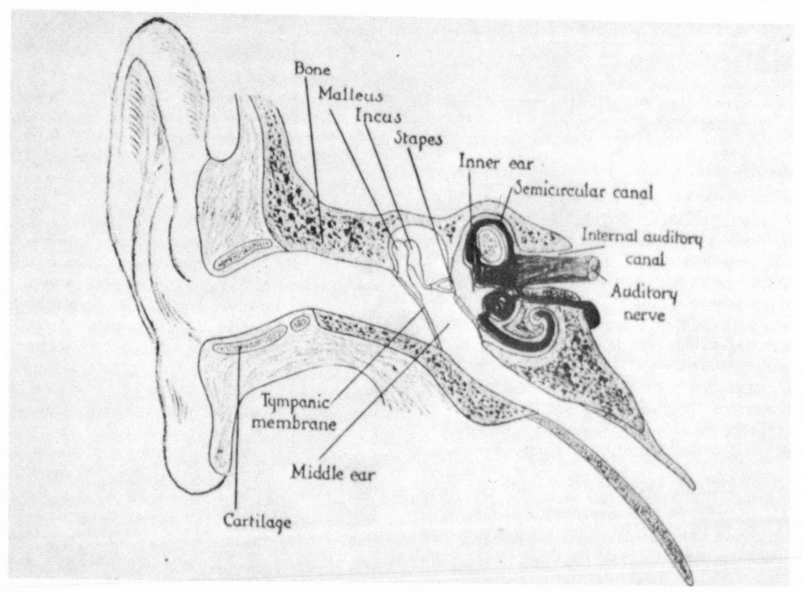

Figure 2

A sound wave collected by the external ear is transmitted to the ear drum, setting into motion the ossicular chain, causing the stapes to move in and out. The endolymphatic fluid thus set in motion acts on the organ of Corti.

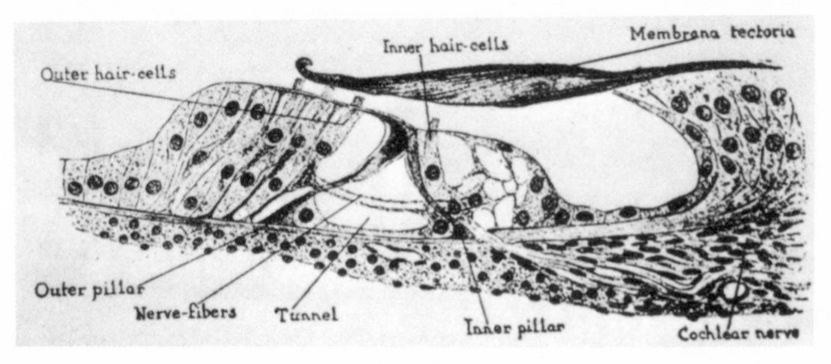

Figure 3
Section of organ of Corti from human cochlea. (x375)

the warning signals, fire alarm, telephone bell? Can they determine the direction from which the voice or sound coming? After several hours exposure to the plant noise, do they notice a temporary loss of hearing? Do they experience a muffled sensation in the ears that interferes with the understanding of speech or other sounds? Do they have "ringing" in the ears? If the answer is "yes" to any of the above questions, a hearing conservation program is needed.

The Subcommittee on Noise of the Committee on Conservation of Hearing and the Research Center, Subcommittee on Noise, has prepared a Guide for Conservation of Hearing in Noise. They state: "To be effective, a hearing conservation program should include: (a) a noise-exposure analysis; (b) provision for control of noise-exposure; (c) measurement of hearing."

The assessment of the noise-exposure is not a function within the limits of this paper. The classification of the noise, whether impulsive (drop hammers, punch presses, explosions) or steady (compressed air, diesel engines, fans, motors, lathes), is clinically important. A single sharp noise (firecracker at ear) is sufficient to produce hearing loss, whereas continuous exposure to some noises may not be expected to produce a hearing impairment. Those wishing information on noise level measurements and measurements of noise-exposure are referred to the Guide described earlier.

Nature of the Clinical Facilities

Ideally, the medical department should have special forms to record the essential history and physical findings related to the auditory function. At the beginning of history-taking, valuable information may be obtained. The physician may get important clues at this time; for example, the hard-of-hearing worker speaks with a loud and strained voice and probably has a sensorineural type of impairment; if his voice is soft, he could have otosclerosis.

On the occasion of a worker presenting the complaint of "sudden deafness," the physician should have an orderly plan of investigation. The case must have a complete "workup," beginning with a careful history of the conditions existing at the time of onset of the "deafness." Certain circumstances associated with the onset may indicate the etiology and the pathologic process. Sudden deafness, with or without vertigo due to some type of vascular lesion, may occur in the middle third of life or after. Virus infection may cause sudden deafnesss. The history usually is that of a "cold" or upper respiratory infection a few weeks before the onset. The deafness is commonly associated with tinnitus and dizziness.

Other viral infections causing deafness are well known: unilateral deafness occasionally occurring after mumps; deafness due to measles, though less common than mumps, usually is bilateral. The virus of rubella attacks the inner ear of the fetus in the first trimester. The Ramsay-Hunt syndrome is thought to be a viral infection sometimes causing deafness, associated with herpes about the ear and facial paralysis.

The record of the physical findings should include a rough diagram of the right and left ear drums. In the event of a perforation, a visual representation of the size, shape and area of involvement can be drawn on the eardrum sketch. The condtion of the ear canals should be noted for size, shape, for presence of cerumen or discharge, and for blood — old or new.

The provision for reporting the preemployment and subsequent audiograms and related tests is probably the most essential part of the record (Table II). Hearing ability measured in the preemployment test establishes an auditory

TABLE I
COMPARATIVE PROVISIONS FOR OCCUPATIONAL HEARING LOSS IN WORKMEN'S COMPENSATION CASES

Jurisdiction	Is Occupational Hearing Loss Compensable?		Basis of Compensation: Schedule In Weeks		Basis of Compensation: Wage Loss % of Body Function Others	Maximum Dollar Amount Payable		Must Employee Leave Work to File Claim		Test Frequencies and Rating Scales for Determining Hearing Loss and Impairment*	Compensation For Tinnitus Basis for Same	Provision For Hearing Aid
	Noise-Induced	Traumatic	One Ear	Both Ears		One Ear	Both Ears	Noise	Traumatic			
Alabama	No	Yes	50	150		$ 1,650	$ 4,950		No	M. E.	No	No
Alaska	Yes	Yes	52	200		1,800	7,000	No	No	AAOO & AMA	Questionable*	Yes
Arizona	No	Yes	85	255		11,000	33,000		No	AAOO & AMA	Yes	Yes
Arkansas	Yes	Yes	40	150		1,400	5,250	No	No	M. E.	Questionable*	Yes
California	Yes	Yes	*Special formula					No	No	*See notes	Yes	Yes
Colorado	No	Yes	35	139		1,408.75	5,594.75		No	AAOO & AMA	No	No
Connecticut	Yes	Yes	52	156		2,860	8,580	No	No	M. E.	Yes	Sometimes
Delaware	Yes	Yes	75	175	Possible	3,750	8,750	No	No	M. E.	—	Yes
Dist. of Columbia	Yes	Yes	52	200		3,640	14,000	No	No	AAOO & AMA	*	
Florida	Yes	Yes	40	150		1,680	6,300	No	No	M. E.	No experience	Yes
Georgia	Yes*	Yes	60	150		2,220	5,550	No	No	M. E.	No	No
Hawaii	Yes	Yes	52	200	Yes	5,850	22,500	No	No	AAOO & AMA	Yes*	Yes
Idaho	Yes	Yes	35	150		1,050	4,500	No	No	AAOO & AMA—M. E.	Yes	Yes
Illinois	Yes	Yes	50	125		*See notes		No	No	M. E.	Questionable	
Indiana	No	Yes	75	200		3,150	8,400		No	M. E.	No	No
Iowa	Yes	Yes	50	175		1,850	6,500	No	No	M. E.	Yes	Yes
Kansas	Yes	Yes	30	110		1,140	4,180	No	No	AMA Formula of Feb. 1947	No	Yes
Kentucky	No	Yes	*			2,080				M. E.	No	
Louisiana	No	Yes				Depends on wages			Yes	M. E.	No	No
Maine	No	Yes	50	100	Possible	1,950	3,900		No	M. E.	No	No
Maryland	No	Yes	75	175		1,875	4,375	No	No	M. E.	*	Yes
Massachusetts	No	Yes	150	400		3,000	8,000		No	M. E.	Yes*	No
Michigan	Yes	Yes						Yes	Yes	M. E.	Yes	Yes
Minnesota	Yes	Yes	55	170		2,475	7,650	No	No	M. E.	Yes	Yes
Mississippi	Yes	Yes	40	150		1,400	5,250	No	No	M. E.		No
Missouri	Yes	Yes	17 44	100 N. 168 T.		765 1,980	4,500 7,560	Yes (6 mo.)	No	AAOO & AMA—M. E.	Yes	No
Montana	No	Yes	40	200		2,000	10,000		No	AAOO & AMA	No	
Nebraska	No	Yes	50	100		1,850	3,700		No	M. E.		No
Nevada	Yes	Yes	85	255		2,000	6,000	No	No	M. E.	Body basis—Yes	Yes
New Hampshire	No	Yes	52	114		2,184	4,788		No		No	Yes
New Jersey	Yes	Yes	60	200		2,400	8,000	No	No	AMA Formula of Feb. 1947	Yes*	No
New Mexico	Yes	Yes	40	150	Yes	1,520	5,700	No	No	M. E.	Questionable	No experience
New York	Yes	Yes	60	150		3,300	8,250	Yes (6 mo.)	No	AAOO & AMA	Yes*	No
North Carolina	No	Yes	70	150		2,450	5,250		No	M. E.		
North Dakota	Yes	Yes	50	200		1,575	6,300	No	No		Yes	Yes
Ohio	Yes	Yes	*Must be total.			1,225	6,125	No	No	M. E.	No	No
Oklahoma	No	Yes	100	200		3,000	6,000		No	M. E.		
Oregon	Yes	Yes	60°	192°		2,790	8,928	No	No		No	Yes
Pennsylvania	Yes	Yes		180*						M. E.		
Rhode Island	Yes	Yes	60	150		1,620	4,050				Possibly*	Yes
South Carolina	No	Yes	70	150		2,450	5,250		No	M. E.	No	
South Dakota	No	No			Yes*					AAOO & AMA		
Tennessee	No	Yes	75	150		2,700	5,400		No	M. E.	No	Yes
Texas	No	Yes		150		?	?		No	AAOO & AMA	Possible	Yes
Utah	Yes	Yes	50	100		2,100	4,200	No	No	AAOO & AMA	No	No
Vermont	No	Yes		215			7,740		No	M. E.	Possible	No
Virginia	Yes	Yes	50	100		1,850	3,700	No	No	M. E.	No	Yes
Washington	Yes	Yes				1,950	6,825	No	No	AAOO & AMA	No	Yes
West Virginia	Yes	Yes	60	180		2,280	6,840	No	No	M. E.	No	Yes
Wisconsin	Yes	Yes	36 50	216 N. 333⅓ T.		1,539 N. 2,137 T.	9,234 N. 14,250 T.	Yes (6 mo.)	No	AAOO & AMA	Possible	Yes
Wyoming	No	Yes	50% Statutory loss of one eye			2,500	5,000		No	M. E.	No	Yes
Canada Provinces	Yes	Yes			Yes*	11.25 per month for life	112.50	Yes	No	*See notes	Seldom	Yes

AAOO *Guide for the Evaluation of Hearing Impairment, a Report of the Committee on Conservation of Hearing: Transactions of the American Academy of Ophthalmology and Otolaryngology, Vol. 63, No. 2, March-April, 1959.

AMA Guides for the Evaluation of Permanent Impairment of Ear, Nose, Throat and Related Structures: JAMA, August 19, 1961, Vol. 177, pp. 489-501.

M.E. Medical Evaluation

This survey was developed and conducted by Mr. Ralph E. Gintz, Director of Workmen's Compensation, Industrial Commission, State of Wisconsin; and by Meyer S. Fox, M.D., Department of Otolaryngology, Mount Sinai Hospital, Milwaukee, Wis.

baseline to provide a record for comparison of subsequent changes in hearing that may occur as a result of accident, age, disease or noise-exposure.

Audiometry

The essentials of good audiometry are followed. The operator of the audiometer is trained in these essentials and personally conducts all tests; i.e., whisper and conversational voice, Weber, Rinne, Schwabach, and finally the audiometric tests. All tests are conducted in a soundproof room. Calibration of the audiometer is checked at the beginning of each day's work. The first tone used is a 1,024 cycle, proceeding in sequence to the highest frequency. In this way fatigue to the ear as a result of a prolonged period of testing the lowest to the highest frequency is avoided. A reasonably loud signal, based upon the apparent hearing loss (to whisper or conversational voice), is given. Attenuation is introduced in 5 db steps until no response is indicated. The maneuver is repeated until uniform response is obtained three times. Interruption of the sound is frequently employed, care being taken to avoid audible or visual recognition of the process. Masking is used in those cases demonstrating a disparity of 30 db or more between the two ears. In no case should the masking intensity reach a level that would unfavorably influence the ear under test.

The otological evidence of the aging process is shown on the audiogram. The loss of the high tones, called "presbycusis," is thought to be caused by degenerative changes occurring in the organ of Corti; nerve cells, including the spiral ganglion; and changes in the central nervous system. These changes begin as early as the fourth decade and are progressive.

The audiometer is limited to the use of a pure tone sound consisting of only one frequency. The nature of the ordinary speech sound is very complex, since it does not consist of a simple sinusoidal wave, but of a superposition of many waves of different frequencies. Therefore, the attempt to arrive at some reliable formula for using the pure tone audiometric chart for estimating the patient's hearing loss for speech has been less than satisfactory. The oldest method is to average the decible losses for the 512, 1,024 and 2,048 frequencies and then to multiply this decible average by 0.8.

Audiologic Testing

The term "deafness" is misleading. Webster defines "deaf" as "wanting or deprived of the sense of hearing either wholly or in part." Actually, the difficulty lies not so much in hearing speech as in understanding it. Often the worker will complain, "I can hear the sound, but I don't get the word."

The incapacity to hear speech correctly can be measured in two ways. The monitored live-voice method can be presented to the patient either through headphones or by "free field" (no headphones used, the examinee hearing the speech directly from a loud speaker). Secondly, the speech test may be presented by recorded speech, using phonetically balanced (PB) monosyllabic words at intesities well above threshold. The PB maximum score (PBM) is the percentage of PB words which the patient can repeat correctly after they been presented at appropriate intensity.

Following the audiologic testing, both vestibular systems must be examined. Vertigo is commonly an associated symptom in cases of "sudden deafness." Nystagmus is one of the most common findings in the vertiginous patient; therefore, the examiner first looks for spontaneous nystagmus. This is best done at the time of the Rhomberg test. While the patient is standing, with eyes open, we look for nystagmus. If none exists, the patient is directed to place the tip of the index finger to the nose; first the right hand, then the left (cerebellar signs). If neither spontaneous nystagmus nor pastpointing are present, we proceed to the postural tests.

Postural Tests

Postural tests are not generally used, although these tests will often make the diagnosis and demonstrate objective evidence of the patient's subjective complaint of dizziness. We conduct the postural tests according to the method of Cawthorne. The patient is seated upon the examination table. He is instructed that he will be placed in various positions and that it is imperative that he keep his eyes open at all times.

Frenzel glasses are used, and the patient is instructed to look directly forward. Spontaneous nystagmus is noted, if present; if absent, the patient's head is held by the examiner and the patient is placed in the supine position with the head to one side (the side indicated as the offending position given in the history). A latency period of ten seconds is permitted for the recognition of the positional nystagmus. If the nystagmus does not occur after ten seconds, the patient is brought upright and again the examiner looks for nystagmus. If absent after ten sconds, the patient is placed in the supine position to the opposite side of the first procedure. If nytagmus is not observed, the patient is returned to the sitting position and again examined for the presence or absence of nystagmus. In the absence

---

Table II, at right, is from the Guide for Conservation of Hearing in Noise, Revised 1964, a Supplement to the Transactions of the American Academy of Ophthalmology and Otolaryngology.

TABLE II
FORM TO RECORD AUDITORY HISTORY AND PHYSICAL FINDINGS

HEARING CONSERVATION DATA CARD NO.________ TYPE OF AUDIOGRAM: REFERENCE AND/OR PRE-EMPLOYMENT ☐ RECHECK ☐ OTHER________

A. IDENTIFICATION

LAST NAME	MIDDLE	FIRST	SEX	DATE OF BIRTH		
			MALE FEMALE	DAY	MO.	YR.

SOCIAL SECURITY NUMBER	COMPANY NUMBER

B. CURRENT NOISE-EXPOSURE

JOB TITLE OR NUMBER	DEPARTMENT OR LOCATION	TIME IN JOB		
		NONE	MOS.	YRS.

NOISE-EXPOSURE

STEADY NOISE	IMPULSE NOISE	PER CENT TIME NOISE ON	EMPLOYEES ESTIMATE OF OWN HEARING
CONTINUOUS ☐	CONTINUOUS ☐	10 20 30 40 50	GOOD
INTERMITTENT ☐	INTERMITTENT ☐	60 70 80 90 100	FAIR
			POOR

C. AUDIOGRAM

TIME SINCE MOST RECENT NOISE-EXPOSURE	DURATION OF MOST RECENT NOISE-EXPOSURE
0-20 MIN, 1 HR, 4-7 HRS, 1 DA 21-50 MIN, 2-3 HRS, 8-16 HRS, 2-3 DAS 4+ DAS	0-20 MIN, 1 HR, 4-7 HRS 21-50 MIN, 2-3 HRS, 7+ HRS

AGE	DATE OF AUDIOGRAM	DAY OF WEEK	TIME OF DAY	EAR PROTECTION WAS EAR PROTECTION WORN? YES NO

RIGHT EAR								LEFT EAR							
250	500	1000	1500	2000	3000	4000	6000	250	500	1000	1500	2000	3000	4000	6000

D. PREVIOUS NOISE-EXPOSURE AND MEDICAL HISTORY

PREVIOUS EMPLOYMENT (LAST 3 JOBS)

TYPE OF WORK	FOR WHOM	HOW LONG

HISTORY

HEAD INJURY (WITH UNCONSCIOUSNESS) ☐
HEARING LOSS IN FAMILY (BEFORE AGE 50) ☐
TINNITUS FOLLOWING NOISE-EXPOSURE R L

STATUS

PERFORATIONS OF DRUMHEAD R L
DRAINAGE FROM EAR R L
MALFORMATION OF EAR R L

RECORD ANY COMMENTS SUBJECT MAKES ABOUT HEARING

TECHNICIAN________
PHYSICIAN________

AUDIOGRAM

TIME SINCE MOST RECENT NOISE-EXPOSURE	DURATION OF MOST RECENT NOISE-EXPOSURE
0-20 MIN, 1 HR, 4-7 HRS, 1 DA 21-50 MIN, 2-3 HRS, 8-16 HRS, 2-3 DAS 4+ DAS	0-20 MIN, 1 HR, 4-7 HRS 21-50 MIN, 2-3 HRS, 7+ HRS

AGE	DATE OF AUDIOGRAM	DAY OF WEEK	TIME OF DAY	EAR PROTECTION WAS EAR PROTECTION WORN? YES NO

RIGHT EAR								LEFT EAR							
250	500	1000	1500	2000	3000	4000	6000	250	500	1000	1500	2000	3000	4000	6000

AUDIOGRAM

TIME SINCE MOST RECENT NOISE-EXPOSURE	DURATION OF MOST RECENT NOISE-EXPOSURE
0-20 MIN, 1 HR, 4-7 HRS, 1 DA 21-50 MIN, 2-3 HRS, 8-16 HRS, 2-3 DAS 4+ DAS	0-20 MIN, 1 HR, 4-7 HRS 21-50 MIN, 2-3 HRS, 7+ HRS

AGE	DATE OF AUDIOGRAM	DAY OF WEEK	TIME OF DAY	EAR PROTECTION WAS EAR PROTECTION WORN? YES NO

RIGHT EAR								LEFT EAR							
250	500	1000	1500	2000	3000	4000	6000	250	500	1000	1500	2000	3000	4000	6000

AUDIOGRAM

TIME SINCE MOST RECENT NOISE-EXPOSURE	DURATION OF MOST RECENT NOISE-EXPOSURE
0-20 MIN, 1 HR, 4-7 HRS, 1 DA 21-50 MIN, 2-3 HRS, 8-16 HRS, 2-3 DAS 4+ DAS	0-20 MIN, 1 HR, 4-7 HRS 21-50 MIN, 2-3 HRS, 7+ HRS

AGE	DATE OF AUDIOGRAM	DAY OF WEEK	TIME OF DAY	EAR PROTECTION WAS EAR PROTECTION WORN? YES NO

RIGHT EAR								LEFT EAR							
250	500	1000	1500	2000	3000	4000	6000	250	500	1000	1500	2000	3000	4000	6000

AUDIOGRAM

TIME SINCE MOST RECENT NOISE-EXPOSURE	DURATION OF MOST RECENT NOISE-EXPOSURE
0-20 MIN, 1 HR, 4-7 HRS, 1 DA 21-50 MIN, 2-3 HRS, 8-16 HRS, 2-3 DAS 4+ DAS	0-20 MIN, 1 HR, 4-7 HRS 21-50 MIN, 2-3 HRS, 7+ HRS

AGE	DATE OF AUDIOGRAM	DAY OF WEEK	TIME OF DAY	EAR PROTECTION WAS EAR PROTECTION WORN? YES NO

RIGHT EAR								LEFT EAR							
250	500	1000	1500	2000	3000	4000	6000	250	500	1000	1500	2000	3000	4000	6000

RESEARCH CENTER, SUBCOMMITTEE ON NOISE

TABLE III*

CHARACTERISTICS OF NYSTAGMUS WITH PERIPHERAL AND CENTRAL LESIONS

	Peripheral	Central
Direction	Constant in different head positions	May vary with different head positions
Onset after assuming new position	Latency period of 3 to 10 seconds	Immediate onset
Amplitude	Usually large	Usually small
Duration	3 to 10 seconds	Much longer than 10 seconds. May continue indefinitely as long as provocative posture is maintained
Fatigability	Nystagmus can only be elicited 1, 2, or 3 times in diminishing intensity	Nystagmus recurs each time provocative posture is assumed

*Schuknecht, H. F.: A Clinical Study of Auditory Damage Following Blows to the Head. Ann. Otol. 59:331, 1950.

of nystagmus, the patient is placed supine with the head lower than the shoulders (midline, head-hanging position). If nystagmus is absent after ten seconds, the patient is returned to the upright position and again examined for nystagmus.

If nystagmus occurs, the intensity, direction or nature, the duration and presence or absence of vomiting, are noted on the record (Table III). If nystagmus is noted in one or more of the positions, that position is held until it subsides. If the nystagmus persists for more than 30 seconds, it will probably continue as long as the critical position is maintained. Should nystagmus persist for 30 seconds, the patient is returned to the upright position, and after a short pause is placed in the position which provoked the nystagmus. If the nystagmus does not reappear, it is recorded as readily fatigable. If the nystagmus reappears with diminished intensity, and on further attempts does not occur at all, it is recorded as gradually fatigable; and if the nystagmus continues with undiminished intensity despite repetition, it is recorded as nonfatigable.

Acoustic neuroma is frequently ushered in by slight dizziness. Unilateral hearing loss progresses gradually to total deafness. The patient falls to the side of the lesion, and nystagmus is toward the affected side. Roentgenograms demonstrate enlargement of the internal acoustic meatus.

Patients with supratentorial brain tumor usually have a persistent sense of imbalance. Vision fails, disks of the optic nerve are choked, and sensory changes occur. Nystagmus is generally related to position, and the patient may vomit if the provocative position is maintained. Nystagmus of a vertical nature strongly suggests brain tumor.

With infratentorial tumor, the Rhomberg test is positive, and pastpointing and adiadochokinesia are associated. Nystagmus may be spontaneous and continuous. If postural, nystagmus will recur and persist each time the offending position is assumed. The gait is ataxic.

Caloric Tests

On completion of the postural tests, the reaction to caloric tests is carried out:

(1) 3 cc ice water irrigated into external auditory canal — 20 seconds. (Cold air may be used instead of water).

(2) Normal response: approximately 30 seconds latent period, nystagmus appears in lateral gaze to opposite side (first degree), then in the midline gaze (second degree), to side of stimulation (third degree), fast component is to the opposite side stimulated. Patient also experiences vertigo. Reaction lasts one and one-half to two minutes. The other ear is then stimulated after about a five-minute interval.

The caloric test determines by minimal stimulation the degree of excitability of each labyrinth. A third-degree response is hyperactive.

The absence of response to 3 cc of ice water suggests severe hypoactivity. When this occurs the stimulus is increased to 6 cc of ice water. Failure to provoke a response to this stimulus indicates a "dead or nonreactive labyrinth." A fracture involving the labyrinth will produce a nonreactive response to maximal caloric stimulation.

The patient with the onset of sudden deafness needs a detailed general physical examination. Is there a peripheral vascular condition, a hemorrhagic defect, a hypertensive diathesis, or some other basis for an inner-ear vascular episode? Serologic virus tests, including serial mumps antibody titers studies, should be made.

Acoustic Trauma

Sudden deafness may result from acoustic trauma. This term is reserved for the impairment of hearing as a result of damage produced by sudden explosive or impulsive sound and/or a head blow capable of producing either skull fracture or serious brain concussion. The worker with skull fracture commonly will have blood in the canal or the middle ear on the affected side. The ear exposed to explosive sound usually will show a ruptured ear drum. The patient will complain of ringing and fullness of the ear. The loss of hearing is immediate and nonprogressive.

Schuknecht has shown that hearing loss following concussion involves the higher frequencies. The frequency most involved is the 4,096 cps, which of itself does not involve speech reception. Dizziness and tinnitus are frequent associated and localizing symptoms. X-ray investigation is in order to demonstrate the presence or absence of basal or temporal bone fracture. When a fracture is visible, there can be little question about the severity of the blow to the head (Table IV).

In some cases of acoustic trauma the hearing loss is temporary. Recovery can be demonstrated by serial audiograms. In other cases, acoustic trauma and/or head injury may produce a permanent hearing loss. If evidence of recovery of hearing is not present after four weeks, the loss is likely to be irreversible. However, for medicolegal reasons one should not attempt to establish the amount of damage until several months have passed.

Sudden deafness may occur during the course of administration of certain drugs, but this is rare. The drugs of known ototoxicity are: dihydrostreptomycin, neomycin, quinine and kanamycin. The onset of the deafness is commonly ushered in by dizziness and tinnitus. The audiogram shows a sensorineural type hearing loss and the cold caloric test gives an absent or diminished response.

A patient complaining of "sudden deafness" may have a functional or psychogenic hearing impairment. Search for an organic basis to explain the etiology of the hearing loss is negative. History will bring out an emotional conflict or a period of stress that immediately preceded the onset of the hearing loss. The functional hearing loss is an unconscious device which the patient uses to escape an intolerable situation. To him his hearing loss has an organic basis as manifested by the tinnitus, which is commonly an associated symptom of "hysterical deafness." His hearing impairment varies with the degree of anxiety. Hearing is restored when the origin of the anxiety is recognized and resolved.

Malingering

Perhaps the most troublesome and most frequent case presenting the complaint of sudden deafness is, lacking a better term, "the malingerer." It has been estimated that of all industrial claims for damage, 90% are possible malingerers and 20% are obvious malingerers.[1]

The malingerer feigns illness or inability to avoid duty or to gain an advantage (financial compensation or escape from military duty). He deliberately fabricates his symptoms. When unobserved, he abandons his symptoms. His hearing test shows an irregular response. He claims "he hears nothing" when the good ear is occluded by the examiner's finger and a shout, loud enough for response by bone conduction, is given. If the good ear is irrigated (in the manner used to remove wax from the ear), he will open his eyes on command even though the irrigation

TABLE IV*
DIFFERENTIAL CLINICAL FINDINGS IN HEARING LOSS

	Labyrinthine Concussion	Longitudinal Fracture	Transverse Fracture
Bleeding from ear	Never	Very common	Rare
Injury to external auditory canal	Never	Occasional	Never
Rupture of drum	Never	Very common	Rare—commonly a hemotympanum
Presence of cerebrospinal fluid	Never	Occasional	Occasional
Hearing loss	All degrees—partial to complete recovery	All degrees—combined type, partial to complete recovery	Profound nerve type—no recovery
Vertigo	Occasional—mild and transient	Occasional—mild to transient	Severe—subsides; nystagmus to opposite side
Depressed vestibular function	Occasional—mild	Occasional—mild	No response
Facial nerve injury	Never	25%—usually temporary	60%—often permanent
X-ray signs	Negative	25%—in squamous and mastoid area	60%—in occiput in pyramid in vertex-mento films

*Schuknecht, H. F.: A Clinical Study of Auditory Damage Following Blows to the Head, Ann. Otol. 59:331-359, 1950.

going on in the good ear is enough for complete masking of that ear.

Batteries of tests exist to unmask the malingerer. There are the reliable but unsophisticated tests just mentioned, the easily executed Stenger's test and the Lombard test; up to the complicated test requiring special equipment, the delayed-feedback test, and the psychogalvanic-skin-resistance test.

Glorig[2] tells us "the practice of labeling a person simulating an illness as a malingerer is being discouraged by many examiners. They point out that inasmuch as simulation of an illness may be either conscious or unconscious, it is wise to be extremely careful before terming a patient a malingerer."

The diagnosis of malingering is a most serious charge and is seldom proved without an actual admission by the patient of "intent to defraud." I share this point of view; however, I believe we have the technical knowledge and the hardware that permits us to identify the "nonorganic" hearing loss that is reported to be caused by an incident or noise exposure at work. If we are to function as responsible physicians, we must be prepared to recognize the person whose deafness is consciously motivated by seeking unwarranted compensation.

Physician's Responsibility

From the point of view of the physician, our duty is clearly the exercise of our skills towards the preservation of hearing. In the words of Shakespeare's Julius Caesar, the plant physician must in effect say to the employee, "Lend me your ear." It is he who must quarterback the team, he who organizes the programs of hearing conservation.

Hearing is one of our most valuable senses. Severe deafness is one of the most distressing afflictions of mankind. The importance of deafness as a cause of disability is not generally appreciated, mainly because the crippling affect is not visible. The tragedies of the loss of hearing are seldom sudden or dramatic; they are "just terribly personal" and more often than not direct the individual down the lonely path to the world of silence. Our opportunity is in the early recognition of a noise-induced loss of hearing. At this time steps can be taken to prevent progress of the loss to the involvement of the speech frequencies.

Our responsibility extends into the area of both management and laborer. Through the overall cooperation and understanding of the noise problem in industry, the program of prevention of hearing loss can be accomplished. The conservation of hearing program must have the support of labor, management and the medical profession if we are to be successful.

References

1. Critical evaluation of methods for testing for non-organic hearing impairment. Minutes of the 2nd meeting of Working Group No. 36, Armed Forces National Research Councils. Committee on Hearing and Bio-Acoustics, Washington, 1959.

2. Glorig, Aram: Audiometry, Principles and Practices. Williams and Wilkins, 1965.

Long-Term Study Relating Temporary and Permanent Hearing Loss

JOSEPH SATALOFF, MD; LAWRENCE VASSALLO, MS; JOSEPH M. VALLOTI, MD; AND HYMAN MENDUKE, PhD

DURING the last 15 years, many studies have been published on the relationship between permanent threshold shift (PTS) and temporary threshold shift (TTS).[1-9] In many of these, subjects with normal hearing were exposed to noises of various intensities, spectra, and durations under essentially "laboratory" conditions and the temporary threshold shifts induced in these normal ears were then compared to the permanent hearing losses of industrial personnel who had been working in noise conditions presumably comparable to the laboratory conditions. Few studies relating TTS and PTS have been conducted on personnel in an actual working environment and fewer still have involved long-term follow-up studies to relate TTS and PTS in the same subjects.

In 1961 we initiated a longitudinal study in a large paper plant in an attempt to answer certain questions concerning the effect of industrial noise on temporary threshold (TT) and on permanent threshold (PT). We were especially interested in determining whether permanent threshold shift due to noise could occur in the absence of temporary threshold shift measured within ten minutes after exposure and whether ears exhibiting some amount of TTS in turn incurred more permanent damage than did ears exhibiting no TTS.

Material and Method

Initially, there were 21 subjects of whom 15 were men and six were women. Table 1 lists the number of subjects in each age group and their years of employment in the paper plant as of 1961. During the three years after 1961, five men and two women left the study due to transfers, illness, etc. To our knowledge, none of these reasons was at all related to hearing or to noise. All subjects were machine operators exposed to reasonably steady noise levels, at the operator's station, of approximately 93 db in the 300-2,400 octave band widths and with an overall maximum level of about 100 db.

Reprint requests to 1721 Pine Street, Philadelphia, Pa 19103 (Dr. Sataloff).

All hearing tests were conducted by trained technicians. Instrument calibration was checked daily on normal ears. To insure that the calibration of the audiometer had not drifted to any great extent (while still staying within specific tolerances), thresholds of nonnoise exposed personnel were taken in 1961 and again in 1964. These thresholds were approximately the same for the two test periods. Subjects were seated in a prefabricated sound proof booth which met the American Standard criteria for background noise in test rooms. In 1961 and 1964 hearing testing was done at the following times:

BW/BV/61 = Before Work, Before Vacation, in the middle of the week in 1961 (16 hours had elapsed since the end of the preceding workday).

AW/BV/61 = After Work, Before Vacation, in the middle of the week in 1961 (within ter minutes after noise exposure).

BW/AV/61 = Before Work, After Vacation of two weeks in 1961.

AW/AV/61 = After Work, After Vacation in 1961 (within ten minutes after exposure).

BW/BV/64 = Before Work, Before Vacation, in the middle of the week in 1964.

BW/AV/64 = Before Work, After Vacation of three weeks in 1964.

The difference between BW/BV/61 and AW/BV/61 and between BW/AV/61 and AW/AV/61 reflected TTS on workdays which had been preceded by rest period durations of 16 hours and two weeks respectively. The threshold observed before work after vacation (BW/AV/61) was considered the best indication of permanent threshold (PT) since there had been a two week rest period from the noise. The difference between AW/BV/61 and BW/AV/61 was considered the best indication of TTS since this is a comparison of thresholds measured after eight hours noise exposure and again after two weeks rest from noise.

In 1964, we were able to repeat hearing measurements before work in the middle of the work week (BW/BV/64) and before work following a three-week vacation (BW/AV/64); unfortunately management considerations made it impossible to repeat the audiograms after work on these days. There was, therefore, no opportunity to verify the daily TTS values as determined in 1961. We did consider, however, the differences (improvements) in thresholds measured BW/BV/64 and BW/AV/64 were evidence of TTS. It was also possible to evaluate the shifts in permanent threshold that had occurred between 1961 and 1964 (BW/AV/64)-(BW/AV/61).

Results and Comment

1961.—The 1961 testing sequence on 21 subjects revealed that there were some daily

BEFORE WORK AFTER VACATION — BEFORE WORK BEFORE VACATION — AFTER WORK BEFORE VACATION

CHRONIC TTS — ACUTE TTS

TOTAL TTS

Relationship of temporary threshold shifts and total threshold shift.

shifts in prevacation thresholds (Table 2, *row 1*) for the frequencies 1,000 to 8,000 cycles per second, almost all of which were significant at the 0.05 level or better. Thresholds measured before and after work after the vacation period (*row 2*) revealed greater degrees of TTS for all frequencies between 1,000 and 6,000 cps, all of which were significant at the 0.01 level. The difference in daily TTS between that observed before vacation and that observed after vacation (*row 3*) was significant only at 3,000 and 4,000 cps. The other differences, while in the same direction for all frequencies except 8,000 cps, were not large enough to be significant with a sample of only 21 subjects.

Improvements in thresholds after a rest period are indicative of the presence of TTS. There is a daily or "acute" TTS which is evident in the comparison of before work and after work thresholds. There is

TABLE 1.—*Number of Subjects in Age Groups*

1961 Age Group	No. of Subjects	1961 Time at Job (Yr)
20–29	5	8, 4, 2, 2, 7
30–39	6	18, 15, 15, 14, 10, 8
40–49	6	17, 16, 16, 25, 19, 32
50–59	3	19, 27, 27
60+	1	31

TABLE 2.—*1961 Results on 21 Subjects*

Row			Frequency in CPS						
			500	1,000	2,000	3,000	4,000	6,000	8,000
1.	Change in threshold from before work/before vacation to after work/before vacation (Acute or daily TTS)	Mean threshold change in db	1.0	3.6	4.7	2.3	3.0	2.2	3.0
		Level of significance	NS	<0.01	<0.01	<0.05	<0.05	0.1 (NS)	<0.05
2.	Change in threshold from before work/after vacation to after work/after vacation (TTS after first day's work after 2 wk. vac.)	Mean threshold change in db	2.5	5.7	7.0	7.3	7.0	4.4	1.4
		Level of significance	0.1 (NS)	<0.01	<0.01	<0.01	<0.01	<0.01	NS
3.	Difference between TTS observed daily and TTS observed on first day of work after vacation	Mean difference in db	1.5	2.1	2.3	5.0	4.0	2.2	−1.6
		Level of significance	NS	NS	NS	<0.01	<0.05	NS	NS
4.	Change in threshold from before work/before vacation to before work/after vacation (Chronic TTS)	Mean threshold change in db	−1.5	−1.7	−2.5	−5.0	−5.5	−0.6	1.0
		Level of significance	NS	NS	<0.05	<0.01	<0.01	NS	NS
5.	Change in threshold from after work/before vacation to after work/after vacation	Mean threshold change in db	0.0	0.2	−0.4	0.0	−1.4	1.5	0.6
		Level of significance	NS	NS	NS	NS	NS	NS	NS

1964 Results on 14 of Original 21 Subjects

Row			Frequency in CPS						
			500	1,000	2,000	3,000	4,000	6,000	8,000
6.	Change in threshold from before work/before vacation to before work/after vacation (Chronic TTS)	Mean threshold change in db	−1.0	−1.7	0.2	−0.8	−1.9	1.7	2.3
		Level of significance	NS	NS	NS	NS	NS	NS	NS
7.	Change in postvacation before work thresholds from 1961 to 1964	Mean threshold change in db	−1.3	−1.0	2.9	6.3	1.9	5.6	7.7
		Level of significance	NS	NS	<0.05	<0.01	NS	<0.05	<0.01

TABLE 3.—*Threshold Changes*

Frequency in CPS	Change in Threshold From Before Work/Before Vacation to After Work/Before Vacation (Acute TTS)	Change in Threshold From Before Work/Before Vacation to Before Work/After Vacation (Chronic TTS)		Total TTS	Change in Threshold From Before Work/After Vacation to After Work/After Vacation
500	1.0	−1.5	=	2.5	2.5
1,000	3.6	−1.7	=	5.3	5.7
2,000	4.7	−2.5	=	7.2	7.0
3,000	2.3	−5.0	=	7.3	7.3
4,000	3.0	−5.5	=	8.5	7.0
6,000	2.2	−0.6	=	2.8	4.4
8,000	3.0	1.0	=	2.0	1.4

also a "chronic" TTS which is evident in the comparison of before work thresholds taken before vacation and then after vacation. Here, the improvement in thresholds resulting from the long rest period shows how much TTS is chronically present. The total TTS is the difference in thresholds after work/before vacation, and before work/after vacation. Relationship of these temporary shifts is shown in the Figure. As shown in Table 3, *column 3,* the "total TTS is quite large, especially between 1,000 and 4,000 cps inclusive. It is of interest to note here that the threshold shift on the first day of work after vacation (Table 3, *column 4*), is almost exactly the same as the "total" TTS (chronic plus acute TTS). Table 2, *row 4,* shows that the changes (improvements) in threshold after a long rest period are significant at 2,000, 3,000, and 4,000 cps.

1964.—Comparisons of thresholds of 14 of the original 21 subjects taken after 16 hours off from work and three weeks off from work (Table 2, *row 6*) showed minimal differences in mean thresholds (no TTS). A comparison of postvacation threshold improvement in 1961, (Table 2, *row 4*) with that in 1964, (*row 6*) points out the apparent cessation of TTS in 1964. However, a comparison of 1961-1964 postvacation threshold (Table 2, *row 7*) indicates changes in permanent threshold at 2,000 cps, 3,000 cps, 6,000 cps, and 8,000 cps, all of which were significant at at least the 0.05 level.

In most of the 14 subjects, the amount of permanent shift was under 10 db and the amount of TTS shown in 1961 was also under 10 db.

There were some subjects who sustained PTS of 10 db or more without showing any evidence of TTS: at 3,000 cps, three subjects had from 10 to 18 db PTS and two had 18 db PTS; at 8,000 cps two had 12 db and one had 30 db PTS. (The values of 12, 18, and 22 db were obtained as averages of the two ears.)

There were also some subjects who showed no PTS but had demonstrated TTS of 10 db or more: at 500 cps one subject had 18 db TTS; at 1,000 cps two had 18 db TTS; at 2,000 cps one had 12 and one had 22 db TTS; at 3,000 cps one had 12 db, one had 15 db, and another had 18 db TTS; and at 4,000 cps one had 12 db TTS and another 15 db TTS.

Relating the permanent threshold shift which occurred in the three year period to that expected to occur during the same period (age and frequency being the variables) it was found that at 1,000 and 4,000 cps the "actual" and "expected" losses were

TABLE 4.—*"Actual" and "Expected" Hearing Losses in Three-Year Period*

Test Frequency (CPS)	Noise Exposed Group Expected Loss (db) *	Actual Loss (db)	Degree of Loss Greater Than "Expected" (db)
1,000	0.5	−1.0	−1.5
2,000	0.8	2.9	2.1
3,000	1.6	6.3	4.7
4,000	1.9	2.7	0.8
6,000	2.8	5.6	2.8
8,000	2.7	7.7	5.0
	Nonnoise Exposed Group		
1,000	0.0	−1.0	−1.0
2,000	0.0	−1.5	−1.5
3,000	1.3	0.9	−2.4
4,000	1.8	1.9	0.1
8,000	2.3	3.8	1.5

* Mean expected loss for the three-year period. The age of each individual is taken into account in arriving at these estimates. For example, a 23-year-old subject is not expected to sustain any change in threshold at 2,000 cps during the three-year period, whereas a 55-year-old subject may have a 2 db deterioration in threshold for 2,000 during the same time span. A mean "expected loss" is arrived at for each frequency through these individual determinations. (The nonnoise exposed group is 85% women with a mean age of 33—thus the slight difference in expected loss for the three years).

approximately the same. Losses at the other frequencies were from 2 to 5 db greater than expected as shown in Table 4.

It appears that at 4,000 cps permanent shift due to noise has reached its maximum and further shifts in threshold will be the result of the aging process. Noise damage appears to be continuing to the surrounding areas, particularly at the 3,000, 6,000, and 8,000 cps sites, but not down to the 1,000 cps area.

There may be some inclination to consider the small shifts as unimportant. They are, however, substantial and significant since they occurred in a relatively short period of three years and were greater than the losses expected due to aging.

The proposal that after ten years exposure, noise of a given level will not cause further permanent shifts in threshold does not appear to apply in this study. As of 1964, 13 of the 14 subjects had ten or more years exposure to their noisy jobs, yet there appears to be continuing trauma even though it is no longer apparent when tested for as a daily phenomenon. The fact that the hearing losses in this study were comparatively mild and could continue to deteriorate may be a factor in considering the proposal that hearing levels remain unchanged after ten years exposure. If the hearing losses were more severe there might not have been further loss in the noise environment.

Conclusions

In 1961 hearing tests were done on 21 subjects working in a reasonably steady state overall noise level of about 100 db with approximately 93 db in the 300-2,400 octave band width. The hearing tests measured hearing threshold levels after eight hours of work, after 16 hours rest, after a two week vacation, and again after eight hours of work. There was evidence of small but significant temporary shifts in threshold when hearing was tested during the work week. After a two week rest period, larger threshold shifts were observed on the first day of work. This was not the result of more trauma resulting from the first day's exposure. It was the result of thresholds returning to their "true" levels with rest and then shifting back to the middle of the week levels after exposure. This indicates that thresholds measured before and after work during the work week reveal only part of the TTS which exists.

In 1964 hearing studies were obtained on 14 of the original subjects, the rest having been separated for reasons other than noise exposure. Hearing thresholds were obtained after a 16 hour rest and after a three week vacation. Whereas in 1961, a two week rest period showed significant improvement in thresholds for 2,000, 3,000, and 4,000 cps (evidence of TTS), there were no similar improvements in thresholds in 1964 after a three week rest period (no evidence of TTS). However, a comparison of 1961 and 1964 postvacation, prework thresholds yielded permanent threshold shifts for 2,000, 3,000, and 6,000 cps greater than is to be expected from the aging process. It appears from these results that TTS no longer is taking place (or at least it doesn't show up as a daily phenomenon) but that permanent shifts due to noise are still taking place.

There were some subjects who showed 10 db or more PTS without any evidence of TTS.

There were also some subjects who showed TTS over 10 db but failed to develop any further PTS.

It would appear that absence of significant TTS measured within ten minutes of a day's exposure to the steady noise described in this study does not rule out the development of progressive hearing loss greater than is to be expected from the aging process.

REFERENCES

1. Wilson, W.H.: Prevention of Traumatic Deafness: A Preliminary Report, *Arch Otolaryng* **37**:757-767, 1943.
2. Wilson, W.H.: Prevention of Traumatic Deafness: Further Studies, *Arch Otolaryng* **40**:52-59, 1944.
3. Gravendeel, D.W., and Plomp, R.: Permanent and Temporary Diesel Engine Noise Dips, *Arch Otolaryng* **74**:405-407, 1961.
4. Kylin, B.: Temporary Threshold Shift and Auditory Trauma Following Exposure to Steady-State Noise, *Acta Otolaryng*, suppl 152, 1960.
5. Ward, W.D.; Glorig, A.; and Sklar, D.L.: Temporary Threshold Shift From Octave Band Noise: Applications to Damage Risk Criteria, *J Acoust Soc Amer* **31**:522, 528, 1959.
6. Glorig, A.; Summerfield, A.; and Ward, W.D.: Observations on Temporary Auditory Threshold Shift Resulting From Noise Exposure, *Ann Otol Rhin Laryng* **67**:824, 1958.
7. Nixon, J.D., and Glorig, A.: Noise-Induced Permanent Threshold Shift at 2000 cps and 4000 cps, *J Acoust Soc Amer* **33**:904-908, 1961.
8. Oppliger, G.C.; VonSchulthess, G.; and Grandjean, E.: The Relation Between Auditory Fatigue and Permanent Noise Deafness, *Acta Otolaryng* **52**:415-428, 1960.
9. Nixon, J.C.; Glorig, A.; and Bell, D.W.: Predicting Hearing Loss From Noise-Induced TTS, *Arch Otolaryng* **81**:250-256, 1965.

Temporary and Permanent Hearing Loss

A Ten-Year Follow-Up

JOSEPH SATALOFF, MD; LAWRENCE VASSALLO, MS;
AND HYMAN MENDUKE, PhD

Two major and related problems persist in the realm of occupational hearing loss: (1) the development of a reliable test to predict an employee's sensitivity to noise exposure, and (2) a clear understanding of the relation between temporary threshold shift (TTS) and permanent threshold shift (PTS).

It seems logical to assume that individuals who sustain habitual TTS due to exposure to intense noise have demonstrated their "susceptibility" to acoustic trauma and will eventually develop permanent hearing damage. On the basis of a postulated direct relationship between TTS and PTS, Wilson,[6] in 1944, suggested a susceptibility test to detect individuals most likely to develop permanent hearing loss as the result of prolonged exposure to intense noise. He exposed personnel to a pure tone for eight minutes and measured the resulting TTS in the octave frequency above the exposure signal. He then predicted that those subjects who developed a significant amount of TTS would develop permanent hearing loss after prolonged exposure to noise, while those who did not show much TTS would not develop any permanent hearing loss from this particular noise; he presented no follow-up data that would permit evaluation of this theory.

Glorig et al[3] measured TTS in personnel exposed for eight hours to certain industrial noise. They expressed the belief that if no TTS (12 db or less at 2,000 cps) is produced after eight hours exposure, no PTS would result from this noise exposure. They suggested also that PTS can be predicted by the use of formulae developed from TTS studies. They presented no follow-up data, but a longitudinal study is reported to be in progress.[5]

About 11 years ago a series of auditory experiments was performed on 105 subjects at the Naval Base in Philadelphia.[4] The chief purpose was to test the validity of Wilson's proposed susceptibility test. This report presents the results of an 11-year follow-up on some of the individuals studied at that time.

In 1951 pure tone audiograms were performed on 105 subjects between 24 and 69 years of age, most being under 50. The testing was done under excellent conditions using a standard technique and an instrument of known calibration. Only subjects with normal hearing or sensorineural loss were included. After establishment of air conduction hearing thresholds, the right ear was exposed to a 2,000 cps tone of 95 db on the audiometer for eight minutes. With a stop watch, thresholds at 4,000 cps were successively obtained at ten seconds, 30 seconds, one minute, two minutes, and three minutes after cessation of the 2,000 cps stimulus. After a rest period the same procedure was performed on the opposite ear. While no conclusions concerning the predictability of PTS through TTS were reached it was found that in subjects with unilateral loss at 4,000 cps the degree of TTS at 4,000 cps was significantly higher in the good ear than in the ear with impaired hearing. There was some evidence that subjects with good hearing at 2,000 cps but bilateral loss at 4,000 cps had a slightly smaller TTS than subjects with normal hearing at both frequencies. Where subjects had bilateral losses at both 2,000 cps and 4,000 cps, TTS became markedly

---

This study was supported by the Foundation for Medical Research in Hearing, Philadelphia.

smaller. The inverse relationship between threshold loss at 4,000 cps and degree of TTS was a group phenomenon but individual variations were marked. Many subjects with good hearing failed to show TTS and some with impaired hearing did show a TTS.[4]

Prior to the start of the study in 1951 most of the men had histories of gunfire exposure and also of working daily in jet engine testing areas with noise levels up to 105 db or more for short periods. These past exposures were probably responsible for the high tone hearing losses found in our personnel at the start of this investigation. Between 1951 and 1962 the noise levels in which these men worked decreased markedly. During this period they were exposed to an average overall daily noise level of about 90 db with a maximum of about 85 db in any of the three octave bands between the frequencies 300 to 2,400 cps. On occasion some of the men were exposed to jet engine noise up to 100 db overall, but ordinarily they wore ear protectors in such situations.

In 1962 we were able to obtain audiograms on 33 subjects of the original 105 studied. We could not ascertain the reasons for nonavailability in each of the 72 "missing" subjects. Of these, 21 would have been past retirement age since they were 55 or older in 1951. We feel confident that the reasons for unavailability of the remaining 49 were unrelated to hearing.

Based on the age distribution 11 years earlier, 65% of the group should have been under 50 in 1962. There was in fact rather good agreement with this proportion since 58% of those available in 1962 were under 50.

We were, unfortunately, unable to reevaluate TTS in 1962. Threshold levels were determined in a standard manner, but with a different audiometer. This required a slight correction at 2,000 cps to match the calibration of the instrument used initially.

One would expect some further loss of hearing in the higher frequencies in the 11-year interval between tests. In the period between 30 and 40 years of age, a 3 db reduction in hearing level at 2,000 cps is expected in a normal non-noise-exposed male population. At 4,000 cps approximately 8 db reduction in hearing threshold level is expected between 30 and 40 years of age. For each advancing decade after 30 years and for each test frequency above 1,000 cps, one can expect slightly increasing shifts in threshold.

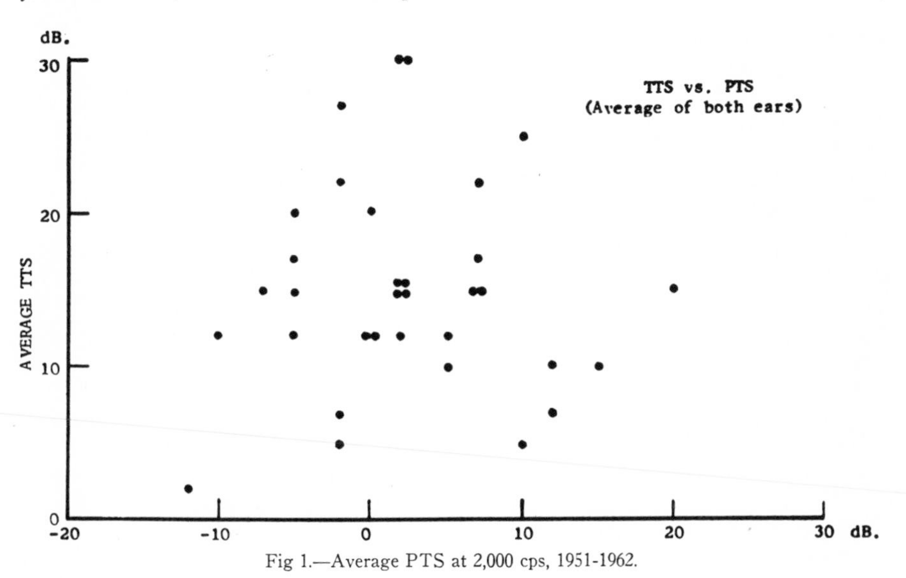

Fig 1.—Average PTS at 2,000 cps, 1951-1962.

It is the usual practice to subtract these "expected" age losses when working with a study of this type.[1,2] In reality it is not definitely known whether the effects of noise are separate from the effects of age or if there is interaction between noise and age. In any case, age correction factors were not used in our study. As will be pointed out later, this made little difference in the final results.

Since threshold shifts due to noise exposure generally occur in the higher frequencies, we will report on the frequencies 2,000, 4,000, and 8,000 cps. Right and left ear thresholds obtained in 1962 at each frequency were averaged and compared with the 1951 averages. The permanent threshold shift at each frequency was then compared with the decibel levels of TTS obtained in the 1951 study.

As seen in the scattergrams, Fig 1, 2, and 3, there is no apparent relationship between PTS and TTS for any of the frequencies between 2,000 and 8,000 cycles. Some individuals who had sustained little or no TTS (10 db or less) showed a PTS of as much as 20 db while other individuals who had sustained as much as 30 db TTS showed little or no PTS.

The observed threshold changes for the 11-year period were generally no greater than would have been expected from the aging process. Corrections for aging leave the relationship, or rather, lack of relationship, between TTS and PTS unaltered.

Originally, this experiment was designed to investigate the relation of TTS to PTS and to explore the validity of Wilson's susceptibility test. In the follow-up study we also had hoped to substantiate the damage risk criteria set forth by Glorig et al.[3] As it turned out, the industrial noise level diminished to 90 db overall, on the average, shortly after our study commenced in 1951 and there was probably insufficient noise exposure to produce grossly measurable amounts of daily TTS. Under these conditions, TTS values had no predictive value for PTS 11 years later. Our findings, therefore, do support the contentions of Glorig et al, who postulated that noise levels of this magnitude are insufficient to produce much TTS and therefore produce no PTS. The noise levels of 90 db overall with a maximum of 85 db in any of the three octave bands between the frequencies 300 to 2,400 cycles did not produce in our longitudinal study any abnormal PTS in the 33 subjects.

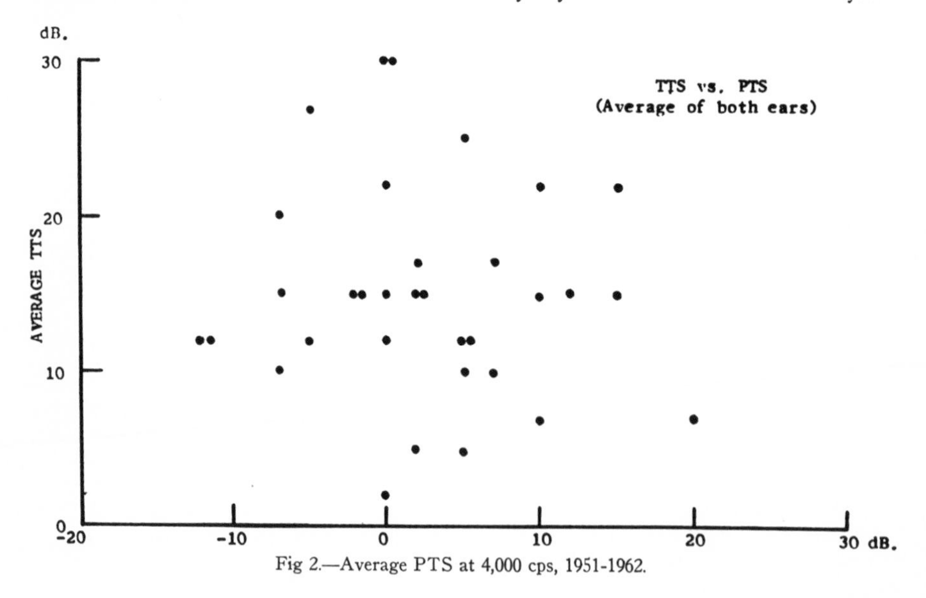

Fig 2.—Average PTS at 4,000 cps, 1951-1962.

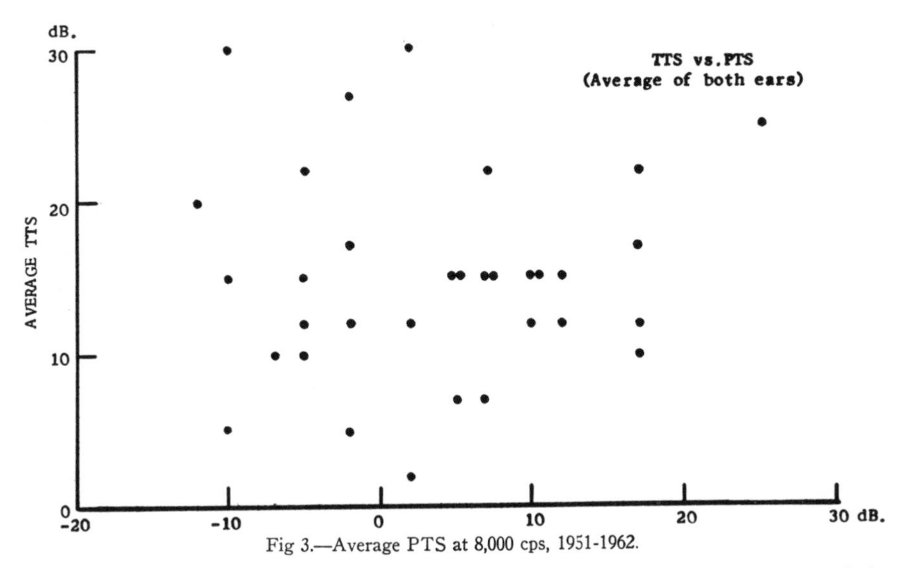

Fig 3.—Average PTS at 8,000 cps, 1951-1962.

This was true of those with normal hearing in 1951 as well as in those with a sensorineural high tone loss produced previously by noise exposure. The hearing losses in half (17) of our subjects were originally diagnosed "sensorineural" due to gunfire exposure or to industrial noise. The remaining 16 had hearing within normal limits. Of the 17, only two showed a deterioration in excess of 10 db at 4,000 cps but not at 8,000 cps; two at 8,000 cps but not at 4,000 cps, and one at both 4,000 and 8,000 cps. These factors, we think, are an interesting by-product of this study in that there appears to be no progressive aspect to noise-induced hearing loss when noise levels are reduced below traumatic levels. Furthermore, these noise levels appear to be safe even to ears that experimentally showed "susceptibility to trauma" as demonstrated by the presence of TTS after exposure to a pure tone of 95 db for eight minutes.

Conclusions

1. There is no apparent relationship between temporary threshold shift (TTS) produced experimentally by exposure to a pure tone and permanent threshold shift (PTS) in employees exposed for 11 years to average daily noise levels of 90 db overall with a maximum of about 85 db in any of the three octave bands between the frequencies 300 to 2,400.

2. Noise-induced hearing loss does not seem to be progressive in personnel who worked for 11 years with reduced daily maximum noise levels of about 85 db in any of the three octave bands between 300 to 2,400. Subjects with normal hearing showed no evidence of noise-induced PTS in such an environment.

REFERENCES

1. Exploratory Subcommittee Z 24-X-2: The Relations of Hearing Loss to Noise Exposure, New York: American Standards Association, Inc., 1954.

2. Glorig, A., and Davis, H.: Age, Noise and Hearing Loss, Ann Otol 70:556-571, 1961.

3. Glorig, A.; Ward, D.; and Nixon, J.: Damage Risk Criteria and Noise Induced Hearing Loss, Arch Otolaryng 74:413-423, 1961.

4. Sataloff, J.; Menduke, H.; and Hughes, A.: Temporary Threshold Shift in Normal and Abnormal Ears, Arch Otolaryng 76:52-54, 1962.

5. Steffen, T.; Nixon, J.; and Glorig, A.: Stapedectomy and Noise, Laryngoscope 73:1044-1060, 1963.

6. Wilson, W. H.: Prevention of Traumatic Deafness: Further Studies, Arch Otolaryng 40:52-59, 1944.

The Environment in Relation to Otologic Disease

JOSEPH SATALOFF, MD AND JOHN A. ZAPP, JR., PhD

Introduction

In a broad sense, man, as an organism, lives and functions in an "environment" which can be described as the aggregate of all external influences affecting the life and development of the organism. The environment consists of the air man breathes, the buildings in which he lives and works, the clothing he wears, the food and water he ingests, etc. The environment has always directed the course of man's development as epitomized in the Darwinian concepts "struggle for existence" and "survival of the fittest." Man, on the other hand, has been able to make profound changes in his environment, and this, in turn, has introduced new factors which act on, and affect, the life and development of man.

Environmental changes are perhaps most marked in the areas which may be called "occupational," and date roughly from the industrial revolution. They are occurring at a more rapid rate than at any previous time in history. The new products and services which have evolved spill over, of course, into the nonoccupational environment, and the question arises as to what effect all these environmental changes may have on the incidence of new and old diseases.

The subject of this article is the environment in relation to otologic disease. There is such a relation and it has been known for a long time. W. Gilman Thompson, MD, in his book *The Occupational Diseases*,[1] published in 1914 and billed as the first of its kind to be published in this country, devotes a little more than three pages to diseases of the ear. Thompson describes effects on the external ear, the tympanum, the middle ear, and the internal ear, which may be characterized as otologic disease of environmental origin.

While Thompson was explicitly aware of the injurious effect of noise on hearing acuity, the increasing noise level of the modern environment, and, particularly, of some work environments, has led to an intensive study of this phenomenon over the past two decades. Much valuable knowledge has been gained thereby, but there is a possibility that we may have come to regard noise-induced hearing loss as the only form of otologic disease of environmental origin.

While not ignoring noise, this article will attempt to direct attention to other forms of otologic disease which are also of environmental origin.

The External Ear

External Otitis.—Irritation and inflammation of the skin of the ear canal is a common finding, particularly in warm and humid climates, and in industrial environments with high temperatures and humidity. The skin of the ear canal is thin and very delicate. Hence, prolonged exposure to moisture and warmth increases the susceptibility of the skin to irritation from chemical and mechanical irritants and leads to excoriation which serves as a fertile base for infection by bacteria or fungi.

Industrial environments which are very dusty (including agriculture) may result in the deposition of dust and grit or of fibrous material on the surface of the skin of the ear canal, and these may irritate because of chemical action (eg, dusts containing lead, zinc, arsenic, mercury, chromates, acids or alkali, etc) or by direct mechanical action when rubbed against the skin. A great many organic chemicals, including disinfectants, medicaments, and cosmetics, have been known to cause external otitis.

Hairs and fibrous materials such as horse hair, kapok, cotton, wool, glass fibers, and even feathers may lodge in the ear canal and initiate irritation and inflammation.

While the ear plug is becoming increasingly common in industry to protect against excessive noise, the improper care of these protectors can be a potent cause of external otitis. Dirty plugs, plugs washed with soap or treated with disinfectants and insufficiently rinsed, plugs too large for the given ear canal, or plugs lubricated with saliva, can readily produce inflammation in the ear canal.

The most common of all causes of external otitis is prolonged exposure of the skin of the canal to water, and the most common exposure is during swimming and showering, both of which have become increasingly popular. Since the ear infection usually becomes apparent several days after the exposure to water, it may be blamed on other causes such as the work environment.

It is not uncommon to encounter allergic external otitis of environmental origin. Many chemicals, drugs, and cosmetic preparations are not only skin irritants but potential skin sensitizers in susceptible individuals. Bacteria and fungi may produce an allergic inflammation, as may, also, certain medicaments used to control them. Kapok, feathers, and animal hairs encountered in the occupational environment or from pillows have been implicated.

In acute external otitis, the skin of the ear canal is generally edematous or excoriated. Tenderness in pronounced around the entire ear, and the pain is aggravated when the patient chews or when he presses against the ear. The patient frequently complains of itching and discomfort. Among the early findings in external otitis is a discharge in the ear canal which may resemble wax (cerumen) except that it often has a bad odor and is actually of a different consistency and appearance. External otitis is frequently aggravated by scratching inside the ear canal.

It is important that external otitis be differentiated from otitis media. This may be difficult unless the eardrum is visible and appears to be normal (not inflamed). The pain in otitis media is usually very deep and is not aggravated by movements of the jaw while chewing. Sneezing and coughing, however, often produce severe pain because of increased pressure in the inflamed middle ear. If the discharge in an ear canal has a stringy mucoid appearance (such as found in the nose during rhinitis), it is almost invariably indicative of otitis media with a perforated eardrum.

Whenever a clear-cut diagnosis is not possible, therapy should be directed to both the external and the middle ear. For external otitis, the medication is principally applied locally to the outer ear; for otitis media it is directed to the middle ear, nasopharynx, and systemically.

Care should be taken to avoid too strenuous effort to introduce an otoscope into the tender and inflamed ear of a patient with external otitis. Such a procedure would be painful and would only aggravate the patient. In many instances, the preliminary diagnosis must be based on the history, superficial examination, and clinical experience until the infection in the external canal subsides and permits a more satisfactory examination of the eardrum.

Of the many successful methods of treating external otitis, one of the best is to insert snugly into the swollen ear canal a large cotton wick soaked in Burow's solution of about 1:10 dilution. The same wick should be kept in place and wetted with the same solution for a period of 24 to 48 hours. The wick should not be allowed to become dry. The Burow's solution changes the pH in the canal and inhibits the growth of some pathogenic organisms while the ear is healing. If the infection persists despite this treatment, other specific drugs such as antibiotics, administered locally and systemically, are indicated.

Ear drops of one sort or another are so frequently used that they warrant comment. In the acute stage of external otitis, the canal is generally so swollen that the drops cannot get into the canal, or, if they do, they seldom contact the affected areas on the superior part. A snugly inserted wick is much more effective. Applying the medication with cotton-tipped applicators to all portions of the canal is effective if it is done frequently and gently.

The use of strong medicaments and excessive manipulation of the swollen ear canal should be avoided, as should overtreatment. These may even replace an irritation due to infection with one due to the treatment itself. In allergic external otitis, steroids combined with antibiotics have proven effective. Antipruritic medication and relief from pain are extremely helpful while the basic cause is being treated. Irrigation of the ear with water frequently delays healing and encourages further infection, and should be avoided. Most patients with external otitis can be ambulatory and continue working while under treatment.

The External Ear

Impacted Wax (Cerumen).—Obstruction of the ear canal due to excessive wax (cerumen) is not uncommon. It is generally precipitated by the accumulation of excessive "dirt" in the ear canal. This may be dust or fibers of occupational origin. The dirt mixes with the cerumen and forms a hardened mass which tends to become enmeshed with the hairs in the ear canal, and, thus, is prevented from falling out by itself.

The wax-secreting glands are situated only in the skin covering the cartilaginous or outer part of the ear canal, and wax is formed only in this outer area. When wax is found impacted more deeply in the bony portion or against the eardrum, it has usually been pushed in there. If the ear canal becomes completely obstructed with wax, the patient experiences a hearing loss.

Interestingly enough, the patient with impacted cerumen often gives a history of sudden, rather than gradual, hearing loss, described as "suddenly going deaf." This experience may occur while chewing or while attempting to remove wax with a finger or some probe. It occurs because the wax is compacted into the narrower portion of the ear canal.

If cerumen becomes lodged against the eardrum, a rushing type of tinnitus is sometimes reported, or the patient may be aware of hearing his own heartbeat. The hearing loss is invariably accompanied by a feeling of fullness, and the loss is generally greater in the lower frequencies. The loss is most often less than 30 db and is rarely greater than 40 db.

A hearing loss should not be attributed merely to the presence of a large amount of cerumen in the ear canal until other causes are ruled out. Even if there is only a pinpoint opening through the cerumen, the patient may hear almost normally. The hearing loss does not become apparent until the ear canal is completely blocked. It is important, therefore, to perform air-conduction audiograms after removal of the wax to be certain that hearing has been fully restored.

If irrigation has been used to remove impacted cerumen, the ear canal should be dried afterward or the small amount of water remaining in the deep pit at the anterior inferior portion of the ear canal may cause a feeling of fullness as well as slight interference with hearing.

The prevention of this particular form of otologic disease consists in measures which will prevent the entry of excessive amounts of dirt or other irritating material into the ear canal. These might include the reduction of gross particulates in the air through better ventilation or housekeeping, improved personal hygiene, and the wearing of clean cotton plugs in the ear to screen out the foreign matter. On the other hand, the wearing of dirty ear plugs may actually introduce dirt into the ear canal and aggravate the condition.

The Tympanum

Ruptured Eardrum.—We speak of an eardrum being ruptured when it is suddenly penetrated by a foreign body like a piece of metal or by an explosion or slap across the ear. Otherwise, a hole in the eardrum is called a perforation rather than a rupture.

Usually a rupture is more irregular and is not immediately accompanied by signs of inflammation.

Most ruptured eardrums that are caused by penetrating objects occur in the posterior portion of the drum because of the curve of the external canal and the slope of the eardrum. Ruptures caused by sudden and intense pressure change, as by a blow to the side of the head or by an explosion, are more frequently in the anterior part of the drum and occasionally in the pars flaccida or upper part.

Whenever an explosion occurs in the proximity of the patient, in addition to a complete examination for other injuries, it is advisable to examine the eardrums in order that proper therapy can be instituted in the event that the eardrum has been ruptured.

The single treatment for ruptured eardrums is systemic antibiotics. Nose blowing should be avoided and, above all, there should be no unnecessary probing in the canal. The introduction of medications into the canal should be avoided to prevent entry of infection into the middle ear. In most instances, the rupture heals spontaneously if infection is prevented. If healing does not occur, it may be necessary to encourage healing by cauterizing the edges of the ruptured eardrum or to perform a myringoplasty at some future date.[2] The hearing loss may be as much as 60 db if the rupture was due to a force severe enough to impair the ossicular chain, but usually the hearing loss is less than 30 db and involves all frequencies.

Spark in the External Ear Canal.—Unfortunately, getting a spark into the external ear canal is not a rare experience in industry where employees are welding, grinding, chipping, or burning. The spark hits the eardrum with devastating effect. Usually, the entire drum is cleanly destroyed, leaving just the handle of the malleus hanging down. Little or no infection accompanies this very painful and traumatic accident.

The hearing loss is usually around 50 or 60 db and affects all frequencies. As in cases of ruptured eardrum, probing the canal and forceful nose blowing should be avoided. The eardrum does not regenerate by itself when such a large amount of it is destroyed. It is essential to graft a new eardrum using either skin or vein.

An accident of this nature is completely preventable. It is essential to protect individuals, not only with safety glasses, but with ear protection as well when they work in industrial areas where free sparks are produced. The use of insert protectors probably is the most advisable means of protection. Absorbent cotton in the ears is of minimal help and inadequate protection. In addition to providing relief from pain, it is important to avoid manipulation in the ear canal or irrigating with water. Antibiotics to prevent infection are essential.

Foreign Body in the Ear Canal.—Hearing loss and fullness in the ear are often the only symptoms produced by a foreign body in the ear canal. It is surprising how long a piece of absorbent cotton or other foreign matter can remain in an ear canal without the patient being aware of its presence. It is only when this foreign body becomes impacted with wax or swollen with moisture that fullness and hearing loss ensue, and medical attention is sought. The hearing loss is due to the occlusion of the canal; it is usually very mild and usually greater in the lower frequencies.

The variety of foreign bodies removed from ear canals ranges from rubber erasers to peas. Most of these cause enough ear discomfort to attract attention before hearing loss becomes prominent, but not always. Caution must always be observed when attempting to remove a foreign body from the ear canal. Usually, special grasping instruments are essential, depending upon the nature of the foreign body. General anesthesia may be advisable unless the foreign body is obviously simple to grasp and remove in one painless maneuver. It is easy to underestimate the difficulty in removing a foreign body and to run into unexpected problems; excessive preparation is better than too little.

The Middle Ear

Acute Otitis Media.—Acute otitis media is an ear infection of comparatively short duration, in contrast with chronic otitis media which persists continuously for many months. If an acute otitis media clears up and, then, because of a persistent anterior perforation in the eardrum, flares up again in a matter of months, this should be considered a recurrent otitis media and not a chronic one.

The common causes of acute otitis media are upper respiratory infections, sinusitis, hypertrophied adenoids, allergies, and improper nose blowing and sneezing. It is notable that all of these causes are external to the ear itself. The otitis media is secondary to a condition prevailing elsewhere in the body, and many of the causative infections are influenced or precipitated by environmental conditions.

Thus, irritant gases, vapors, or particulate matter in the air may cause irritation and inflammation of the nasopharynx which makes it more susceptible to invasion by ubiquitous pathogenic bacteria. Prolonged exposure to chilling climatic conditions may also contribute to a lowering of the resistance to infection, as may, also, nutritionally inadequate diets.

Many patients with anterior perforations whose ears have been dry for months, or even years, may get water in them, or blow their noses improperly during an acute rhinitis, and, in this way, the ear becomes reinfected. In such cases, the otorrhea is usually stringy and mucoid and comes from the area of the eustachian tube.

To prevent otitis media, patients should be cautioned to refrain from improper nose blowing and sneezing. Too forceful blowing while pinching both nares causes build-up of pressure in the nasopharynx; this pressure may force small amounts of infected mucus through the eustachian tube into the ear with resultant otitis media.

While excellent results have been achieved with medication in preventing otitis media, some cases still progress to chronic otitic infections. One of the principal causes for this is failure to use adequate doses of antibiotics. In many cases of middle ear infection, a much higher blood level of antibiotic is necessary than is generally recognized; the reason being that the infection has become walled off and can be reached only by very high blood levels of drug. If adequate antibiotics are prescribed to most patients with otitis media in the early stage, it can be cleared up before severe pain and disability develop. Early treatment can prevent discomfort and loss of work time for employees, and the employees can be treated on an ambulatory basis without loss of work. If the employee reports to the medical department at a time when his eardrum is bulging, a myringotomy is advisable and relief will be immediate. The use of intensive antibiotics along with oral decongestants and nose sprays are effective measures against otitis media.

Aerotitis Media.—The condition known as aerotitis media has been associated with work in caissons, with deep water diving, and with air travel. The condition is now associated most frequently with rapid descent in an airplane.

If changes in atmospheric pressure occur, equalization of pressure between the middle ear and the external ear occurs through the eustachian tube, but if congestion exists in the eustachian tube, equalization of pressure between the middle and outer ear does not occur. When the atmospheric pressure increases rapidly, as in the descent of an airplane, and the eustachian tube is blocked, the pressure on the outside of the tympanum exceeds that on the inside and the eardrum is pushed in towards the middle ear and the promontory of the cochlea. This causes sudden pain, fullness, and hearing loss. When the atmospheric pressure decreases rapidly as in the rapid ascent of an airplane or of a diver, and the eustachian tube is blocked, the higher pressure in the middle ear pushes the eardrum outward, resulting in tinnitus, vertigo, and hearing loss. In the event of a sudden decompression, hemorrhages and air emboli can occur in the middle ear as well as elsewhere in the body.

The prevention of aerotitis media is best accomplished by the avoidance of rapid changes in pressure on the part of those suffering from congestion of the eustachian tubes because of upper respiratory infection or allergy. If such pressure changes cannot be avoided, the use of oral decongestants and nasal sprays or inhalants may produce a temporary decongestion of the eustachian tube and facilitate equalization of pressure between the middle and outer ears.

If aerotitis media has occurred, vigorous chewing movements may induce a temporary opening of the eustachian tubes and bring relief. If the patient is seen soon after the symptoms develop, he can be relieved immediately by a myringotomy followed by politzerization. In all cases, hearing should return to normal when the eardrum is restored to its customary position.

Allergic Manifestations.—Sometimes, a clinical picture of otitis media is the result of an allergic reaction. Lewis [3] and Proetz [4] describe cases of otitis media attributed to food allergies which apparently produced edema of the middle ear. Edema of the eustachian tubes is not an uncommon finding in allergic rhinitis or hay fever and may also simulate the symptoms of otitis media.

The Inner Ear

Acoustic Trauma.—In order to differentiate sudden hearing loss due to accidental and brief exposure to intense noise from the gradual loss caused by prolonged exposure over many months and years, we restrict the term acoustic trauma to the former and call the latter noise-induced or occupational hearing loss.[6]

In acoustic trauma, the patient is usually exposed to a very intense noise of short duration like an explosion or rifle shot. This causes immediate hearing loss accompanied by fullness and ringing in the ear. If the cause is an explosion, there may also be a rupture of the eardrum and disruption of the ossicular chain. If this occurs, generally, a conductive hearing loss is immediately caused without much serious nerve damage developing because the middle ear defect now serves as a protection for the inner ear.

Following acoustic trauma to the inner ear, the patient usually notes that his fullness and ringing tinnitus subside, and his hearing improves. Generally, the hearing returns to normal. Most of us have been exposed to gunfire at one time or another and experienced some temporary hearing loss and then have had our hearing return to normal. In some cases, however, a degree of permanent hearing loss remains. The amount of loss depends on the intensity and duration of the noise and the sensitivity of the ear. Usually, the permanent loss is very mild and consists only of a high-tone dip. If the noise is very intense and the ear is particularly sensitive, the loss may be greater and involve a broader range of frequencies. The milder cases of hearing loss involve only one ear, usually the one closer to the gun or the source of the noise. If the noise is very intense, and the hearing loss moderate, then, usually, both ears are affected to an almost identical degree or perhaps one slightly more than the other. It is hardly possible (as a result of exposure to intense noise) to have a nerve deafness in one ear greater than 60 db in all frequencies with normal hearing in the other ear.

Because there is almost always some degree of temporary hearing loss or fatigue in acoustic trauma, the amount of permanent damage cannot be established until several months after the exposure. In the interim, the individual must be free of other exposure to intense noise that might aggravate the hearing loss. The audiometric patterns in acoustic trauma are similar to those in noise-induced hearing loss, but the history is different and, probably, the manner in which the permanent hearing loss is produced is also different.

After the temporary hearing loss subsides and only permanent loss remains, the hearing level stabilizes and there is no further progression in the hearing loss, according to most investigators.

Noise-Induced Hearing Loss.—If acoustic trauma is an otologic disease of acute onset, noise-induced hearing loss is its chronic counterpart. It has been known for many years that prolonged exposure to high levels of environmental noise can cause a permanent hearing impairment.[1]

Thompson noted that hearing loss caused by prolonged exposure to noise of high pitch was common among loom tenders, spinners, railway engineers, and might even occur in telegraphers. The reverberating noise experienced by boiler makers and structural riveters had resulted in the syndrome named, by the workers themselves, "boilermaker's deafness." It was known that the cause of this kind of hearing loss was due to damage in the inner ear and that the loss was more pronounced for high than for low notes. It was known, also, that high-pitched, rather than low-pitched, noises were most often responsible for the hearing loss. It was known that the hearing loss immediately after exposure to the noise was greater than after a period of rest away from the noise.

What is most interesting, however, is the attitude of the time toward this loss. Thompson states: "Unfortunately, there seems to be no remedy for this hazard, for not alone the ears, but the temporal and other cranial bones are set in acute vibration, and, if a man must work inside a boiler or a gun turret, he has to accept the consequences."

Since 1914, a great amount of quantitative information has been developed about noise-induced hearing loss. This has occurred: (1) because of the greatly increased noise levels of many occupational and nonoccupational environments and, hence, the increasing magnitude of the problem; and (2) because of the evolution of a social attitude which no longer agrees that a worker exposed to high noise levels should have to accept the consequences.

It is not enough to say that prolonged exposure to high noise levels results in permanent hearing loss. We have to define: how long is prolonged; how high is high; noise levels with respect to the sound frequency spectrum; noise levels with respect to continuity or discontinuity of exposure; what we mean by hearing loss, in operational terms.

All of the answers are not in, by any means. The American Academy of Ophthalmology and Otolaryngology, however, published through its Subcommittee on Noise in Industry, in 1957, a "Guide for Conservation of Hearing in Noise" which gives the following summary of what is generally accepted by the medical profession[7]:

Noise-induced hearing loss depends upon noise levels and exposure time. Any attempt to assess the need for hearing conservation must take account of both.

The effect of continuous exposure to steady noise may depend on the way the sound energy is distributed in the noise. Early noise-induced hearing losses are usually confined to the frequencies around 4000 cycles per second. As the exposure lengthens the losses spread to lower frequencies whose audibility is more directly involved in the understanding of speech. Data on noise-induced hearing loss, both temporary and permanent, indicate that the losses occur at frequencies above those that characterize the exposure sounds. Since the most important frequencies to be protected are in the range 500 to 2000 cps inclusive, it follows that the 300-600 and 600-1200 cps bands deserve our major attention if we are trying to protect man's hearing for speech.

At the present time our knowledge of the relations of noise-exposure to hearing loss is much too limited for us to propose "safe" amounts of noise-exposure. We can, however, point to certain noise levels that indicate when it is advisable to initiate conservation of hearing programs. These levels will not be general because a different leve is needed for different types of noise and different schedules of exposure. The hearing conservation level that we now specify tentatively applies only to years of exposure to broad-band steady noises with relatively flat spectra. *It does not apply to short exposures, and above all, it does not apply to impact noises or narrow band noises. This tentative hearing conservation level is stated as follows:*

If the sound energy of the noise is distributed more or less evenly throughout the eight octave bands, and if a person is to be exposed to this noise regularly for many hours a day, five days a week for many years, then: if the noise level in either the 300-600 cycle band or the 600-1200 cycle band is 85 db, the initiation of noise-exposure control and tests of hearing is advisable. The more the octave band levels exceed 85 db the more urgent is the need for hearing conservation.

It is apparent that the above guide defines the limits of only one kind of environmental noise exposure among many possible kinds that might produce a hearing loss. Efforts continue to expand our knowledge of the limits of other kinds of noise exposure which might produce hearing loss.[8,10]

The effects of noise on hearing are of two kinds, temporary and permanent. The temporary effect is known as temporary threshold shift (TTS) and is experienced immediately after exposure to an intense noise. Recovery begins within minutes after the exposure ceases but may not be complete for hours, days, or even months under certain circumstances. If a person is to be put into a noisy environment and it is anticipated that he will thereafter be exposed to this environment many hours a day for many years, it is believed that the temporary threshold shift experienced at the beginning of the exposure can be used to predict the magnitude of the permanent threshold shift that may be obtained after many years of exposure.

Environmental noise produces its effects in the organ of Corti in the inner ear.[11] Sound-induced motion of the fluid in the cochlea induces shearing and bending movements of the hair cells in the organ of Corti which, in turn, result in electrical stimuli transmitted by the auditory nerve. Prolonged and excessive noise eventually produces deterioration and, finally, destruction of the hair cells, and thus disrupts the sound transmission mechanism. The damage occurs first to hair cells associated with the perception of frequencies higher than 2,000 cps, and thus does not interfere with the perception of speech.

Later, the damage spreads to lower frequencies in the auditory range and loss for the hearing of speech becomes apparent.

From the social point of view, the preservation of hearing for speech is of the greatest importance, and in most workman's compensation situations, compensation is paid only for hearing loss in the speech frequencies, 500, 1,000, and 2,000 cps. The fact that initial noise-induced hearing loss occurs at frequencies above 2,000 cps provides a powerful tool for the detection of the condition and for preventive action before the loss spreads to the speech-hearing frequencies.

The pure tone air-conduction audiogram, if taken properly, provides an accurate map of hearing acuity for a range of selected frequencies. The frequencies monitored by audiometry should cover the range 500-6,000 or 8,000 cps. Early noise-induced hearing loss is usually centered about the 4,000 cps frequency, and, thus, is detected as a 4,000 cycle (or thereabouts) dip or notch. A deepening and spreading of this notch as exposure continues is predictive of eventual loss in the hearing frequencies, and should trigger the initiation of a hearing conservation program.

Noise-induced hearing loss, excluding temporary threshold shift, is irreversible. It cannot be cured, therefore, it must be prevented. One method of prevention is the reduction of the environmental noise level to safe limits, if this be possible. If it is not possible, the individual ears must be separated from the environmental noise by sound-attenuating barriers: (1) at the sound source; (2) somewhere between the individual and the sound source; or (3) at or in the ears (ear muffs or ear plugs). Reduction of the time of exposure (intermittency) also helps in the prevention of hearing loss.

Conservation of hearing programs in industry are no longer uncommon and have proved highly effective in the prevention of occupational hearing loss.[12]

Meniere's Disease.—One could hardly call Meniere's disease an environmental one; however, the attacks of dizziness so often present in this condition are certainly of great concern to industry. It is necessary, therefore, that we have an understanding of Meniere's disease to be able to diagnose it and differentiate it from other conditions causing vertigo and deafness.[13]

The characteristic findings in Meniere's disease are generally clear cut. The typical history is marked by a sudden onset of dizziness, hearing loss and a seashell-like tinnitus in one ear. The room seems to spin or the patient feels himself to be spinning. Everything goes around, and the tinnitus is severely aggravated during the attack. Sometimes, nausea and even vomiting may occur. These symptoms may last for a period of minutes or several days and then disappear. The same symptoms may then recur at varying time intervals. After several attacks, the hearing loss and tinnitus may persist. The hearing loss may not be severe, but voices begin to sound tinny and muffled on the telephone, and it becomes difficult for the patient to follow a conversation because of inability to distinguish between different words that have related sounds. In addition, there is distortion of sound and a reduced threshold of discomfort for loud noises.

The vertigo is of a specific type that involves some sort of motion. It is "subjective" when the patient has a sensation of moving or "objective" when things move about the patient. Usually, the motion is described as rotary, especially during the acute attack. Occasionally, between acute attacks, patients have a mild feeling of motion whereby they seem to fall to one side and can not keep their balance. Less often, there is a strange up-and-down or to-and-fro motion. Other types of sensation such as a feeling of faintness or weakness or seeing spots before the eyes, or just vague "dizziness" should not be attributed necessarily to Meniere's disease. There must also be some evidence of hearing loss being, or having been, present for a diagnosis to be Meniere's disease. It is true that in the very early phases, hearing may be lost only during the acute attack and that it may be normal between attacks, but even in such cases, the patient will recall having experienced deafness, fullness, and roaring in at least one ear. If these subjective symptoms are absent and no hearing loss is present, a diagnosis of Meniere's disease should be made with utmost caution, and further studies should explore the possibility of some other likely cause for the dizziness.

Patients are more frequently concerned about the vertigo and tinnitus than about their hearing loss. For an individual who works on scaffolding or drives a car or truck, a sudden unexpected attack of vertigo beyond his control is indeed a serious problem. These patients are invariably apprehensive and gravely concerned. As a matter of fact, nervous tension is a most prominent symptom in patients with Meniere's disease. Most of them seem to be much improved and less subject to attacks when they are relaxed and free of tense situations. People with Meniere's disease generally should not undertake tasks where a sudden attack of dizziness could cause injury to themselves or others.

The ocean-roaring or seashell-like tinnitus is also a matter of grave concern. When individuals are in quiet surroundings or when they are under tension, the tinnitus may become so alarming to them that they are often willing to undergo any type of surgery, even if it means the loss of all their hearing in the affected ear, as long as the noise can be made to disappear. As yet, we have no specific, reliable procedure to control tinnitus and consistently preserve hearing.

The hearing loss generally is not as disturbing to the patient as the other two symptoms, but when it happens to a housewife or an executive who uses the telephone frequently, it becomes an important issue.

The audiologic findings in a typical early case of Meniere's disease show low-tone hearing impairment with reduced bone conduction. If the bone conduction were not reduced, it would be conductive instead of sensory hearing loss. There is no air-bone gap since both air conduction and bone conduction are reduced to the same degree. The patient's ability to discriminate in the bad ear is so reduced that he can distinguish only 40% of the speech that he hears at ordinary levels of loudness. Furthermore, if speech is made louder, (from 50 to 70 db) the patient distinguishes even less (contrary to expectations). This brings out one of the most important features of Meniere's disease; distortion. Distortion explains many of the symptoms; not only the reduced discrimination but the tinny character that speech assumes in the patient's ears and his ability to discriminate between words on the telephone.

Another interesting phenomenon in patients with Meniere's disease is called a lower threshold of discomfort and is manifested by the patient's complaint that loud noises bother him. This difficulty is partially related to distortion but principally to the phenomenon of recruitment. Recruitment is a telltale audiologic finding in the diagnosis of inner ear deafness and particularly Meniere's disease. It stems from an abnormally rapid increase in the sensation of loudness and, when present to a marked degree, permits us to classify a case with reduced bone conduction as being sensory and to localize the damage to the inner ear. In the absence of recruitment, the localization is uncertain and the condition must be considered sensorineural.

Meniere's disease is believed to be due to a disturbance which increases the quantity of fluid in the cochlea. It is not obviously of environmental origin except for a relatively few cases that have been demonstrated to be due to allergy. Atkinson [14] found that a minority of the patients with Meniere's disease in his series showed a positive reaction to an intracutaneous test with histamine, which he believed differentiated them from those whose Meniere's disease was on a nonallergic basis. Other authors have reported cases of Meniere's disease due to specific food allergies.[15]

Direct Head Trauma.—While it may be stretching a point to consider traumatic head injury as environmental in origin, it must be conceded that the environment in which men work, and particularly travel by automobile, is often conducive to impact injuries. A share of these result in blows to the head which may, in turn, have otologic manifestations.

Conductive type of hearing loss following head injury may be due to a simple disruption of the ossicular chain without damage to the eardrum or the rest of the middle ear. This is rare. More often, the conductive hearing loss is due to a longitudinal fracture of the temporal bone which causes bleeding into the middle ear and a tear in the eardrum. Occasionally, the ossicular chain is also disrupted in this type of injury and there may also be a temporary facial paralysis and a cerebrospinal otorrhea. In longitudinal fractures, x-rays show a fracture line extending in the middle cranial fossa from the outer ear

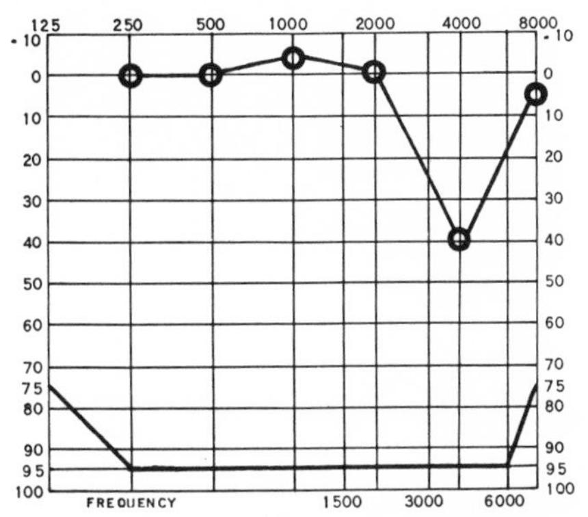

Fig 1.—Dip at 4,000 cps due to direct head trauma.

inwards towards the foramen magnum. This line is parallel to the superior petrosal sinus. We should also note that bone conduction is also reduced. This can be due to the blood in the middle ear (hemotympanum), in which case, the bone conduction returns to normal when the pathology resolves or it may be due to damage to the inner ear, cochlear concussion, which may resolve in part or remain permanently affected.

A sensorineural type of hearing loss due to direct trauma can occur even without evidence of bone fracture. However, it usually accompanies a longitudinal temporal bone fracture. Interestingly enough, a sensory hearing loss may be found in the ear on the

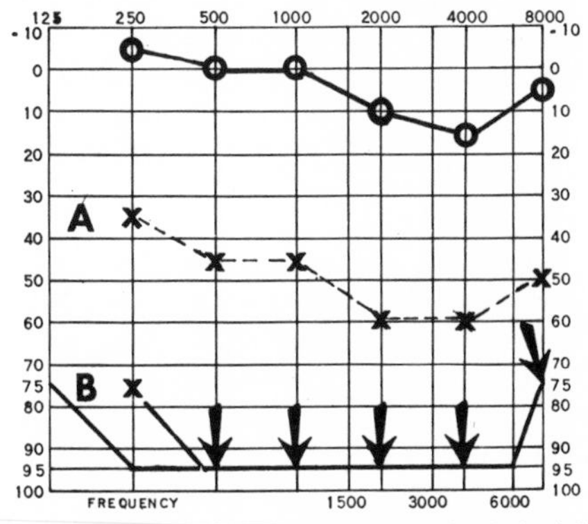

Fig 2.—Curve *A* indicates thresholds of the left ear without masking noise in the right ear (shadow curve). Curve *B* shows that, with the right ear masked, there is no hearing in the left ear for the output limits of the audiometer.

	Labyrinthine Concussion	Longitudinal Fracture	Transverse Fracture
Bleeding from ear	Never	Very common	Rare
Injury to ext auditory canal	Never	Occasional	Never
Rupture of drum	Never	Very common	Rare, commonly a hemotympanum
Presence of cerebrospinal fluid	Never	Occasional	Occasional
Hearing loss	All degrees, partial to complete recovery	All degrees, combined type, partial to complete recovery	Profound nerve type, no recovery
Vertigo	Occasional, mild and transient	Occasional, mild and transient	Severe, subsides; nystagmus to opposite side
Depressed vestibular function	Occasional, mild	Occasional, mild	No response
Facial nerve injury	Never	25%, usually temporary	60%, often permanent
X-ray signs	Negative	25%, in squamous and mastoid area	60%, in occiput in pyramid in vertex-mento films

side of the head opposite to that where the injury was sustained (contrecoup). The hearing loss has practically the same characteristics as that produced by exposure to intense noise. If the head injury is comparatively mild, only the hair cells of the basilar end of the cochlea are affected and the hearing loss is like that shown in Fig 1. If the injury is more severe, a greater area in the cochlea is affected and even the nerve itself may also be damaged.

Another cause of sensorineural deafness is transverse fracture of the temporal bone. The fracture line on x-ray is perpendicular to the superior petrosal sinus and the long axis of the temporal bone. A severe blow to the back of the head can produce this type of fracture. The accompanying hearing loss is generally very severe and often total. It is caused by fracture into the vestibule of the inner ear and destruction of the cochlea. Blood fills these areas and can be seen through an intact eardrum (in contrast to the torn eardrum caused by longitudinal fracture).

In many cases, the facial nerve is also damaged, causing a complete facial paralysis including the forehead on one side. Cerebrospinal fluid may also fill the middle ear and drain out of the eustachian tube especially if the eardrum is intact. Vertigo and nystagmus are invariably present after a transverse fracture of the temporal bone, and the hearing loss is generally permanent.

A purely neural type of hearing impairment can occasionally occur with transverse fracture of the temporal bone when the fracture includes the internal meatus and crushes the auditory nerve. Facial paralysis is also present, generally, but may clear up.

The inner ear is so well protected that a blow to the head must be quite severe to produce hearing loss. When there is evidence of fracture, there is little question about the severity of the blow. When there is no visible fracture, however, and hearing loss is present, it may be difficult to determine, in some cases that come to litigation, whether the hearing loss was caused by the injury or was previously present. In such cases, it is well to recall that if the patient did not exhibit any period of unconsciousness, the chances are that the blow was not severe enough to produce cochlear concussion with significant permanent hearing loss. Vertigo and tinnitus must be associated with such damage and both are likely to persist for extended periods after the injury.

We must also bear in mind that a portion of the conductive and sensorineural hearing loss resulting from trauma may be reversible and that it is necessary to wait at least several months before appraising the degree of permanent hearing loss.

A few cases have been reported in which the patients did not become aware of their hearing loss until several weeks after the injury, and the hearing loss seemed to progress. This type of loss is extremely unusual and, while it is possible, no proof is available that the injury is the sole etiology.

It is unjustifiable to assume that, merely because a patient sustained a head injury, any hearing loss present was caused by the injury. If the hearing impairment is conductive, there must be evidence of damage to the eardrum, or the audiologic findings must indicate some ossicular chain defect. This latter can be very readily confirmed by elevating the eardrum and examining the ossicular chain. If the hearing loss is sensory or sensorineural, then the audiological findings must fit the characteristic patterns that have been established for these types of hearing impairment. Fig 2 shows a case in which an individual claimed a hearing loss as the result of a head injury. Because this was a subtotal unilateral loss with normal hearing in the other side, every effort was made to try to explore this case very carefully. It was found, after much investigation, that the patient had had mumps many years ago and that his hearing loss was due to mumps labyrinthitis rather than to the injury. Upon establishing a better understanding with the patient, he freely admitted the situation.

Schuknecht has crystallized the differential diagnosis of hearing loss due to head injuries.

Presbycusis.—Presbycusis is that loss of hearing acuity which is associated with aging.[16] It is found in the general population of those who are not specifically exposed to noisy job environments. It is not at all clear, however, that some presbycusis is not the result of lifetime exposure to the general noise level of the environment. In any event, the average person shows no loss of hearing for speech in early presbycusis. The losses are first evident in frequencies greater than 2,000 cps and tend to become more severe with increasing frequency. In calculating compensation for noise-induced hearing loss, it is often customary to make allowance for that part of the loss due to presbycusis.

Drugs.—Certain drugs, particularly of the mycin series, and certain antibiotics have been known to produce, as side effects, injury to the auditory nerve. This has resulted in permanent hearing loss—in some patients almost total—and should be considered in the prescription of these drugs.

The Environment and Conservation of Hearing

The historic concern of medical science for the prevention as well as the cure of disease extends to the conservation of hearing. Since hearing loss is easier to prevent than to cure, it is fitting that great emphasis be placed on conservation of hearing programs.

In recent years, a great amount of work has been devoted to the problem of understanding and preventing noise-induced hearing loss, a phenomenon of environmental origin and of increasing importance as the noise levels of working environments and, indeed, of the general living environment increase. We have indicated, however, that there are other causes of hearing loss, also associated with the environment, which should not be neglected. For this reason, we include, in this section, a brief discussion of classification of hearing impairment and related causes.

The phenomenon of hearing involves an auditory pathway, along which vibrations in the air are transformed into mechanical vibrations in a bony chain, thence to waves in a fluid, thence to electrical impulses in the auditory nerve. A disruption of the pathway at any point will result in a hearing loss. If a pathologic condition occurs in the external ear canal, the middle ear, or the eustachian tube, it may produce interference with the mechanical vibrations of the tympanum or the ossicular chain leading to the oval window. Hence, the hearing loss produced, if any, is called a conductive hearing loss. If the pathologic condition is in the cochlea, auditory nerve, or higher brain centers, the interference will be in the sensory or neural components of the auditory pathway, or a mixture of the two, and the resultant hearing loss is called sensorineural.

The common causes of conductive hearing loss are rupture of the eardrum due to explosion or traumatic injury by a spark:

aerotitis media due to a pressure differential between the external and middle ear; infections of the external and/or middle ear; impacted cerumen; and otosclerosis. As has been shown, all but the last of these can be of environmental origin and are amenable to good hygiene measures.

In sensorineural hearing impairment, the damage lies medial to the stapes footplate, ie, in the inner ear or auditory nerve. The otologist is now able, in many instances, to distinguish the specific site of damage in the various cases of sensorineural hearing loss. Common causes of sensorineural hearing impairment are: noise-induced deafness, acoustic trauma, presbycusis, direct head trauma, eighth nerve tumor, drug reactions, Meniere's disease, and certain systemic allergies and infections. Mixed hearing loss, having both conductive and sensorineural components, can also occur.

While there are many distinguishing features of conductive and sensorineural hearing impairment, the most widely applicable tool for distinguishing between them is bone conduction audiometry. In conductive hearing loss, the bone conduction audiogram is usually good and much better than the air conduction audiogram. In sensorineural deafness, the audiometric thresholds by air conduction and by bone conductions are at approximately the same level.

In brief, there are many causes of hearing impairment which can be associated with the environment in which man lives and works. The physician must be aware of, and understand, all of them if he is to carry out an intelligent program for the conservation of hearing.

REFERENCES

1. Thompson, W. G.: Occupational Disease, New York: D. Appleton-Century Company, 1914.
2. Shambaugh, G. E., Jr.: Surgery of the Ear, Philadelphia: W. B. Saunders Co., 1959.
3. Lewis, E. R.: Ann Otol 38:185, 1929.
4. Proetz, A.: Ann Otol 40:67, 1931.
5. Jones, M. F.: Ann Otol 47:910, 1938.
6. Davis, H.: Hearing and Deafness, S. R. Silverman, ed., New York: Holt Rhinehart & Winston, Inc., 1960.
7. Subcommittee on Noise in Industry: Guide for Conservation of Hearing in Noise, Los Angeles: American Academy of Ophthalmology and Otolaryngology, 1957.
8. Eldridge, D.: Armed Services Technical Information Agency, No. AD245980, 1960.
9. Reference deleted.
10. Guild, S.: J Lab Clin Med 4:153-180, 1919.
11. Sataloff, J.: Industrial Deafness, New York: McGraw-Hill Book Co. Inc, 1957.
12. Schuknecht, H. F.: Meniere's Disease, Trans Amer Laryng Rhinol Otol Soc 1963.
13. Atkinson, M.: Observations on Etiology and Treatment of Ménière's Syndrome, JAMA 116: 1753-1760, 1941.
14. Urbach, E., and Gottlieb, P. M.: Allergy, 2nd ed, New York: Grune & Stratton, Inc., 1946.
15. Schuknecht, H. F.: Presbycusis, Laryngoscope, 65:404-419, 1955.

LOUDNESS ADAPTATION AS A FUNCTION OF FREQUENCY, INTENSITY, AND TIME

WARREN H. TEICHNER AND ERNEST SADLER

A. INTRODUCTION

The decline in loudness of pure tones under sustained stimulation has received considerable attention in the past few years. Although they have not supported their conclusions with specific experimental citations, some reviewers (9, 17) have noted that loudness probably does not adapt with continued stimulation.

In 1929, von Bekesy measured the decrease in loudness of tones of various frequencies and intensities and found that it varied greatly as a function of the intensity of the stimulating tone but bore only a slight relationship to frequency. Presenting a long-continued tone to one ear and a second tone of 0.2-second duration to the other ear just after the first tone ended, von Bekesy instructed the subjects to adjust an attenuator so as to produce the same loudness in the second tone as the terminal portion of the first one. He reported that as the duration of the first tone was increased, the intensity of the second tone had to be decreased. Because of the extremely short duration of the second tone, it seems unlikely that the subjects actually manipulated an attenuator to equate the two tones for loudness. Rather it would seem that this manipulation was done by the experimenter and that the subjects gave verbal or other signalling responses. To permit recovery between test runs, von Bekesy found it necessary to follow each presentation with a quiet period of 3 to 8 minutes depending on the duration of the first one.

In 1930, using essentially the same technique, Wood (16), as described in Geldard, also found that as the duration of the first or standard tone increased, the intensity level of the comparison tone necessary to provide a loudness match decreased. Hood (4), in 1950, described a technique permitting successive measurements which differed from that of both von Bekesy and Wood in that the two tones were on simultaneously. In Hood's method, simultaneous tones were first presented to each ear for a brief period (10 to 20 seconds). The intensity of one tone, the standard, was fixed, while the intensity of the other tone, the comparison, was variable. The subject attenuated the comparison stimulus until the two tones appeared equally loud. Two or three matches were made in this manner. Then without interruption the standard tone alone was continued while the comparison tone was presented only at periodic intervals for short durations during which the subject again equated the tones for loudness. The difference between the intensity level required for a loudness match initially and the intensity level required after the standard tone had been on for some time was taken as the amount by which the response of the test ear had declined under sustained stimulation. This decline has been called "perstimulatory fatigue" or "adaptation" by various investigators.

Using this procedure, Hood measured adaptation for various intensities and frequencies. He concluded that adaptation was essentially complete in about 3.5 minutes, irrespective of intensity, and that the maximum amount of adaptation increased as a function of the intensity of the standard tone. At high intensities, adaptation amounted to as much as 50 db.

In view of the considerable degree of adaptation shown by Hood's results, Egan (2) attempted to determine if the amounts of adaptation were a function of the method used to equate the tones for loudness. He compared an equal-loudness method, in which the subject was instructed to make the tones equally loud, and the median-plane localization method, in which the subject was told to adjust the sound until only one tone was heard, this being localized in the median plane. He had subjects match identical frequencies and also match frequencies that differed from one another by 5 to 200 cps. The frequency difference was expected to prevent the subject from localizing the sound in the median plane, and thus force him to make a true equal-loudness balance. Egan concluded from the results that the two methods did not differ in the amount of adaptation measured. He also found greater amounts of adaptation for higher frequencies.

Egan and Thwing (3) found that considerably less adaptation was measured by von Bekesy's method, in which the comparison tone comes on after the standard tone has ceased, than by Hood's method, in which the loudness balance is made while both tones are on.

Jerger (6), using the median-plane localization method, extended the study of adaptation to pure tones by using a wider range of frequencies and intensities than had been used before. In view of Egan's (2) results, it is of some interest that at high frequencies (2000 to 8000 cps) a number of Jerger's subjects were unable to localize the stimulus in the median plane during the period of sustained stimulation. In these cases, they were told to make an equal-loudness match. Jerger found that the time at which maximum adaptation occurred was a function of both frequency and intensity. At frequencies up to 500 cps, maximum adaptation seemed to be reached after 3 minutes of stimulation, but in the range from 1000 to 8000 cps, only adaptation with the very low intensities showed any tendency to stabilize at a maximum value. At the higher intensities, 70 to 90 db, adaptation continued to increase even after 5 minutes. This result is not in agreement with Hood's conclusion that the point of maximum adaptation is reached at about 3.5 minutes, irrespective of intensity.

In another study by Jerger (1956), adaptation was measured for a 5000 cps tone at 80 db over a 10-minute period of sustained stimulation. The point of maximum adaptation was not reached until the 7th minute, a finding in close agreement with Carterette's (1) results for adaptation to sustained thermal noise, and removed from Hood's (4) finding of 3.5 minutes.

There is also wide variation as to the maximum amount of adaptation which occurs. Most investigators (2, 3, 6, 13, 14) reported between 20 to 25 db while Hood (4) reported up to 50 db and Palva (10), using a von Bekesy type attenuator, only 8 db.

Small (11) points out that a likely cause of these wide differences lies in the wide variation in psychophysical procedures which have been used as well as in other methodological techniques. The purpose of this study, therefore, was to reinvestigate the question by designing an experiment intended to overcome possible artifacts resulting from procedures of previous studies.

Perhaps the most basic point among studies reporting adaptation is the question of whether the subject actually matched the tones for loudness or whether his response was due to some other kind of judgment. Egan (2) found no difference in amount of adaptation reported when the subject was instructed to match loudness and when he was instructed to localize in the median plane. However, Jerger's (6) finding that some of his subjects were unable to localize in the median plane suggests the possibility of ambiguous reporting in at least some of Egan's subjects. The finding that subjects were able to match tones of different frequency suggests that they can make a loudness match, but not that they actually do so. Further indication that the subject is in fact matching loudness might be gained by requiring the subject to make judgments of loudness ratios as well as equal loudness judgments.

If he is able to do both and if both show adaptation, then considerable support might accrue to the genuineness of loudness as his judgmental response and to the adaptation phenomenon as well. This was the technique of the present study, to use equal-loudness and twice-as-loud judgments.

A second point concerns the similarity of the standard and comparison tones during the period when the subject makes his judgment. It would seem reasonable that the greater the physical difference between these two, the more difficult the subject's judging task would be. Thus, the most acceptable procedures would seem to be those which minimized the differences between tones. The two tones should not only be identical in frequency, but also as nearly alike as possible in both physical and apparent intensity. Thus, the common procedure of presenting a highly attenuated comparison tone may permit the possibility of confounding physical and subjective intensity in the subject's judgment at the instant of stimulation as well as successively during the period of adjustment. If adaptation is a function of intensity then this procedure would have the advantage of minimizing adaptation of the test ear during the adjustment period. On the other hand, it has a disadvantage in that it requires the subject to compare successive pairs of tones of widely different subjective intensity. Moreover, one might expect that the time required for the adjustment would also be longer than if the tone were presented at its original intensity and to the extent that adaptation depends on time the test ear itself might adapt considerably. In fact, adjustment times of 15 seconds have been the most widely used. Related disadvantages of this procedure are that the subject is likely to overshoot in making his adjustment and if a step-attenuator is used, as has usually been the case, possibly important amounts of overshooting are difficult to avoid. Further, rather than acting as a control, the procedure of randomly attenuating the comparison tone before each match (2, 3, 6, 13) might complicate the situation by introducing a variable reaction time in the subject's judgment (12).

Little use has been made of the procedure of presenting the subject with a comparison tone identical in intensity (and frequency) to the standard. The primary objection to this procedure is that if adaptation is a function of intensity, such a procedure might be expected to lead to more rapid adaptation of the test ear; and, consequently, unless very rapid judgments could be made, this should result in less apparent adaptation than would be obtained by presenting an attenuated comparison tone. However, such a procedure might permit extremely rapid judgments, and if the subject used a continuously variable attenuator he could make rapid adjustments of the comparison tone. The latter consideration led to the adoption of this technique in the present experiment.

Another important experimental point concerns the difference in sensitivity of the two ears. As far as can be determined, no previous investigator has reported using subjects with ears matched in sensitivity. The usual procedure has been to use the difference in intensity obtained from an initial loudness match and one obtained after a period of continued stimulation. This procedure is acceptable only on the assumption that ears of different sensitivity adapt along the same time function. In view of the disagreement among studies reported, a safer procedure would appear to be to select subjects, in advance, with equally sensitive ears. This procedure, which was taken in the present study, permitted direct use of the difference between the standard and comparison ears at any point in time.

Also of importance is the duration of the recovery interval between stimulations. This has varied from one experiment to another, from 20 seconds to 8 minutes depending on the duration of the standard tone. Because of the uncertain nature of loudness adaptation, the length of the interval required for recovery would appear to be critical. Some consideration was given to this problem by running a preliminary experiment in which subjects were placed under the combinations of frequency and intensity used in the present study. Stimulation periods were 0.25, 0.50, 1.0, 2.0, and 5.0 minutes and the corresponding recovery times were 1.0, 1.0, 3.0, 5.0, and 10.0 minutes. The results obtained under these conditions did not differ from the results obtained when the recovery times were equal to the stimulation period plus 30 seconds. Therefore, times equal to the stimulation period plus 30 seconds were used in the present study.

B. Method

1. *Subjects*

The subjects were four female undergraduate students. These students were selected from 35 volunteers. The basis for selection was an agreement between the two ears within 2 db at absolute threshold of the frequencies used in the experiment. The subjects were paid for their participation. These subjects had previously participated in an experiment of 3-month duration which involved continuous binaural stimulation and had also been given preexperimental practice in the judgments required for the present study. Thus, they were considered to be highly skilled subjects at the start of the present study.

2. *Apparatus*

The signal was generated by a Hewlett-Packard audio oscillator and led to a step-up audio transformer. At this point the signal was divided into a standard and comparison channel. The channel for the standard tone went to the experimenter's continuously variable attenuator (General Radio), and from there directly to one earphone. The comparison channel was passed through an Atcotrol timer to a 2 db step attenuator (General Radio), then to the subject's continuously variable attenuator (General Radio). At this point it was further divided with one channel going to the subject's other earphone and another channel to a General Radio Graphic Level Recorder. PDR-10 earphones, mounted in MX-41/AR cushions, were matched in impedance; and all impedances within the system were appropriately matched. Calibrations were made with a Hewlett-Packard vacuum-tube voltmeter, the Graphic Level Recorder, and a General Radio Sound Level Meter. The subjects sat in a sound-absorbent chamber; the experimenter monitored the experiment from an outside control area permitting observation through a one-way window.

3. *Procedures*

Prior to each experimental session, the subject was required to rest in the sound room for 5 minutes before donning the earphones. The signal for donning the earphones and for other phases of the session was a silent flashing light controlled by the experimenter.

Two frequencies, 3500 cps and 5000 cps, were used, each at two intensities, 85 and 95 db, re 0.0002 dynes per square centimeter. Each of these four combinations was presented to the standard ear for 0.25, 0.50, 1.0, 2.0, and 5.0 minutes, and then was also presented to the test ear for judgment. During judgment periods, therefore, both ears were stimulated simultaneously. For each of the 20 intensity-frequency-time combinations, the subject made an equal-loudness and a twice-as-loud judgment, setting the attenuator as rapidly as possible. These 40 observation conditions made up an experimental series. The experimental series was replicated three times for each subject, with each replication consisting of a different random presentation of the combinations. Sessions were limited to one hour, with no more than two sessions for one subject in any one day. A complete replication took approximately five sessions.

The following is an example of an actual run: a 5000 cps tone at 95 db was transmitted to one ear (standard tone). After 2 minutes the same tone was also transmitted to the other ear (comparison tone). Immediately upon hearing the comparison tone, the subject, by means of his attenuator, manipulated the intensity level of the comparison tone until it sounded equal in loudness to the standard tone. Upon completion of the match, both tones were discontinued by the experimenter. A continuous graphic record was kept of the subject's attenuator settings. The interval between runs was equal to the period of stimulation by the standard tone plus 30 seconds.

C. Results

The amount of adaptation in db, re 0.0002 dynes per square centimeter, for the equal-loudness matches, was obtained as the difference between the

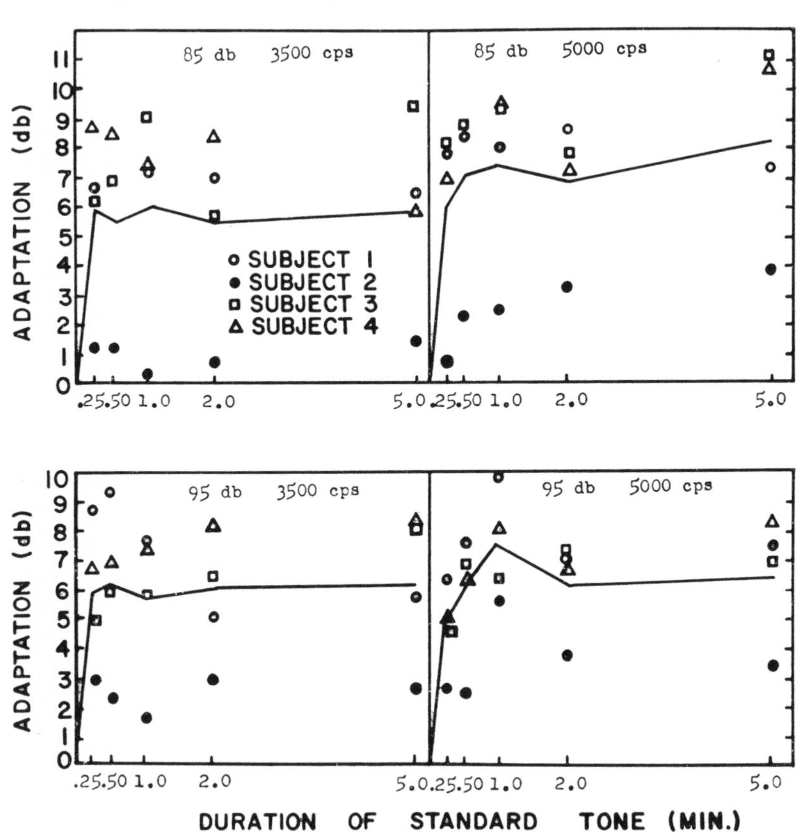

FIGURE 1
Mean and Individual Adaptation for Equal-Loudness Judgments

final intensity setting of the comparison tone and the intensity of the standard tone. A less direct method was necessary in the case of the twice-as-loud judgments. The sone values of the standard tones were computed and then doubled. These doubled sone values were then reconverted to db. The final intensity settings of the comparison tones were then subtracted from the reconverted db values to give the amount of adaptation. The intensity values of the comparison tones used in computing adaptation were the geometric means of the intensity settings of the three replications.[2]

A mean adaptation curve was derived for each experimental condition by obtaining the geometric mean of all the subjects at each condition. Figure 1 shows these curves for the equal-loudness matches with the individual subject values, and Figure 2 shows the mean curves for the twice-as-loud judgments, again with the individual subject values. Inspection of the mean curves shows an initial increase under all conditions after which the curves do not suggest any further systematic changes. There was a greater amount of adaptation with 5000 cps than with 3500 cps, irrespective of the match that was made. The individual curves indicate surprisingly little between-subject variability as compared with the results of Jerger (6), who found that subjects under the same condition differed in adaptation by as much as 40 db. In contrast, subjects in this study differed at the most by 9 db and, in general, by only 4 or 5 db. Inspection of these figures also shows that the mean curves for the equal-loudness matches are flatter than those for the twice-as-loud matches, even though the range of between-subject variability is greater for the equal-loudness matches. In Figure 2, the trend of the mean curves between 2 and 5 minutes suggests the possibility of a recovery from adaptation, which is not exhibited by the curves in Figure 1.

An analysis of variance was performed to evaluate variation in amount of adaptation as a function of the type of judgment made and intensity, frequency, and duration of the standard tone. The only significant effect was that of frequency. Since no difference in amount of adaptation from .25 to 5.0 minutes was demonstrated, t tests were performed to see if the amounts of adaptation at 15 seconds of stimulation differed significantly from no adaptation. Table 1 summarizes the results of these t tests, showing the geometric mean of amounts of adaptation after .25 minutes of stimulation and the level of significance of each mean. Inspection of Table 1 suggests that less adap-

[2] The data were also analyzed in terms of arithmetic means. These were not found to differ by more than rounding errors from the geometric means. However, since decibles are on a logarithmic scale, geometric means are reported. It should be noted, though, that the more variable the subject is in his repetitions of the same judgment, the more greatly the arithmetic and geometric means may differ. Thus, their agreement in the present study reflects the favorable reliability of the data.

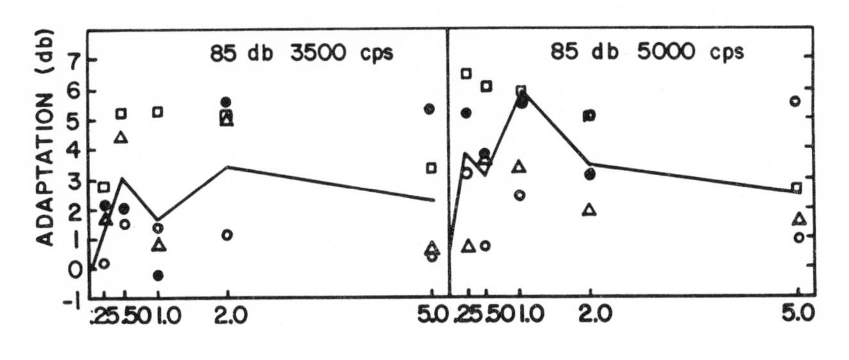

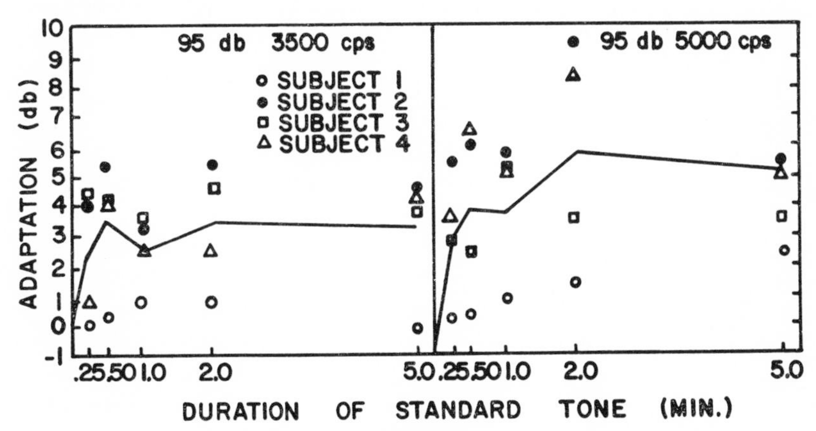

FIGURE 2
MEAN AND INDIVIDUAL ADAPTATION FOR TWICE-AS-LOUD JUDGMENTS

TABLE 1
TABLE OF t TESTS BETWEEN DIFFERENCES IN ADAPTATION BEFORE STIMULATION AND AFTER .25 MINUTES OF STIMULATION

Judgment made	Stimulus condition	Geom. mean adaptation (db)	t
equally loud	85 db 3500 cps	5.66	3.56**
equally loud	85 db 5000 cps	5.94	3.41**
equally loud	95 db 3500 cps	5.82	4.69***
equally loud	95 db 5000 cps	4.70	6.02***
twice as loud	85 db 3500 cps	1.64	2.93*
twice as loud	85 db 5000 cps	3.79	3.03*
twice as loud	95 db 3500 cps	2.36	2.07
twice as loud	95 db 5000 cps	3.08	3.35**

* $p < .05$.
** $p < .025$.
*** $p < .01$.

tation was obtained with the equal-loudness judgments than with the twice-as-loud judgments. This suggestion, however, was not upheld by the analysis of variance as no significant difference was found in amount of adaptation as a function of the kind of judgment made. Table 1 does show that with one exception—the twice-as-loud judgment at 3500 cps, 95 db—the mean amount of adaptation obtained at each condition was greater than zero.

D. Discussion

The results of the present study show that a significant amount of loudness adaptation occurs within .25 minutes of stimulation. This does not support those reviewers (9, 17) who have stated that such adaptation probably does not occur. The present finding that subjects were able to make reliable judgments of both equally and twice-as-loud supports the notion that loudness was actually the basis of the judgment in this experiment. The amount of adaptation obtained, however, was very small compared to most previous studies. This may have been expected as a consequence of the method used; that is, presentation of an unattenuated comparison tone should produce more adaptation of the test ear than should presentation of an attenuated comparison tone. A compensating factor was the decrease in the time required to make the match. The mean time for all conditions was 5 seconds.

Some previous studies (6, 13) have reported that adaptation is independent of frequency above 1000 cps. In contrast, the present study found a significant frequency effect above 1000 cps as more adaptation occurred at 500 cps than at 3500 cps.

The finding that adaptation was not affected by intensity is also in disagreement with previous results. This too might have been due to the use of the unattenuated test tone. That is, if adaptation were complete within 15 seconds, differences in adaptation produced by 10 db differences in stimuli might be difficult to measure. Either a stimulus duration of less than .25 minutes or a difference in stimulus intensity of more than 10 db might result in a significant intensity effect. This problem deserves further study.

As an explanation for the results of this experiment, the possibility exists that contraction of the tympanic muscles may be partially or wholly responsible for the adaptation obtained. Luescher (7), in his observations on the stapedius muscle, reported an increase in duration of muscle contraction with an increase in intensity of the stimulating tone, with the high intensities giving a steady contraction that lasted throughout the presentation of the tone up to 10 or 15 seconds. This agrees well with the maximum adaptation time obtained in the present experiment.

Metz (8) found that these contractions rapidly attained a maximum, remained at maximum for several seconds, and then, although still being stimulated, slowly relaxed. This process appears to be a reflex adaptation and not a muscular fatigue, for Metz found that if the tone were stopped and then immediately presented again, the contraction reappeared with full magnitude. Jepsen (5) reported that these muscle reflexes occurred only when the stimulating tone was at a high intensity level. According to Wever and Lawrence (15), the amount of protection afforded by contraction of these muscles varies greatly with frequency, with the greater protection being afforded for the lower frequencies. This finding is in opposition to that of the present study in that more adaptation was obtained at 5000 cps than at 3500 cps. On the other hand, the findings of Luescher (7), Metz (8), and Jepsen (5) do not provide specific disagreement with the lack of significant intensity effect found in the present study. That is, the 10 db difference between the tones used may be within the difference limen of the tympanic muscles. However, as noted above, that is a question deserving further study. The relationship between the response of the tympanic muscles to continued stimulation and the present data is certainly suggestive. However, as noted by Small (11), a variety of theoretical approaches are possible.

E. Conclusions

The method used did not provide results in agreement with previous studies. In contradiction to these studies, intensity did not affect the amount of adaptation obtained; frequencies above 1000 cps did affect adaptation; adaptation was complete within .25 minutes and the amount of adaptation obtained was relatively small. It was demonstrated, however, that loudness was the basis of the judgments made and that the amount of adaptation is independent of whether the subject makes equal-loudness settings or double-loudness settings.

References

1. Carterette, E. C. Perstimulatory fatigue for continuous and interrupted noise. *J. Acoust. Soc. Amer.*, 1955, **27**, 103-111.
2. Egan, J. P. Perstimulatory fatigue as measured by heterophonic loudness balances. *J. Accoust. Soc. Amer.*, 1955, **27**, 111-120.
3. Egan, J. P., & Thwing, E. J. Further studies on perstimulatory fatigue. *J. Acoust. Soc. Amer.*, 1955, **27**, 1225-1226.
4. Hood, J. D. Studies in auditory fatigue and adaptation. *Acta Otolaryng.*, 1950, Suppl. No. 92.
5. Jepsen, O. The threshold of the reflexes of the intra-tympanic muscles in a normal material examined by means of the impedance method. *Acta Otolaryng.*, 1951, **39**, 406-408.
6. Jerger, J. F. Auditory adaptation. *J. Acoust. Soc. Amer.*, 1957, **29**, 357-363.

7. LUESCHER, E. Die Funktion des Musculus Stapedius beim Menschen. *Z. Hals-Nasen-Ohrenheilk.*, 1929, **23**, 105-132.
8. METZ, O. Studies on the contraction of the tympanic muscles as indicated by changes in the impedance of the ear. *Acta Otolaryng.*, 1951, **39**, 397-405.
9. OSGOOD, C. E. Method and Theory in Experimental Psychology. New York: Oxford Univ. Press, 1953.
10. PALVA, T. Studies on per-stimulatory adaptation in various groups of deafness. *Laryngoscope,* 1955, **65**, 829-847.
11. SMALL, A. M., JR. Auditory adaptation. In J. Jerger (Ed.), *Modern Developments in Audiology*. New York: Academic Press, 1963.
12. TEICHNER, W. H. Recent studies of simple reaction times. *Psychol. Bull.,* 1954, **51**, 128-149.
13. THWING, E. J. Spread of perstimulatory fatigue of a pure tone to neighboring frequencies. *J. Acoust. Soc. Amer.,* 1955, **27**, 741-748.
14. VON BEKESY, G. Experiments in Hearing. New York: McGraw-Hill, 1960.
15. WEVER, E. G., & LAWRENCE, M. Physiological Acoustics. Princeton: Princeton Univ. Press, 1954.
16. WOOD, A. G. Unpublished Master's thesis, University of Virginia, Charlottesville, 1930 (cited in Geldard, F. A., *The Human Senses*). New York: Wiley, 1953.
17. WOODWORTH, R. S., & SCHLOSBERG, H. Experimental Psychology. New York: Holt, 1954.

INDEX